The Biochemistry of the Stilbenoids

The Biochemistry of Natural Products Series

Series editors: Professor J.B. Harborne, *University of Reading*, *Reading, UK*; Herbert Baxter, *Executive editor*, *London, UK*

Books in this series are devoted to the evaluation and presentation of current knowledge of all classes of naturally occurring substances, with emphasis on their metabolic and ecological biochemistry. Natural product research is of world-wide importance and the invited authors are internationally recognized authorities who are leaders in their discipline. Individual volumes will be devoted to particular classes of metabolites and subjects will be selected for their scientific significance, topicality and economic importance. The high standard of the series will be ensured by appropriate review procedures. The editors are confident that these significant and timely volumes will make a major contribution to advancing our knowledge of the biochemistry of naturally occurring organic compounds.

Forthcoming titles

Porphyrins: Analytical chemistry, clinical biochemistry, toxicology and pharmacology
F. DeMatteis and C.K. Lim

Biochemistry of Cardiac Glycosides
S. Malcolm

Polyacetylenes and Related Compounds
H. Greger

Non-protein Amino Acids
E.A. Bell and R.J. Nash

Biochemistry of Steroidal Alkaloids
J.G. Roddick

Biochemistry of Tropane Alkaloids
R.J. Robins, N.J. Walton, B. Drager, R.B. Barlow and W.J. Griffin

Biochemistry of the Quinilizidine Alkaloids
M. Wink

The Biochemistry of the Stilbenoids

John Gorham

School of Biological Science
University of Wales, Bangor, UK

with contributions from

Motoo Tori and Yoshinori Asakawa

Faculty of Pharmaceutical Sciences
Tokushima Bunri University
Japan

CHAPMAN & HALL
London · Glasgow · Weinheim · New York · Tokyo · Melbourne · Madras

Published by Chapman & Hall, 2–6 Boundary Row, London SE1 8HN, UK

Chapman & Hall, 2–6 Boundary Row, London SE1 8HN, UK

Blackie Academic & Professional, Wester Cleddens Road, Bishopbriggs, Glasgow G64 2NZ, UK

Chapman & Hall GmbH, Pappelallee 3, 69469 Weinheim, Germany

Chapman & Hall USA, One Penn Plaza, 41st Floor, New York NY 10119, USA

Chapman & Hall Japan, ITP-Japan, Kyowa Building, 3F, 2–2–1 Hirakawacho, Chiyoda-ku, Tokyo 102, Japan

Chapman & Hall Australia, Thomas Nelson Australia, 102 Dodds Street, South Melbourne, Victoria 3205, Australia

Chapman & Hall India, R. Seshadri, 32 Second Main Road, CIT East, Madras 600 035, India

First edition 1995

Typeset in 10/12 Palatino by Colset Pte Ltd, Singapore
Printed in Great Britain by TJ Press (Padstow), Cornwall

ISBN 0 412 55070 9

A catalogue record for this book is available from the British Library

Library of Congress Catalog Card Number: 94–68997

∞ Printed on permanent acid-free text paper, manufactured in accordance with ANSI/NISO Z39.48–1992 and ANSI/NISO Z39.48–1984 (Permanence of Paper).

Contents

Preface

This volume brings together most of the published literature on stilbenes, bibenzyls, phenanthrenes, 9,10-dihydrophenanthrenes and related compounds. Interest in these compounds has increased in recent years, and the number of papers concerned with their distribution, chemistry, physiology and pharmacology has increased accordingly. The number of types of compound produced by the stilbenoid branch of the phenylpropanoid–polymalonate pathway has also increased to include spirans, phenanthropyrans, etc. Recent investigations into the molecular biology of stilbene synthesis have increased our understanding of this pathway and its relationship to that of the flavonoids. The transfer of stilbene synthase activity into tobacco offers the promise of introducing new phytoalexins into crop plants. The details of the later stages of stilbenoid biosynthesis are, however, largely unexplored.

The liverworts are now known as sources of a large number of stilbenoids, and uniquely of bis(bibenzyl)s. These structures have all been discovered in the 35 years since the identification of lunularic acid as a dormancy factor in *Lunularia cruciata*.

While some taxa are known to produce a bewildering array of stilbenoid compounds, there is still much to be discovered. This is illustrated by the recent reports of resveratrol and piceatannol accumulation in the Poaceae. Their physiological properties are also receiving greater attention. The range has increased from antimicrobial activity to interactions with enzyme activity and the growth of plant and animal cells. The stilbenoids offer considerable opportunities for further research in a variety of fields, from taxonomy to pharmacology.

Bangor
1994

J. Gorham

1

Introduction

The stilbenoids (Figure 1.1) are a group of phenolic compounds derived from cinnamic acid (via the shikimic acid pathway) and three acetate units (from malonyl coenzyme A). The first part of the pathway is common to both stilbenoids and flavonoids, the two biosynthetic routes diverging at the point of cyclization of a styryl-3,5,7-triketoheptanoic acid (Figure 1.2). A C-acylation produces a chalcone and subsequent modifications lead to the flavonoids (aurones, flavones, flavonols, flavanones, dihydro-flavonols, dihydrochalcones, isoflavonoids, biflavonoids, anthocyanins, tannins, etc.). An aldol condensation of the same intermediate polyketide produces a stilbene-2-carboxylic acid that is usually an unstable or bound intermediate in pathways to a range of structures collectively known as the stilbenoids (bibenzyls (dihydrostilbenes), bis(bibenzyl)s, stilbenes, phenyldihydro-isocoumarins, phenanthrenes, 9,10-dihydrophenanthrenes and their derivatives). Stilbene- and bibenzyl-2-carboxylic acids are not commonly encountered in plants, with a few notable exceptions (lunularic acid, notholaenic acid, gaylussacin). Both the stilbenoid and flavonoid pathways occur together in liverworts and higher plants. A more detailed account of the biosynthesis of stilbenoids, as far as it is known, can be found in Chapter 5, together with methods for chemical synthesis. The number of possible structures based on the few basic stilbenoid rings is very large, although biosynthetic pathways favour certain substituents at particular carbon atoms. The number of stilbenoid compounds found in plants is now in excess of 300, compared with just over 100 listed in 1980 (Gorham, 1980).

To reduce confusion it is worth making a few points about the nomenclature and structural drawing of stilbenoids. While the terms 'bibenzyl' and 'stilbene' are convenient for the parent hydrocarbons shown in Figure 1.1, the strict rules of the code of chemical nomenclature of the International Union of Pure and Applied Chemists define these compounds as 1,2-diphenylethane and 1,2-diphenylethene, respectively. This is of

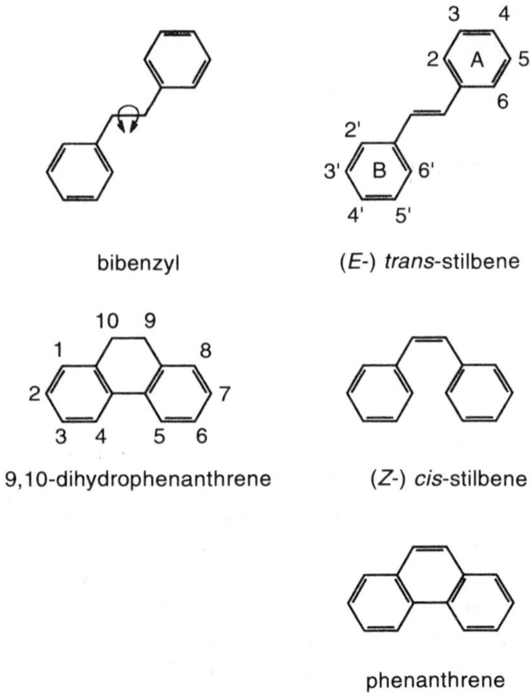

Figure 1.1 Basic structures of the stilbenoids.

practical significance in that derivatives of these hydrocarbons are often indexed in abstracting journals as substituted diphenyl-ethanes and -ethenes.

The conventional system of numbering the atoms of these compounds is shown in Figure 1.1. Since the hydrocarbons are symmetrical, there is no logical reason for starting the numbering in any one of the two benzene rings of the hydrocarbons. The naturally occurring stilbenes and bibenzyls are not symmetrical in the biosynthetic sense, since one ring (A) is derived from a (substituted) phenylpropanoid molecule and the other (B) from the condensation of three molecules of malonyl coenzyme A. Thus the B ring frequently retains the 3,5-substitution pattern (as in pinosylvin, **143a**), while the A ring will initially have the substitution pattern of the phenylpropanoid precursor (no substituents for cinnamoyl coenzyme A, 4-substituted for coumaryl coenzyme A, etc.). These patterns are subject to considerable modification after cyclization. For many highly substituted stilbenes it is not immediately apparent which ring is, in the biosynthetic sense, ring A or ring B. Since radio-tracer experiments have not been performed to determine the origins of the rings of most of these compounds,

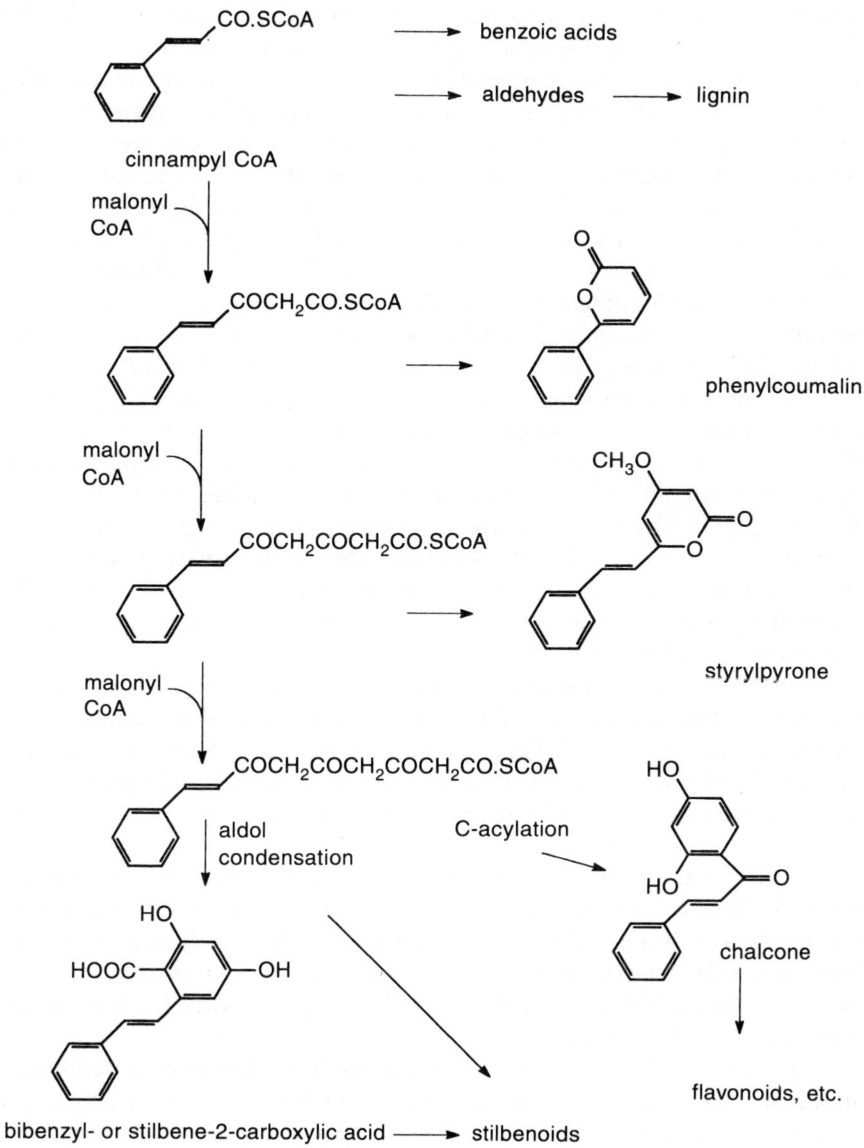

Figure 1.2 The phenylpropanoid–polymalonate pathway.

it is simpler to adopt the convention of numbering the substituents so that the numerical sum of the substituents is as small as possible, and so that there are equal or fewer number-prime substituents on the second ring to plain numbered substituents on the first ring. Where there is still a possibility of choice, smaller substituents (hydroxyls) are given smaller numbers than larger substituents (methoxyls, etc.). Resveratrol, to take a simple example, is thus 3,4′-dihydroxy-5-methoxystilbene rather than 3′,4-dihydroxy-5′-methoxystilbene or 4′,5-dihydroxy-3-methoxystilbene. Some authors have numbered the atoms in stilbene from 1 to 14, with atoms 7 and 8 being the ethene bridge. Gnetifolin E (**163**), for example, was described as isorhapontigenin-12-*O*-D-glucopyranoside by Lin *et al.* (1992).

The same problem of symmetry applies to the phenanthrenes and 9,10-dihydrophenanthrenes, and the same rules can be applied. Since the carbon atoms of phenanthrenes are numbered from 1 to 10 (Figure 1.1), the synonyms can be even more confusing. To take a simple example again, the orchid 9,10-dihydrophenanthrene orchinol (**273**) could be 5,7-dimethoxy-2-hydroxy-9,10-dihydrophenanthrene rather than 2,4-dimethoxy-7-hydroxy-9,10-dihydrophenanthrene, except that the sum of the substituent numbers is 14 in the former name and 13 in the latter. The literature contains several examples of the same compound being numbered differently by different authors.

A problem in the diagrammatic representation of bibenzyls arises from the free rotation (unless sterically hindered) about the ethane bond, shown by the circular arrow in Figure 1.1. Since drawings of bibenzyl structures are usually two-dimensional, there are two possibilities – the planar *trans* or *cis* configurations. In the case of stilbenes the ethane bond does not allow free rotation and there are only two possible configurations. Most naturally occurring stilbenes have the *trans*- (*E*-) configuration, but *cis*- (*Z*-) stilbenes have also been found. Isomerization between the two forms may be induced by strong heat or light. Rotation about the aryl-acyl bonds also means that the larger substituents on the benzene rings may be represented in one of two ways for each ring. A total of four drawings are possible for a stilbene substituted in both rings.

Among the earliest reports of compounds now classified as stilbenoids in plants were those of Hesse (1899) on rhubarb and Asahina and coworkers (Asahina and Asano, 1929, 1930a, b, 1931; Asahina and Miyake, 1916; Asahina and Ueno, 1916) on hydrangenol and phyllodulcin from *Hydrangea*. The early work on stilbenoids was reviewed by Tsukida (1954), Billek and Kindl (1962a) and Billek (1964), and more recently by Gorham (1980, 1989). The distribution of stilbenoids in bryophytes was covered in reviews by Asakawa (1982, 1984, 1988, 1990a, b), Gorham (1990) and Mues and Zinsmeister (1988). Other major reviews have focused on the biosynthesis of stilbenoids (Kindl, 1985) and their role in resistance of wood to fungal attack (Hart, 1981).

SPERMATOPHYTA

Angiospermae

Dicotyledons

Magnoliales
Laurales
Piperales
Aristolochiales
Rafflesiales
Nymphaeales
Illiciales
Nelumbonales
Ranunculales
Papaverales
Sarraceniales
Trochodendrales
Cercidiphyllales
Eupteleales
Didymelales
Hamamelidales
Eucommiales
Urticales
Barbeyales
Casuarinales
Fagales
Betulales
Balanopales
Myricales
Juglandales

Leitneriales
Caryophyllales
Polygonales
Plumbaginales
Theligonales
Dilleniales
Paeoniales
Theales
Violales
Passiflorales
Cucurbitales
Begoniales
Capparales
Tamaricales
Salicales
Ericales
Diapensiales
Ebenales
Primulales
Malvales
Euphorbiales
Thymelaeales
Saxifragales
Rosales

Fabales
Connarales
Nepenthales
Podostemales
Myrtales
Hippuridales
Rutales
Sapindales
Geraniales
Polygalales
Cornales
Celastrales
Rhamnales
Oleales
Santalales
Eleagnales
Proteales
Dipsacales
Gentianales
Polemoniales
Scrophulariales
Lamiales
Campanulales
Calycerales
Asterales

Monocotyledons

Alismales
Hydrocharitales
Potamogetonales
Triuridales
Liliales
Iridales
Zingiberales
Orchidales
Juncales
Cyperales

Bromeliales
Commelinales
Eriocaulales
Restionales
Poales
Arecales
Cyclanthales
Arales
Pandanales
Typhales

Gymnospermae

Coniferales
Taxales
Cycadales
Ginkgoales
Gnetales

PTERIDOPHYTA

Psilotales
Lycopodiales
Isoetales
Equisetales
Filicales

BRYOPHYTA

Musci
Anthocerotae
Hepaticae

Figure 1.3 Distribution of stilbenoid-producing taxa. Taxa in bold type are known to contain stilbenoids.

Stilbenoids are widely distributed in the plant kingdom (Figure 1.3), although some structures are characteristic of particular plant families. They are found in liverworts but not in mosses (Chapter 2), in some ferns (Chapter 2), in gymnosperms (Chapter 3) and in many dicotyledonous angiosperms (Chapter 4), but until recently in only a few monocots other than orchids. In many plants stilbenoids are found in high concentrations in wood (*Pinus, Eucalyptus*) and in storage tissues (*Dioscorea, Tamus, Rheum* and orchid pseudobulbs). Our present knowledge of the distribution of stilbenoids in plants can only be regarded as fragmentary, given the number of taxa which have been systematically investigated, and the rate at which new reports have appeared in the literature in recent years. This is particularly true for stilbenoids that are minor constituents of plants which synthesize a large number of phenolic compounds. Improved techniques for detection and identification (Chapter 6), and for the synthesis of reference compounds (Chapter 5) have greatly assisted the search for stilbenoids. The development of instrumentation for analysis has also contributed substantially to our ability to determine chemical structures of natural products. Without the aid of the 400-MHz nuclear magnetic resonance (NMR) spectrometer the chemistry of bis(bibenzyl)s, for example, could not have been established so easily. A number of stilbenoids have been identified following investigations of the biological activities of plant extracts. There is an increasing literature on the antibiotic and other activities of stilbenoids, but relatively few systematic investigations of biological activity have been published (Chapter 7).

The stilbenoids are a group of compounds that are attracting increasing interest because of their biological activities, their taxonomically related distribution and the biochemistry and molecular biology of their pathway.

2

Stilbenoids in bryophytes and pteridophytes

2.1 INTRODUCTION

Except for an unusual brominated stilbenoid, 3,3′-dibromo-4,4′,5,5′-tetrahydroxybibenzyl, from the alga *Polysiphonia urceolata* (Kursata, Amiya and Nakano, 1976), stilbenoids in lower plants are confined to liverworts and one genus of ferns. The classification of bryophytes is in some cases contentious, but generally three main classes or divisions are recognized, the Hepaticae (liverworts), Musci (mosses) and Anthocerotae (hornworts). The small genus *Takakia*, discovered in the 1950s and containing only two incompletely studied species, is in some ways intermediate between leafy liverworts and mosses and has been placed in a fourth division, the Takakiophyta. Since they were only known as female gametophytes until the discovery of male plants in the late 1980s, their final taxonomic position is still uncertain. Stilbenoids have not been detected in mosses or hornworts, and *Takakia* has not been fully investigated chemically.

Within the Hepaticae six orders are recognized (Gradstein, 1990). The subclass Jungermanniidae contains three orders: the Calobryales (containing about ten species of *Haplomitrium* (= *Calobryum*), the Jungermanniales (leafy liverworts, about 3500 species), and the Metzgeriales (simple thalloid liverworts, about 600 species). The other subclass, the Marchantiidae, also contains three orders: the Monocleales (containing one, or perhaps a few, thalloid species with a long cylindrical capsule opening by a single slit), the Marchantiales (about 200 more complex thalloid and mainly perennial species), and the Sphaerocarpales (about 20–25 short-lived thalloid species with reduced sporophytes). Some authorities also split the Treubiales (about seven species) from the Metzgeriales. Stilbenoids have been found in all six orders of liverworts (Table 2.2, below).

Some stilbenoids isolated from liverworts, especially the bis(bibenzyl)s, are very unusual compounds. Although many bibenzyls have been isolated not only from bryophytes but also from other classes of the plant kingdom, bis(bibenzyl)s are restricted to the liverworts, and recognized as chemical markers for the Hepaticae. Bibenzyl-2-carboxylic acids are also unusual compounds which are characteristic of liverworts and ferns of the genus *Notholaena* (Table 2.1).

Although some detailed reviews have appeared recently (Asakawa, 1982, 1984, 1988, 1990b; Gorham 1990; Mues and Zinsmeister, 1988; Zinsmeister, Becker and Eicher, 1991), the structure and distribution of the compounds listed in Tables 2.2 and 2.3 will be discussed briefly in this chapter.

2.2 BRYOPHYTE BIBENZYLS

A number of different bibenzyls have been isolated not only from the order Jungermanniales but also from the Marchantiales. These bibenzyls are all products of the malonate–shikimate pathway. Structures were elucidated by both synthesis and spectroscopic methods. The main synthetic method is the Wittig reaction between corresponding benzaldehydes and the Wittig reagents derived from benzyl halides with various functional groups, followed by hydrogenation to produce dihydrostilbenoids (see Chapter 5). The position of the oxygen functions were determined both by NMR decoupling and by experiments using the nuclear Overhauser effect (NOE) (Chapter 6). Normally one ring is less substituted or in most cases without any functions, while the other is usually highly substituted. In this review bryophyte bibenzyls are classified into four groups:

1. Group A: simple bibenzyls – bibenzyls having oxygen functions on the aromatic rings, including the methylenedioxy derivatives;
2. Group B: bibenzyls having prenyl or geranyl substituents regardless of the position of the substitution, but not including the cyclized derivatives;
3. Group C: bibenzyls having one more additional heterocyclic ring, including phthalides, but excluding the methylenedioxy derivatives;
4. Group D: other than aromatic groups, namely precursors to the bibenzyls and hybrids between aromatics and terpenoids.

2.2.1 Group A (simple bibenzyls)

Radula species in particular are rich sources of bibenzyl derivatives (Asakawa, 1982; Asakawa *et al*., 1982). 3-Hydroxybibenzyl (**2**), 3-hydroxy-4-methoxybibenzyl (**3**) and 4-hydroxy-3-methoxybibenzyl (**4**) have been detected in *R. frondescens* (Asakawa and Inoue, 1987), together with 3-methoxybibenzyl (**1**) which has also been found in *R. buccinifera*

Table 2.1 Stilbene-2-carboxylic acids in ferns of the genus *Notholaena*

Species	*Compounds*	*Authors*
Notholaena dealbata	4′,5-dimethoxy-3-hydroxybibenzyl-2-carboxylic acid (**notholaenic acid**)	Wollenweber and Favre-Bonvin (1979); Wollenweber (1982)
	3,4′-dimethoxy-5-hydroxybibenzyl-2-carboxylic acid (**isonotholaenic acid**)	Wollenweber *et al.* (1993)
N. nivea var. *nivea*	3,4′-dimethoxy-5-hydroxybibenzyl-2-carboxylic acid (**isonotholaenic acid**)	Wollenweber *et al.* (1993)
N. chilensis	3,4′-dimethoxy-5-hydroxybibenzyl-2-carboxylic acid (**isonotholaenic acid**)	Wollenweber *et al.* (1993)
N. limitanea var. *limitanea*	3,4′-dimethoxy-5-hydroxybibenzyl-2-carboxylic acid (**isonotholaenic acid**)	Wollenweber *et al.* (1993)
N. limitanea var. *mexicana*	4′,5-dimethoxy-3-hydroxybibenzyl-2-carboxylic acid (**notholaenic acid**)	Wollenweber and Favre-Bonvin (1979); Wollenweber (1982)
	3,4′-dimethoxy-5-hydroxybibenzyl-2-carboxylic acid (**isonotholaenic acid**) (one accession)	Wollenweber *et al.* (1993)

Table 2.2 Stilbenoids in the Hepaticae

	Compounds	*Species*	*Authors*
BIBENZYLS			
Group A			
1	3-methoxybibenzyl	*Frullania dilatata* var. *anomala*	Asakawa *et al.* (1981a)
		Radula buccinifera	Asakawa *et al.* (1982)
		Radula frondescens	Asakawa and Inoue (1987)
		Radula javanica	Asakawa, Kondo and Tori (1991)
2	3-hydroxybibenzyl	*Radula frondescens*	Asakawa and Inoue (1987)
3	3-hydroxy-4-methoxybibenzyl	*Radula frondescens*	Asakawa and Inoue (1987)
4	4-hydroxy-3-methoxybibenzyl	*Radula frondescens*	Asakawa and Inoue (1987)
5	3,5-dimethoxybibenzyl	*Radula javanica*	Asakawa, Kondo and Tori (1991)
6	Lunularin* (3,4′-dihydroxybibenzyl)	*Dumortiera hirsuta*	Gorham (1977a,b)
		Marchantia berteroana	Asakawa and Campbell (1982)
		Marchantia chenopoda	M. Tori *et al.* (unpublished)
		Marchantia paleacea var. *diptera*	Asakawa *et al.* (1984)
		Marchantia polymorpha	Asakawa *et al.* (1984, 1990)
		Ricciocarpos natans	Wurzel and Becker (1990)
		Frullania bonincola	Asakawa *et al.* (1981a)
		Frullania dilatata	Asakawa *et al.* (1981a)
		Fruilania tamarisci	Asakawa *et al.* (1981a)
7	4-hydroxy-3′-methoxybibenzyl	*Plagiochila stephensoniana*	Asakawa and Campbell (1982)
		Radula frondescens	Asakawa and Inoue (1987)
		Frullania davurica	Asakawa *et al.* (1981a)

8	3,4′-dimethoxybibenzyl	*Frullania davurica*	Asakawa *et al.* (1981a)
		Frullania falciloba	Asakawa *et al.* (1987a)
		Frullania tamarisci subsp. *obscura*	Asakawa *et al.* (1981a)
		Plagiochila stephensoniana	Asakawa and Campbell (1982)
		Radula buccinifera	Asakawa *et al.* (1982)
9	3-hydroxy-4′-methoxybibenzyl	*Frullania davurica*	M. Toyota and Y. Asakawa (unpublished)
10	2,3,4′-trihydroxybibenzyl	*Pellia endiviifolia*	Asakawa, Takikawa and Aratani (1976)
11	2,4′-dihydroxy-3-methoxybibenzyl	*Frullania brotheri*	Asakawa *et al.* (1981a)
12	Pellepiphyllin (2-hydroxy-3,4′-dimethoxybibenzyl)	*Pellia neesiana*	Benesová and Herout (1970)
13	4-hydroxy-3,4′-dimethoxybibenzyl	*Frullania davurica*	M. Toyota and Y. Asakawa (unpublished)
14	3-hydroxy-4,5-methylenedioxybibenzyl	*Radulla javanica*	Asakawa *et al.* (1982); Asakawa, Kondo and Tori (1991)
15	3-hydroxy-4,3′-dimethoxybibenzyl	*Frullania falciloba*	Asakawa *et al.* (1987a)
16	3′-methoxy-3,4-methylenedioxybibenzyl	*Frullania falciloba*	Asakawa *et al.* (1987a)
		Frullania ericoides	Asakawa, Matsuda and Cheminat (1987)
17	3′-methoxy-3,4-methylene-dioxybibenzyl	*Trocholejeunea sandvicensis*	Asakawa (1984)

Table 2.2 (*Contd*)

Compounds		Species	Authors
18	Lunularic acid* (2-carboxy-3,4′-dihydroxybibenzyl)	*Ricciocarpos natans (M)*	Kunz and Becker (1992)
		Marchantia polymorpha (M)	Abe and Ohta (1983); Ohta *et al.* (1984a,b)
		Marchantia polymorpha	Adam and Becker (1994)
		Jungermannia sublata	Abe and Ohta (1983); Ohta *et al.* (1984a,b)
		Lophocolea heterophylla	Abe and Ohta (1983); Ohta *et al.* (1984a,b)
		Calypogeia tosana	Abe and Ohta (1983); Ohta *et al.* (1984a,b)
19	5,4′-dihydroxybibenzyl-2-*O*-6-D-glucopyranoside	*Ricciocarpos natans*	Kunz and Becker (1992)
20	4′-hydroxy-5-methoxy-3,4-methylenedioxybibenzyl	*Plagiochila chacabucensis*	Asakawa and Inoue (1984b)
21	3,4,4′-trimethoxy-2-methoxy-carbonylbibenzyl	*Plagiochila spinulosa*	Rycroft (1990)
22	3-hydroxy-3′-methoxy-4,5-methylenedioxy bibenzyl	*Frullania amplicrania*	Asakawa *et al.* (1981a)
23	5,3′-dimethoxy-3,4-methylene-dioxybibenzyl	*Frullania ericoides*	Asakawa, Matsuda and Cheminat (1987)
24	3,4-dihydroxy-5,3′dimethoxy-bibenzyl	*Schistochila balfocerii* var. *bekkei*	Gradstein, Matsuda and Asakawa (1985)

25	3,4;3′,4′-bismethylenedioxy-bibenzyl	*Frullania parvistipula*	Asakawa, Matsuda and Cheminat (1987)
26	5,3′,4′-trihydroxybibenzyl-2-*O*-6-D-glucopyranoside	*Ricciocarpos natans*	Kunz and Becker (1992)
27	2-carboxy-3,4′-dihydroxy-bibenzyl-3′-*O*-6-D-glucopyranoside	*Ricciocarpos natans*	Kunz and Becker (1992)
28	4′-hydroxy-3,3′-dimethoxy-4,5-methylenedioxybibenzyl	*Frullania bonincola*	Asakawa, Matsuda and Cheminat (1987)
29	3,4,5,3′,4′-pentamethoxybibenzyl	*Frullania serratta*	Asakawa *et al.* (1991c)
30	3,3′-dihydroxy-4,5;4′,5′-bismethylenedioxybibenzyl	*Frullania parvistipula*	Asakawa, Matsuda and Cheminat (1987)
31	3,4,5,3′,4′,5′-hexamethoxy-bibenzyl (brittonian A)	*Frullania brittoniae* subsp. *truncata*	Asakawa, Takikawa and Aratani (1976)
32	3,4,5,3′-tetramethoxy-4′,5′-methylenedioxybibenzyl (brittonin B)	*Frullania brittoniae* subsp. *truncata*	Asakawa, Takikawa and Aratani (1976)
		Frullania serratta	Asakawa *et al.* (1991c)

Group B

33	3-hydroxy-4-(3-methyl-2-butenyl) bibenzyl	*Radula oyamensis*	Asakawa *et al.* (1982)
34	3,5-dihydroxy-4-(3-methyl-2-butenyl)bibenzyl	*Radula complanata*	Takikawa, Tori and Asakawa (1989)
		Radula buccinifera	Asakawa *et al.* (1982)
35	3,5-dihydroxy-4-(2,3-epoxy-3-methylbutyl)bibenzyl	*Radula complanata*	Asakawa *et al.* (1982)

Table 2.2 (*Contd*)

Compounds		Species	Authors
36	3-hydroxy-5-methoxy-4-(3-methyl-2-butenyl)bibenzyl	*Radula oyamensis*	Asakawa *et al.* (1982)
37	2-carboxy-3,5-dihydroxy-4(3-methyl-2-butenyl)bibenzyl	*Radula complanata*	Asakawa *et al.* (1978)
38	3,4′-dihydroxy-4-(3-methyl-2-butenyl)bibenzyl	*Radula oyamensis*	Asakawa *et al.* (1982)
39	3-hydroxy-4′-methoxy-4-(3-methyl-2-butenyl)bibenzyl	*Radula oyamensis*	Asakawa *et al.* (1982)
40	4′-hydroxy-3-methoxy-4-(3-methyl-2-butenyl)bibenzyl	*Radula oyamensis*	Asakawa *et al.* (1982)
		Radula kojana	Asakawa *et al.* (1991b)
41	3,5-dihydroxy-2-(3-methyl-2-butenyl)bibenzyl	*Radula complanata*	Asakawa *et al.* (1991a)
		Radula perrottetii	Asakawa *et al.* (1991a)
		Radula kojana	Asakawa *et al.* (1991b)
		Radula frondescens	Asakawa and Inoue (1987)
		Radula voluta	Asakawa and Inoue (1987)
42	5-hydroxy-3-methoxy-2-(3-methyl-2-butenyl)bibenzyl	*Radula kojana*	Asakawa *et al.* (1991b)
43	3-hydroxy-5-methoxy-2-(3-methyl-2-butenyl)bibenzyl	*Radula kojana*	Asakawa *et al.* (1991b)
44	2-(2,3-epoxy-3-methylbutenyl)-3,5-dihydroxybibenzyl	*Radula kojana*	Asakawa *et al.* (1991b)

45	3-hydroxy-2-(3-hydroxy-3-methylbutyl)-5-methoxybibenzyl	*Radula kojana*	Asakawa *et al.* (1991b)
46	Perrottetin A	*Radula perrottetii*	Asakawa *et al.* (1991a)
47	Perrottetin B	*Radula perrottetii*	Asakawa *et al.* (1991a)
48	Perrottetin C	*Radula perrottetii*	Asakawa *et al.* (1991a)
49	3,5-dihydroxy-2-methoxycarbonyl-6-(3-methyl-2-butenyl)bibenzyl-	*Radula perrottetii*	Asakawa *et al.* (1991a)
50	3,4-dihydroxy-4-(3,7-dimethylocta-2,7-dienyl)bibenzyl	*Radula oyamensis*	Asakawa *et al.* (1982)
51	2-geranyl-3,5-dihydroxybibenzyl	*Radula complanata*	Asakawa *et al.* (1982)
		Radula tokiensis	Asakawa *et al.* (1982)
		Radula oyamensis	Asakawa *et al.* (1982)
		Radula kojana	Asakawa *et al.* (1991b)
		Radula frondescens	Asakawa and Inoue (1987)
		Radula voluta	Asakawa and Inoue (1987)
		Radula javanica	Asakawa, Kondo and Tori (1991)
52	4-geranyl-3,5-dihydroxybibenzyl	*Radula javanica*	Asakawa, Toyota and Takemoto (1978)
		Radula constricta	Takikawa, Tori and Asakawa 1989)
		Radula tokiensis	Asakawa *et al.* (1981b)
53	2-geranyl-3,5,4′-trihydroxybibenzyl	*Radula frondescens*	Asakawa and Inoue (1987)
54	4-geranyl-3,5,4′-trihydroxybibenzyl	*Radula javanica*	Asakawa, Toyota and Takemoto (1978)

Table 2.2 (*Contd*)

	Compounds	*Species*	*Authors*
Group C			
55	2(*R*)-6-hydroxy-2-isopropenyl-4-(2-phenylethyl)-2,3-dihydrobenzofuran	*Radula perrottetii*	Asakawa *et al.* (1991a)
56	Perrottetin D	*Radula perrottetii*	Asakawa *et al.* (1982, 1991a)
		Radula voluta	Asakawa and Inoue (1987)
57	6-hydroxy-4-(2-phenylethyl)-benzofuran	*Radula kojana*	Asakawa *et al.* (1991b)
58	Tylimanthin A	*Tylimanthus urvilleanus*	Asakawa and Inoue (1984a)
59	Tylimanthin B	*Tylimanthus urvilleanus*	Asakawa and Inoue (1984a)
60	Tylimanthin C	*Tylimanthus urvilleanus*	Asakawa and Inoue (1984a)
61	7-hydroxy-2,2-dimethyl-5-(2-phenylethyl)chromene	*Radula kojana*	Asakawa *et al.* (1991b)
		Radula perrottetii	Asakawa *et al.* (1991a)
62	7-methoxy-2,2-dimethyl-5-(2-phenylethyl)chromene	*Radula kojana*	Asakawa *et al.* (1991b)
63	7,8-dihydroxy-2,2-dimethyl-5-(2-phenylethyl)chromene	*Radula perrottetii*	Asakawa *et al.* (1991a)
64	5-hydroxy-2,2-dimethyl-7-(2-phenylethyl)chromene	*Radula kojana*	Asakawa *et al.* (1991b)
65	6-carboxy-5-hydroxy-2,2-dimethyl-7-(2-phenylethyl)chromene	*Radula kojana*	Asakawa *et al.* (1991b)
66	2(*S*)-7-hydroxy-2-methyl-2-(4-methyl-3-pentenyl)-5-(2-phenylethyl)chromene	*Radula kojana*	Asakawa *et al.* (1991b)

67	2(*S*)-6-carboxy-7-hydroxy-2-methyl-2-(4-methyl-3-pentenyl)-chromene	*Radula kojana*	Asakawa *et al.* (1991b)
68	Radulanin I	*Radula javanica*	Asakawa *et al.* (1991a)
69	Radulanin J	*Radula javanica*	Asakawa, Kondo and Tori (1991)
70	Radulanin K	*Radula javanica*	Asakawa, Kondo and Tori (1991)
71	Radulanin A	*Radula complanata*	Asakawa (1982); Asakawa *et al.* (1991a)
		Radula javanica	Asakawa, Kondo and Tori (1991)
		Radula voluta	Asakawa and Inoue (1987)
		Radula buccinifera	Asakawa *et al.* (1982)
72	Radulanin B	*Radula tokiensis*	Asakawa *et al.* (1981b)
73	Radulanin C	*Radula javanica*	Asakawa, Toyota and Takemoto (1978)
		Radula buccinifera	Asakawa *et al.* (1982)
74	Radulanin D	*Radula javanica*	Asakawa, Toyota and Takemoto (1978)
75	Radulanin E	*Radula javanica*	Asakawa, Toyota and Takemoto (1978)
76	Radulanin F	*Radula complanata*	Asakawa *et al.* (1978)
77	Radulanin G	*Radula javanica*	Asakawa, Toyota and Takemoto (1978)
78	Radulanin H	*Radula complanata*	Asakawa *et al.* (1982)
		Radula javanica	Asakawa *et al.* (1982)
79	4′-hydroxyradulanin H	*Radula complanata*	Takikawa, Tori and Asakawa (1989)
80	Radulanin L	*Radula complanata*	Asakawa *et al.* (1991a)
81	Radulanolide	*Radula complanata*	Asakawa *et al.* (1982)
82	Balantiolide	*Balantiopsis rosea*	Asakawa *et al.* (1986)

Table 2.2 (*Contd*)

Compounds		Species	Authors
83	5,6-dimethoxy-3-(4′-methoxyphenylmethyl)phthalide	*Frullania falciloba*	Asakawa, Takikawa and Tori (1987)
84	5,7-dimethoxy-3-(3′-methoxy-4′,5′-methylenedioxyphenylmethyl)phthalide	*Trocholejeunea sandvicensis*	Asakawa (1984)
85	5,7-dimethoxy-3-(3′,4′,5′-trimethoxyphenylmethyl)-phthalide	*Trocholejeunea sandvicensis*	Asakawa (1984)
Group D			
86	Prelunularic acid	*Marchantia polymorpha*	Ohta *et al.* (1983, 1984a,b)
87	4-(*p*-methoxyphenyl)cyclohex-2-en-1-one	*Plagiochila longispina*	Siegel *et al.* (1991)
88	Spinuloplagin A	*Plagiochila spinulosa*	Rycroft (1990)
89	Spinuloplagin B	*Plagiochila spinulosa*	Rycroft (1990)
BIS(BIBENZYL)S			
Type A			
90a	Marchantin A	*Marchantia polymorpha*	Asakawa *et al.* (1983a); Asakawa (1984)*
		Wiesnerella denudata	Asakawa *et al.* (1991c)

91	Marchantin B	*Marchantia polymorpha*	Asakawa *et al.* (1983a); Asakawa (1984)*
		Wiesnerella denudata	Asakawa *et al.* (1991c)
92	Marchantin C	*Marchantia polymorpha*	Asakawa *et al.* (1983a); Asakawa (1984)*
		Dumortiera hirsuta	M. Toyota and Y. Asakawa (unpublished)
		Reboulia hemisphaerica	H.-C. Wei and C.-L. Wu (personal communication)
		Monoclea forsteri	Toyota, Nagashima and Asakawa (1988)
		Monoclea gottschei subsp. *neotropic*	Spörle *et al.* (1989)
		Plagiochila sciophyla	Hashimoto *et al.* (1987)
		Schistochila glaucescens	Tori, Msuya and Asakawa (1990)
93	Marchantin D	*Marchantia polymorpha*	Asakawa *et al.* (1983a); Asakawa (1984)*
94	Marchantin E	*Marchantia polymorpha*	Asakawa *et al.* (1983a); Asakawa (1984)*
95	Marchantin F	*Marchantia polymorpha*	Asakawa *et al.* (1983a); Asakawa (1984)*
96	Marchantin G	*Marchantia polymorpha*	Asakawa *et al.* (1983a); Asakawa (1984)*
97	Marchantin H	*Plagiochasma intermedium*	Tori (1988); Tori *et al.* (1985)
		Marchantia polymorpha	Asakawa, Okada and Perold (1988)
		Marchantia diptera	Wu (1990)
		Plagiochila sciophila	Hashimoto *et al.* (1987)

Table 2.2 (*Contd*)

Compounds		*Species*	*Authors*
98	Marchantin I	*Riccardia multifida*	Tori (1988); Tori *et al.* (1985)
99	Marchantin J	*Marchantia polymorpha*	Asakawa *et al.* (1990)
100	Marchantin K	*Marchantia polymorpha*	Asakawa *et al.* (1990)
101	Marchantin L	*Marchantia polymorpha*	Asakawa *et al.* (1990)
102	Marchantin M	*Mannia subpilosa*	Wei and Wu (1991)
103	Marchantin N	*Mannia subpilosa*	Wei and Wu (1991)
	Marchantin P	*Marchantia chenopoda*	Tori, Aoki and Asakawa (1994)
104	Marchantinquinone	*Mannia subpilosa*	Wei and Wu (1991)
Type B			
105	Isomarchantin C	*Marchantia polymorpha*	Asakawa *et al.* (1987)
		Marchantia palmata	Asakawa *et al.* (1987)
Type C			
106	Pakyonol	*Mannia fragrans*	Huneck *et al.* (1988)
107	Neomarchantin A	*Schistochila glaucescens*	Tori, Masuya and Asakawa (1990)
		Monoclea forsteri	Toyota, Nagashima and Asakawa (1988)
108	Neomarchantin B	*Schistochila glaucescens*	Tori, Masuya and Asakawa (1990)

Type D			
109a	Riccardin B	*Riccardia multifida*	Asakawa *et al.* (1983a,b)
Type E			
110a	Riccardin A	*Riccardia multifida*	Asakawa *et al.* (1983a,b)
		Blasia pusila	T. Hashimoto and Y. Asakawa (unpublished)
111	Riccardin C	*Reboulia hemisphaerica*	Asakawa and Matsuda (1982)
		Marchantia polymorpha	Asakawa, Okada and Perold (1988); Asakawa *et al.* (1990)
		Marchantia palmata	Asakawa *et al.* (1987)
		Monoclea forsteri	Toyota, Nagashima and Asakawa (1988)
		Ricciocarpos natans	Kunz and Becker (1992)
		Blasia pusila	T. Hashimoto and Y. Asakawa (unpublished)
112	Riccardin D	*Marchantia tosana*	Asakawa *et al.* (1984)
		Monoclea forsteri	Toyota, Nagashima and Asakawa (1988)
113	Riccardin E	*Monoclea forsteri*	Toyota, Nagashima and Asakawa (1988)
		Monoclea gottschei subsp. *neotropic*	Spörle *et al.* (1989)
114	Riccardin F	*Marchantia tosana*	Asakawa *et al.* (1984)
	Riccardin G	*Marchantia chenopoda*	Tori, Aoki and Asakawa (1994)

Table 2.2 (*Contd*)

Compounds		*Species*	*Authors*
Type F			
115	Isoriccardin C	*Marchantia polymorpha*	Asakawa *et al.* (1987)
		Marchantia palmata	Asakawa *et al.* (1987)
Type G			
116	Plagiochin A	*Plagiochila sciophila*	Hashimoto *et al.* (1987)
117	Plagiochin B	*Plagiochila sciophila*	Hashimoto *et al.* (1987)
118	Plagiochin C	*Plagiochila sciophila*	Hashimoto *et al.* (1987)
119	Plagiochin D	*Plagiochila sciophila*	Hashimoto *et al.* (1987)
Type H			
120	Isoplagiochin A	*Plagiochila fruticosa*	T. Hashimoto and Y. Asakawa (unpublished)
121	Isoplagiochin B	*Plagiochila fruticosa*	T. Hashimoto and Y. Asakawa (unpublished)

Type I

122	Perrottetin E	*Radula perrottetii*	Asakawa *et al.* (1991a)
			Toyota *et al.* (1985)
		Radula kojana	Asakawa *et al.* (1991b)
		Monoclea forsteri	Toyota, Nagashima and Asakawa (1988)
		Plagiochila sciophila	Hashimoto *et al.* (1987)
		Jungermannia comata	Nagashima, Toyota and Asakawa (1990)
		Marchantia polymorpha	Asakawa *et al.* (1987)
		Monoclea gottschei subsp. *neotropica*	Spörle *et al.* (1989)
		Nardia subclavata	M. Toyota and Y. Asakawa (unpublished)
		Pellia endiviifolia	Hashimoto *et al.* (1991a)
123	Perrottetin F	*Radula kojana*	Asakawa *et al.* (1991b)
		Radula perrottetii	Asakawa *et al.* (1991a)
			Toyota *et al.* (1985)
		Lunularia cruciata	Hashimoto *et al.* (1991b)
124	Perrottetin G	*Radula perrottetii*	Asakawa *et al.* (1991a)
			Toyota *et al.* (1985)
125	Perrottetin E 11′-methyl ether	*Pellia endiviifolia*	Hashimoto *et al.* (1991a)
126	14-hydroxyperrottetin E 11′-methyl ether	*Pellia endiviifolia*	Hashimoto *et al.* (1991a)

Table 2.2 (*Contd*)

Compounds		*Species*	*Authors*
127	7′,8′-dehydroperrottetin F	*Lunularia cruciata*	Hashimoto *et al.* (1991a)
	(unnamed perrottetin-type)	*Jubula japonica*	Toyota and Asakawa (1993)
	14-hydroxyperrottetin E	*Pellia epiphylla*	Cullmann, Adam and Becker (1993)
	14′-hydroxyperrottetin E	*Pellia epiphylla*	Cullmann, Adam and Becker (1993)
	14,14′-dihydroxyperrottetin E	*Pellia epiphylla*	Cullmann, Adam and Becker (1993)
Others			
128a	Paleatin A	*Marchantia paleaceae* var. *diptera*	Hashimoto *et al.* (1991a)
129	Paleatin B	*Marchantia paleaceae* var. *diptera*	Hashimoto *et al.* (1991a)
130a	Cruciatin	*Lunularia cruciata*	T. Hashimoto and Y. Asakawa (unpublished)
	Pusilatins A, B, C, D	*Blasia pusila*	Hashimoto *et al.* (1991a)
PHENANTHRENES			
135	3-hydroxy-2,5-dimethoxy-phenanthrene	*Marchantia tosana*	Asakawa *et al.* (1984)
136	2-hydroxy-3,6-dimethoxy-phenanthrene	*Marchantia paleaceaa* var. *diptera*	Asakawa *et al.* (1984)

137	2-hydroxy-3,7-dimethoxy-phenanthrene	*Marchantia polymorpha*	Asakawa *et al.* (1987)
138	5-hydroxy-3,4-dimethoxy-9,10-dihydrophenanthrene	*Riccardia jackii*	Asakawa (1982)
139	2-hydroxy-3,4,7-trimethoxy-9,10-dihydrophenanthrene	*Plagiochila spinulosa*	Connolly (1985)
140	3,4,7-trimethoxy-9,10-dihydro-phenanthrene	*Plagiochila spinulosa*	Connolly (1985)
	2,3-dimethoxy-7-hydroxy-phenanthrene	*Marchantia polymorpha*	Adam and Becker (1994)
	2,7-dihydroxy-3-methoxy-phenanthrene	*Marchantia polymorpha*	Adam and Becker (1994)
	3,3′-dimethoxy-2,2′,7,7′-tetra-hydroxy-1,1′-biphenanthrene	*Marchantia polymorpha*	Adam and Becker (1994)
	3-methoxy-2,2′,3′,7,7′-penta-hydroxy-1,1′-biphenanthrene	*Marchantia polymorpha*	Adam and Becker (1994)
	2,2′,3,3′,7,7′-hexahydroxy-1,1′-biphenanthrene	*Marchantia polymorpha*	Adam and Becker (1994)
	jakinenol	*Riccardia jakii*	
PHENYLDIHYDROISOCOUMARIN			
	3-(3,4-dihydroxyphenyl)-8-hydroxyisocoumarin	*Marchantia polymorpha*	Adam and Becker (1994)

* see Gorham (1977a) for further species

(Asakawa *et al.*, 1982) and *R. javanica* (= *R. variabilis*) (Asakawa, Kondo and Tori, 1991). 3,5-Dimethoxybibenzyl (**5**) and 3-hydroxy-4,5-methylenedioxybibenzyl (**14**) have been detected in *Radula javanica* (Asakawa *et al.*, 1982, Asakawa, Kondo and Tori, 1991). 4-Hydroxy-3′-methoxybibenzyl (**7**) is widely distributed in the Jungermanniales (Asakawa, 1982) and has been found in *Plagiochila* (Asakawa and Campbell, 1982; Asakawa and Inoue, 1984b; Lorimer, Perry and Tangney, 1993) and one *Radula* species (Asakawa and Inoue, 1987). 4-Hydroxy-3′-methoxybibenzyl (**7**) and 3,4′-dimethoxybibenzyl (**8**) had already been detected in *Frullania davurica* by gas chromatography–mass spectroscopy (GC–MS) (Asakawa, 1982; Asakawa, Matsuda and Cheminat, 1987). Further investigation of the ether extract of *F. davurica* resulted in isolation of these two compounds (Asakawa, Matsuda and Cheminat, 1987). 2,4′-Dihydroxy-3-methoxybibenzyl (**11**) was isolated from *Frullania brotheri* (Asakawa *et al.*, 1981a). 3,4′-Dimethoxybibenzyl (**8**) has also been found in *Frullania* (Asakawa, Matsuda and Cheminat, 1987), *Gonzylanthus, Leucolejeunea, Riccardia* species (Asakawa, 1982), two *Plagiochila* species (Asakawa and Campbell, 1982; Asakawa and Inoue, 1984b), three *Radula* species (Asakawa *et al.*, 1982; Asakawa and Inoue, 1987) and *Frullania falciloba* (Asakawa, Takikawa and Tori, 1987). A similar bibenzyl, 3-hydroxy-4,3′-dimethoxybibenzyl (**15**) has been isolated from *Frullania falciloba* (Asakawa, Takikawa and Tori, 1987) together with 3′-methoxy-3,4-methylenedioxybibenzyl (**16**) which had previously been detected in four *Frullania* species (Asakawa, 1982). Lunularin (**6**), 3′-methoxy-3,4-methylenedioxybibenzyl (**16**) and 4′-hydroxy-3,3′-dimethoxy-4,5-methylenedioxybibenzyl (**28**) were detected in *Frullania bonincola* (Asakawa, 1982). *F. ericoides* produces a new bibenzyl, 5,3′-dimethoxy-3,4-methylenedioxybibenzyl (**23**), together with **16** (Asakawa, Matsuda and Cheminat, 1987). Brittonins A and B (**31** and **32**) were found in *Frullania brittoniae* subsp. *truncatifolia* (Asakawa, Takikawa and Aratani, 1976). *Pellia endiviifolia* produces 2,3,4′-trihydroxybibenzyl (**10**) (Asakawa, Takikawa and Aratani, 1976), while *Pellia neesiana* biosynthesizes the dimethylated derivative **12** (Benesová and Herout, 1970).

Lunularic acid (**18**), a dormancy-inducing bibenzyl derivative (Valio, Burdon and Schwabe, 1969), is widely distributed in liverworts (Huneck and Pryce, 1971; Gorham, 1977a, b; Pryce, 1971a, b, 1972a, b). This bibenzyl-2-carboxylic acid was also isolated from an axenic culture of *Ricciocarpos natans* (Kunz and Becker, 1992) and suspension cultures of *Marchantia polymorpha, Jungermannia subulata* and *Lophocolea heterophylla* grown in Murasige and Skoog's-2 medium (Abe and Ohta, 1983; Ohta *et al.*, 1984b). Calluses of *M. polymorpha* and *Calypogeia tosana* also produce lunularic acid. Lunularin (**6**) is the decarboxylated derivative of **18** (Pryce, 1972b) and has been isolated from or detected in *Dumortiera hirsuta* (Gottsegen *et al.*, 1988), *M. berteroana* (Asakawa and Campbell, 1982), *Marchantia chenopoda* (M. Tori, Aribiyanti and Y. Asakawa, unpublished results), *M. paleacea* var. *diptera* (Asakawa *et al.*, 1984), *M. polymorpha* (Asakawa *et al.*, 1984, 1990), *Ricciocarpos natans* (Wurzel and Becker, 1990) and a large number of other liverworts (Gorham, 1977a).

3′-Methoxy-3,4-methylenedioxybibenzyl (**17**) was found in *Trocholejeunea sandvicensis* (Asakawa, 1984) and 3-hydroxy-3′-methoxy-4,5-methylenedioxybibenzyl (**22**) was isolated from *Frullania amplicrania* (Asakawa *et al.*, 1981a). Two bibenzyls possessing two methylenedioxy groups have been isolated from *Frullania parvistipula* and their structures established as 3,4;3′,4′-bismethylenedioxybibenzyl (**25**) and 3,3′-dihydroxy-4,5;4′,5′-bismethylenedioxybibenzyl (**30**) on the basis of spectral data and by their total synthesis (Asakawa, Matsuda and Cheminat, 1987). A new bibenzyl, 3,4,5,3′,4′-pentamethoxybibenzyl (**29**) has been isolated from *Frullania serrata* (Asakawa *et al.*, 1991c), together with brittonin B (**32**) (Asakawa, Takikawa and Aratani, 1976). Three bibenzyls, 3,4,4′-trimethoxy-2-methoxycarbonylbibenzyl (**21**), 4′-hydroxy-5-methoxy-3,4-methylenedioxybibenzyl (**20**) and 3,4-dihydroxy-5,3′-dimethoxybibenzyl (**24**) have been obtained from *Plagiochila spinulosa* (Rycroft, 1990), *P. chacabucensis* (Asakawa and Inoue, 1984b) and *Schistochila balfourii* var. *bekkei* (Gradstein, Matsuda and Asakawa, 1985), respectively. These three compounds, **30–32**, are the most highly substituted in this group.

The methanol extract of an axenic culture of *Ricciocarpos natans* was fractionated to give three new bibenzyl glycosides: 5,4′-dihydroxybibenzyl-2-*O*-β-D-glucopyranoside (**19**), 5,3′,4′-trihydroxybibenzyl-2-*O*-β-D-glucopyranoside (**26**) and 2-carboxy-3,4′-dihydroxybibenzyl-3′-*O*-β-D-glucopyranoside (**27**) (Kunz and Becker, 1992), together with two previously known glycosides, salidoroside, β-(3,4-dihydroxyphenyl)ethyl-*O*-β-D-glucopyranoside, lunularic acid (**18**) and a cyclic bis(bibenzyl), riccardin C (**111**). The substitution pattern of the two benzene rings and the location of the glucose in the three glucosides (**19**, **26**, **27**) were confirmed by a combination of NOE and long-range ^{13}C–^{1}H correlation spectroscopy (COSY). This is the first report of bibenzyl glycosides in bryophytes, although their aglycones are widely distributed in liverworts.

2.2.2 Group B (prenyl or geranyl bibenzyls)

Radula species are also rich sources of bibenzyls with prenyl or geranyl groups. Twelve such bibenzyls from three *Radula* species were listed in the review by Asakawa (1982). In addition to 3-hydroxy-5-methoxy-4-(3-methyl-2-butenyl)bibenzyl (**36**) and 2-geranyl-3,5-dihydroxybibenzyl (**51**) (corrected from the previously assigned structure 4-geranyl-3,5-dihydroxybibenzyl (**52**), see above), five new bibenzyls, 3-hydroxy-4-(3-methyl-2-butenyl)bibenzyl (**33**), 3,4′-dihydroxy-4-(3-methyl-2-butenyl) bibenzyl (**38**), 3-hydroxy-4′-methoxy-4-(3-methyl-2-butenyl)bibenzyl (**39**), 4′-hydroxy-3-methoxy-4-(3-methyl-2-butenyl)bibenzyl (**40**) and 2-geranyl-3,5-dihydroxybibenzyl (**51**) have been detected in *Radula oyamensis* (Asakawa *et al.*, 1982). *Radula buccinifera* also produces the previously known **1**, **8**, **34**, radulanin A (**71**) and radulanin C (**73**) (Asakawa *et al.*, 1982). Thirteen new bibenzyl derivatives (**40–45**, **51**, **57**, **61**, **62**, **64**, **66** and **67**) have been isolated from *Radula kojana*, together with two bis(bibenzyl)s, perrottetin E (**122**) and perrottetin F (**123**) (Asakawa *et al.*, 1991b). The 2-prenylbibenzyl **41** has been detected in *R. frondescens* and *R. voluta* (Asakawa and Inoue, 1987). The carboxylic acid (**37**) was isolated as the methyl ester dimethyl ether for convenience of separation from *Radula complanata* (Asakawa *et al.*, 1978).

The prenylated bibenzyls found in *Radula* species are very similar to those isolated from *Cannabis sativa* (Crombie, Crombie and Firth, 1988), who reported that the structure of the presumed **52** from *Radula variablis* (= *R. javanica*), also isolated from *R. complanata*, *R. oyamensis* and *R. tokiensis* (Asakawa *et al.*, 1982), should be revised to **51** on the basis of the ^{1}H NMR spectra of synthetic **52** (*p*-cannabigerol) and **51** (*o*-cannabigerol) (Crombie, Crombie and Firth, 1988). The 4-geranyl derivative **52** was identical with that isolated from *Helichrysum umbraculigerum* (Compositae) (Bohlmann and Hoffmann, 1979). 2-Geranyl-3,5-dihydroxybibenzyl (**51**) has been detected in *R. frondescens*, and *R. voluta* (Asakawa and Inoue, 1987). 2-Geranyl-3,5,4′-trihydroxybibenzyl (**53**) has been detected in *R. frondescens* (Asakawa and Inoue, 1987), while 4-geranyl-3,5-dihydroxybibenzyl (= *p*-cannabigerol) (**52**) has been found in *Radula constricta* (Takikawa, Tori and Asakawa, 1989), *R. tokiensis* (Asakawa *et al.*, 1981a) and *R. javanica* (Asakawa *et al.*, 1978b). The 4′-hydroxylated derivative **54** was isolated from *R. javanica* (Asakawa *et al.*, 1978b), and the double-bond isomer **50** was detected by GC–MS in the crude extracts of *R. oyamensis* (Asakawa *et al.*, 1982).

Asakawa (1982) reported that five prenyl bibenzyls, 3,5-dihydroxy-4-(3-methyl-2-butenyl)bibenzyl (**34**), perrottetins A–D (**46′**, **47′**, **48′**, **56′**) and a prenyldihydrochalcone (**49′**) had been isolated from *R. perrottetii* and their structures determined based on 60-MHz NMR spectrometry. However, these structures were revised to **41**, **46**, **47**, **48**, **56** and **49** by high-resolution NMR spectrometry and synthesis of the corresponding methyl ethers of the natural bibenzyls (Asakawa *et al.*, 1991a). The structure of perrottetin A was revised to **46** because of the NOE difference spectrum of the corresponding trimethyl ether and the total synthesis of **46**. Structures **47′** and **48′** originally assigned to perrottetin B and C were revised to **47** and **48** because of chemical transformations of **47** into the trimethyl ether of **46**.

The compound originally assigned structure **49′** from *R. perrottetii* (Asakawa *et al.*, 1982) was later revised to **49** as a result of NOE experiments and chemical transformations. The presence of a conjugated methoxycarbonyl group (1650 cm^{-1}; δ 171.9, s, 52.2 q) was confirmed by infra red (IR) and ^{13}C NMR spectra. The NOE between H-1″ and H-1′ of **49** was observed but no NOE between the OMe group and any other proton was detected. On the other hand, the dimethyl ether of **49** exhibited NOEs between H-1″ and H-1′ as well as H-4 and both OMe-3 and OMe-5. These data, the chemical shift of the C-4 resonating at high field (δ 102.3 in **49**) and the similarity of the NMR spectrum of **49** to that of **41**, forced revision to 3,5-dihydroxy-2-methoxycarbonyl-6-(3-methyl-2-butenyl)bibenzyl (**49**).

An earlier review by Asakawa (1982) also mentioned the isolation of what was thought to be 3,5-dihydroxy-4-(3-methyl-2-butenyl)bibenzyl (**34**) and its monomethyl ether **36** from European *Radula complanata* (Asakawa *et al.*, 1982). Further fractionation of the crude extract of *R. complanata* yielded 3,5-dihydroxy-2-(3-methyl-2-butenyl)bibenzyl (**41**) and a new bibenzyl, radulanin L (**80**) as well as the epoxide **35** (Asakawa *et al.*, 1991a). The retention times of **41** and **34** in the gas chromatograph (GC) were the same and the mass spectra were almost identical. The chemical shifts of H-2 and H-6, C-2 and C-6, and C-3 and C-5 in **34** are the same, since it is symmetrical at C-2. However, the corresponding protons and carbons in **41** show different chemical shifts (Asakawa *et al.*, 1991b; Crombie, Crombie and Firth, 1988). Attempted demethylation of synthetic **36** and the dimethyl ether of **34** using boron tribromide (BBr_3) or sodium thioethoxide (EtSNa) in HAMPA afforded only chroman and the deprenylated products in place of **34** (Asakawa *et al.*, 1991a).

2.2.3 Group C (with one more heterocyclic ring)

On the basis of the ^{1}H NMR spectrum, structure **56a′** was originally proposed for perrottetin D from *R. perrottetii* (Asakawa *et al.*, 1982) and *R. voluta* (Asakawa and Inoue, 1987). The NOEs between H-1′ and H-5, and H-5 and OMe-4 were detected in its dimethyl ether **56b′**. Although the alternative structure **56a′** was not excluded, new results came from its total synthesis (Asakawa *et al.*, 1991a). The absolute configuration at C-2 was established as (*R*), based on the positive Cotton effects exhibited by **56**, its dimethyl ether and its dihydro-derivative (Asakawa *et al.*, 1991a). The structure of **65** was suggested by the spectroscopic data of the methyl ether of **64**. This was confirmed by preparation from 3-hydroxy-5-methoxy-4-(3-methyl-2-butenyl)bibenzyl (**36**). The structure assigned to **65** was supported by NOE difference spectra of its methyl ester methyl ether and comparison of the spectral data with **64** and its methyl ether.

The absolute configuration of C-2 in the new bibenzyls, **66** and **67** from *R. kojana*, was shown to be (*S*) by the presence of the positive Cotton effects at 262 nm ($\Delta\epsilon$ +0.6) and 250.5 nm ($\Delta\epsilon$ +1.3), respectively (Asakawa *et al.*, 1991b). Crombie, Crombie and Firth (1988) have synthesized (±)-*o*-cannabichromene (**66**) and its *p*-isomer by base-catalyzed condensation of 3,5-dihydroxybibenzyl with citral. The spectral data of **66** from the liverwort was identical with those of synthetic *o*-cannabichromene.

Two new bibenzyls 2(*R*)-6-hydroxy-2-isopropenyl-4-(2-phenylethyl)-2, 3-dihydrobenzofuran (**55**) and 7,8-dihydroxy-2,2-dimethyl-5-(2-phenylethyl)chromene (**63**) have been isolated from *Radula perrottetii*, along with the previously known bibenzyls (**41**, **49**, **56** and **61**) (Asakawa *et al.*, 1991a) and the bis(bibenzyl)s, perrottetins E (**122**), F (**123**) and G (**124**). The absolute configuration of **55** at C-2 was (*R*) because of the positive Cotton effect of its methyl ether (Asakawa *et al.*, 1991a). Compound **55** is presumably derived from the coexisting prenyl bibenzyl **41**. The deisopropenylated derivative of **55** having a benzofuran moiety, **57**, has been isolated from *R. kojana* (Asakawa *et al.*, 1991b).

Tylimanthus urvilleanus produces similar dihydrobenzofuran derivatives; tylimanthins A (**58**), B (**59**) and C (**60**), based on GC–MS data (Asakawa and Inoue, 1984a).

Radulanin A (**71**) has been isolated from *R. javanica* (Asakawa, Kondo and Tori, 1991) and *R. complanata* (Asakawa, 1982; Asakawa *et al.*, 1991a). It was also detected in *R. voluta* (Asakawa and Inoue, 1987). Radulanins B (**72**), C (**73**), D (**74**), E (**75**), F (**76**) and G (**77**) were all isolated from *Radula* species and characterized as bibenzyl derivatives having seven-membered rings with oxygen functions (Asakawa, 1982). *R. complanata* also produces radulanin H (**78**) and 3,5-dihydroxy-4-(2,3-epoxy-3-methylbutyl)bibenzyl (**35**) (Asakawa *et al.*, 1982). Confirmation for structure **71** was provided by the ^{1}H and ^{13}C NMR spectral data and a two-dimensional COSY as well as the NOE difference spectra of its methyl ether. The structure assigned to **78** was supported by the NOE difference spectra of its methyl ester methyl ether, the presence of a hydrogen-bonded carboxyl group (IR, 1650 cm^{-1}) and similarity of the NMR spectra of the related compounds. The structure of radulanin L (**80**) isolated from *R. complanata* has been established by an extensive NMR study (Asakawa, Kondo and Tori, 1991). The presence of a 1,2-disubstituted benzene ring and the locations of the functional groups in the two benzene rings were evident from the coupling patterns in the ^{1}H NMR spectra and the NOE difference spectra of **80** and its dimethyl ether (Asakawa *et al.*, 1991a).

As described earlier, *Radula javanica* (= *R. variabilis*) produces several bibenzyls possessing a dihydrooxepin skeleton (Asakawa, 1982). Further fractionation of the methanol extract of *R. javanica* has yielded three unusual new cyclopropanochroman derivatives, radulanin I (**68**), J (**69**) and K (**70**), together with the previously known radulanins A (**71**) and H (**78**) and 2-geranyl-3,5-dihydroxybibenzyl (**51**) (Asakawa *et al.*, 1991b). The presence of 3-methoxybibenzyl (**1**) and 3,5-dimethoxybibenzyl (**5**) was also detected by GC–MS (Asakawa *et al.*, 1991b). The presence of a

1,1,2-trisubstituted cyclopropane ring and two protons on the carbon bearing the ether oxygen of **68** and **69** was deduced from the ^{1}H and ^{13}C NMR spectra. Methylation of **68** gave a methyl ether that was identical to **69**. The substitution pattern and the position of each functional group on the benzene ring of **69** was established by the NOE difference spectra. Comparison of the spectral data of **68** and radulanin K (**70**) showed that the compounds differed only with respect to the carboxylic group, leading to two possible structures for radulanin K. The structure **70** was favoured, as the closely related compounds **68** and **69** occur in the same species. Radulanin J (**69**) and radulanin K (**70**) are optically active, with positive optical rotations and negative Cotton effects (230 nm in **69** and 263 nm in **70**). On the other hand, radulanin I (**68**) and its methyl ether exhibited negative optical rotations and positive Cotton effects (237 nm in **68** and 230 nm in **69**). The absolute value of the optical rotations and the amplitudes of the circular dichroism (CD) curves of **68** and **69** were smaller than those of radulanin J (**69**) which suggested that **68** might be a mixture of enantiomers in a different ratio. However, the ^{1}H NMR spectrum of **68** exhibited the signals of only one enantiomer in the presence of a chiral shift reagent, therefore the difference in optical rotations and the amplitudes of the CD curves had to be ascribed to a small impurity in **68**. The presence of both (*R*) and (*S*) cyclopropanochroman derivatives in *R. javanica* is noteworthy and also represents the first isolation of a cyclopropanochroman from natural sources.

Phthalides are also included in this group and will be discussed briefly. Radulanolide (**81**) was isolated from *R. complanata* as a crystal being the only compound having one or more seven-membered ring as well as a five-membered lactone (Asakawa *et al.*, 1982). The structure was deduced by a combination of the IR, MS, ^{1}H NMR (decoupling and NOE experiments) and ^{13}C NMR spectra. Balantiolide (**82**) was isolated from *Balantiopsis rosea* and its structure established by spectroscopic data (Asakawa *et al.*, 1986). Methylated phthalide (**83**) has been identified as one of the constituents of *Frullania falciloba* and determined as 5,6-dimethoxy-3-(4′-methoxyphenylmethyl)phthalide on the basis of two-dimensional NMR techniques (Asakawa, Takikawa and Tori, 1987). Two phthalides, **84** and **85**, were found in *Trocholejeunea sandvicensis* (Asakawa, 1984). The absolute configurations of these phthalides have not yet been determined.

2.2.4 Group D (others)

A labile aromatic compound, **86**, named prelunularic acid, possesses a 'prearomatic' structure and has been isolated from the ethanol extract of suspension-cultured cells of *Marchantia polymorpha* (Ohta *et al.*, 1983, 1984a, b; Ohta, Katoh and Takeda, 1990). The ultraviolet (UV) spectrum of **86** at pH 12 indicated time-dependent changes with a clear set of isosbestic points at 247, 284 and 307 nm, while the final spectrum was identical with that of lunularic acid (**18**) (wavelength maxima 238 nm ($\log_{\epsilon}$ 4.25) and 293 nm (3.71)) after 160 min at the same pH, suggesting that **86** was directly converted to lunularic acid (**18**) under basic conditions. The absolute configuration of **86** was established to be (*S*) by application of the dibenzoate chirality rule to the bis(p-dimethylamino)benzoate derived from **86** through reduction and acylation. Prelunularic acid is also present in intact *M. polymorpha*. Compound **86** is the first example of an intermediate possessing a 'prearomatic' structure in the phenylpropanoid–polymalonate biosynthetic pathway.

An Ecuadorian liverwort, *Plagiochila longispina*, produces an unusual aromatic compound whose structure was established as 4-(*p*-methoxyphenylethyl)cyclohex-2-en-1-one (**87**) by a combination of UV (237 nm), IR (1660, 1640 cm^{-1}), MS (*m/z* 121, 109, 95), ^{1}H and ^{13}C NMR spectrometry (Siegel *et al.*, 1991). The C-4 position represents a stereogenic centre. However, no optical activity was observed, indicating that **87** must be racemate.

Two very unusual hybrids, named spinuloplagin A and B (**88** and **89**), between a bibenzyl and a fusicoccane diterpenoid, have been isolated from Scottish *Plagiochila spinulosa* and their structures established by X-ray analyses (Rycroft, 1990). The ^{1}H NMR spectrum of **88** showed very broad signals and the two-dimensional NMR spectra gave only limited information. The spectra of **89** were very similar to those of **88**, but the differences in the ^{13}C NMR signals were found around the aromatic ether substituents ($\Delta\delta$ 0.5–3.2 ppm). These substances were found together with the bibenzyl **21** mentioned previously. This is the first example of a hybrid compound between a terpenoid and a bibenzyl in bryophytes or other plants. The absolute configuration is not known.

2.3 BIS(BIBENZYL)S

The first and most abundant bis(bibenzyl) in *Marchantia polymorpha* and other similar liverworts is marchantin A (**90a**). This compound was found in the Hepaticae in the early 1980s by Asakawa's group. Since it does not show an immediate spot on the thin-layer chromatography (TLC) plate on heating with sulphuric acid spraying, although a dark spot was detected under the UV light, the existence was almost ignored until a red spot was found the following morning. The quantity of **90a** isolated from the crude extracts was roughly 10%. The ^{1}H NMR spectrum of **90a** showed only a large hill in the 60- or 90-MHz NMR spectrum. The 400-MHz NMR spectrum had enough resolution to analyse the fine splittings due to couplings of the four aromatic rings. Marchantin A and other bis(bibenzyl)s were studied in some detail in Asakawa's laboratories because of its novel structure and interesting biological activities (see Chapter 7).

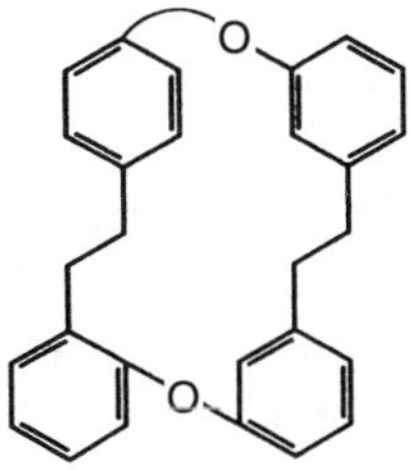

Type A (marchantin)

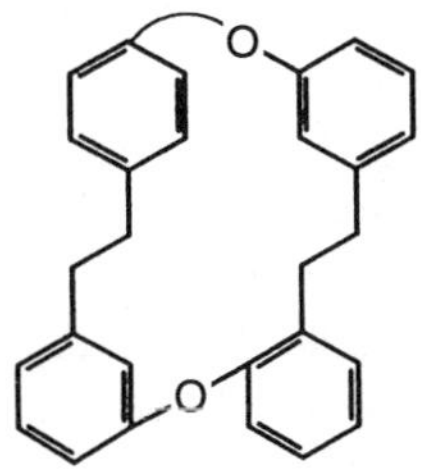

Type B (isomarchantin)

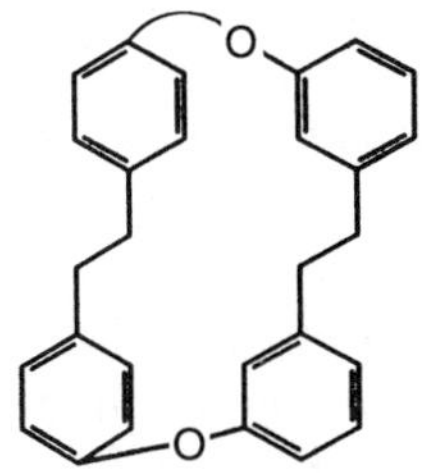

Type C (neomarchantin)

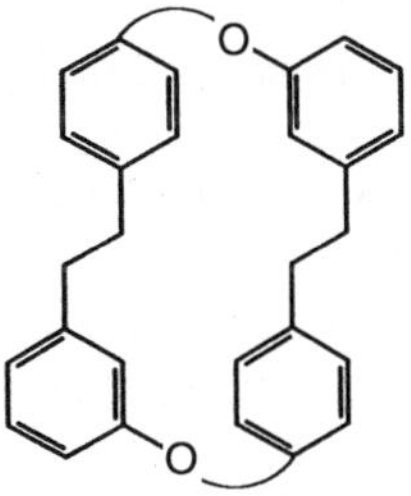

Type D (riccardin B)

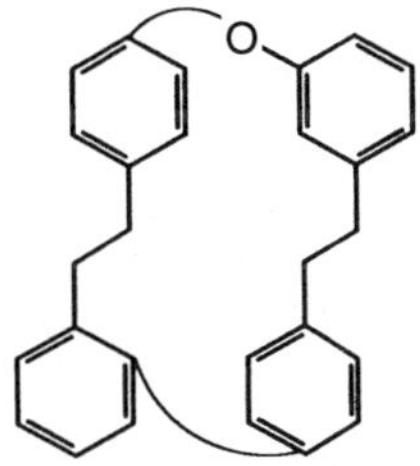

Type E (riccardin)

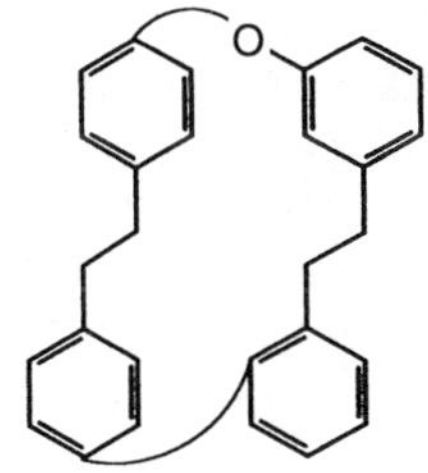

Type F (isoriccardin)

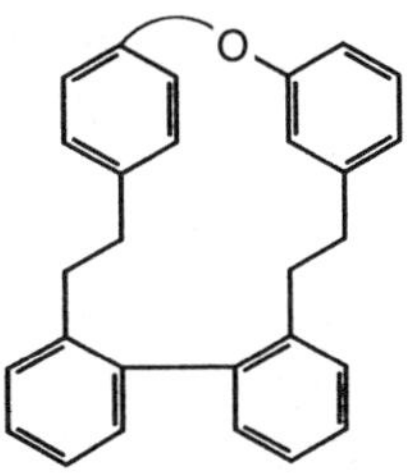

Type G (plagiochin)

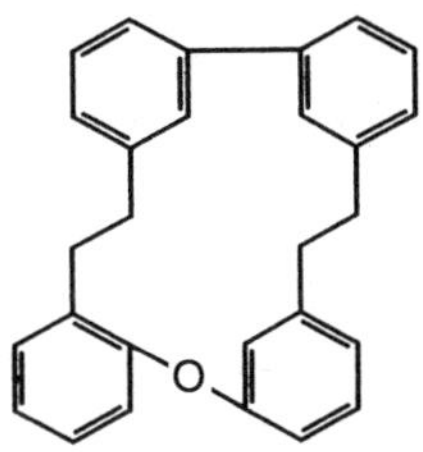

Type H (isoplagiochin)

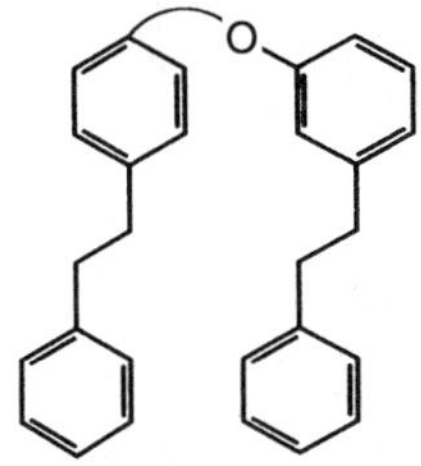

Type I (perrottetin)

Figure 2.1 Nine classes of bis(bibenzyl)s so far isolated from liverworts.

Bis(bibenzyl)s so far reported are classified into nine types on the basis of different modes of cyclization (Figure 2.1).

2.3.1 Type A

The Hepaticae are rich sources of bis(bibenzyl) derivatives. The first reports of novel bis(bibenzyl) derivatives appeared in 1983: riccardin A

(**110a**), isolated from *Riccardia multifida* (Metzgeriales) (Asakawa *et al.*, 1983a, b) and marchantin A (**90a**), isolated from *Marchantia polymorpha* (Marchantiales) (Asakawa *et al.*, 1983b). They have unusual macrocyclic bis(bibenzyl) structures which have so far been found only in the Hepaticae.

The yield of marchantin A (**90a**) depends on the species. For example, pure marchantin A has been isolated in 100–120 g quantities from Japanese *M. paleaceae* var. *diptera* (2 kg of dried material), whereas the total extract of German *M. polymorpha* contained *c.* 20% of marchantin A in weight (Asakawa *et al.*, 1990).

Since the discovery of marchantin A, many congeners have been isolated from the same species and other similar species obtained in different places. *Marchantia polymorpha* collected in Japan or in France produced marchantins B, C, D, E, F and G (**91–96**) (Toyota, 1987). Structures of these marchantins B–G (**91–96**) have been established in a similar manner to that of marchantin A (Chapter 6).

Marchantin A (**90a**) and the related compounds have been isolated from French (Asakawa *et al.*, 1984), German (Asakawa *et al.*, 1990), Indian (Asakawa *et al.*, 1987), Malaysian (A. Tori, Aribiyanti and Y. Asakawa, unpublished results) and South African *M. polymorpha* (Asakawa, Okada and Perold, 1988), Ecuadorian *M. chenopoda* (M. Tori, Aribiyanti and Y. Asakawa, unpublished results), *M. plicata* (Nagashima *et al.*, 1991), Japanese *M. paleacea* var. *diptera* and *M. tosana* (Asakawa *et al.*, 1983a, 1984), as shown in Table 2.2. Marchantins A, B and C were found not only in the thallus (female, male and sterile), but also in the capsule with peduncle of *M. polymorpha* and *M. paleacea* var. *diptera*, while marchantins of *M. tosana* were found only in its sterile thallus (Asakawa *et al.*, 1983a). Marchantins A and B were also isolated from Malaysian *Wiesnerella denudata* belonging to Conocephalaceae (Marchantiales) (Asakawa *et al.*, 1991c), while marchantin C (**92**) has been found in *Dumortiera hirsuta* (Marchantiaceae) (M. Toyota and Y. Asakawa, unpublished results) and *Reboulia hemisphaerica* (H.-C. Wei and C.-L. Wu, personal communication). The last species also produces marchantin C di- (**80**) and monomethyl ether (T. Hashimoto and Y. Asakawa, unpublished results). It is interesting that marchantin C is distributed not only in the Marchantiales and the Monocleales (*Monoclea forsteri* (Toyota, Nagashima and Asakawa, 1988) and *M. gottschei* subsp. *neotropica* (Spörle *et al.*, 1989)), but also in the Jungermanniales, such as *Plagiochila sciophila* (= *P. acanthophylla* subsp. *japonica*) (Hashimoto *et al.*, 1987) and *Schistochila glaucescens* (Tori, Masuya and Asakawa, 1990).

^{1}H- and ^{13}C NMR spectra of marchantins and their derivatives have been completely assigned using NOE difference, ^{13}C–^{1}H COSY and long-range proton selective decoupling (LSPD) spectra, which permitted derivation of the structures of marchantin H (**97**) from *Plagiochasma intermedium* (Marchantiales) and marchantin I (**98**) from *Riccardia multifida* (Metzgeriales) (Tori *et al.*, 1985). The former bis(bibenzyl) has also been isolated from South African *Marchantia polymorpha* (Asakawa, Okada and Perold, 1988), *M. diptera* (Wu, 1990) and *Plagiochila sciophila* (= *P. acanthophylla* subsp. *japonica*) (Hashimoto *et al.*, 1987). Marchantin P has recently been isolated from *Marchantia chenopoda* by Tori, Aoki and Asakawa (1994).

A large collection of *Marchantia polymorpha* in Germany furnished three new bis(bibenzyl)s, marchantins J (**99**), K (**100**) and L (**101**), together with known marchantins A–E (**90a–94**) and G (**96**) (Asakawa *et al.*, 1990). Marchantin J (**99**) constitutes the first record of an ethoxylated compound from bryophytes.

Taiwanese *Mannia subpilosa* produces a cyclic bis(bibenzyl) derivative, marchantinquinone (**104**) as well as marchantin M (**102**) and marchantin N (**103**) (Wei and Wu, 1991). Analysis of ^{1}H and ^{13}C NMR spectra of **104** as well as NOE and spin decoupling experiments showed that **104** was a marchantin-type cyclic bis(bibenzyl) with one phenolic hydroxyl group at C-1′ and a 1,2-disubstituted quinone in ring B. The presence of a quinone was also evidenced by the UV (373 nm) and IR absorption bands (1652 and 1672 cm^{-1}) and by a comparison of the ^{13}C NMR spectrum of **104** with that of 2-methyl-1,4-benzoquinone and 2-methoxy-5-methyl-1, 4-benzoquinone. Further support for the structure was provided by the presence of intense fragment ions at m/z 211 (60%) and 107 (35%) in the mass spectrum corresponding to double benzylic cleavage and the stable *p*-hydroxytropylium ion. Within the family Aytoniaceae (= Grimaldiaceae) of the Marchantiales, *Mannia, Plagiochasma* and *Reboulia* species produce common macrocyclic bis(bibenzyl)s. Synthetic studies of marchantin A (**90a**) (Kodama *et al.*, 1988), B (**91**) (Thi Thu Ha *et al.*, 1991), H (**97**) (Thi Thu Ha *et al.*, 1991) and I (98) (Dienes *et al.*, 1989) have been recorded.

2.3.2 Type B

Isomarchantin C (**105**) has been isolated from Indian *Marchantia polymorpha* and *M. palmata* (Asakawa *et al.*, 1987). There is only one example belonging to this class.

2.3.3 Type C

Neomarchantins A (**107**) and B (**108**) were isolated from *Schistochila glaucescens* (Tori, Masuya and Asakawa, 1990). Neomarchantin A (**107**) corresponds to the desmethyl derivative of pakyonol (**106**) (Huneck *et al.*, 1988). Neomarchantin A (**107**) has also been isolated from *Monoclea forsteri* (Toyota, Nagashima and Asakawa, 1988) and *M. gottschei* subsp. *neotropica* (Spörle *el al.*, 1989). Neomarchantin B (**108**) had one more hydroxyl group (at position 11) than **107**.

Mannia fragrans collected in North Korea elaborates not only cuparane-type sesquiterpenoids but also a cyclic bis(bibenzyl), pakyonol (**106**), whose structure was elucidated by analysis of NMR spectral data (Huneck *et al.*, 1988). Unfortunately, **106** was not derivatized into the corresponding dimethyl ether. The direct comparison of **106** with **107** has not yet been realized. The structure of **106** was confirmed by total synthesis (Nógrádi *et al.*, 1990) and conformational studies (Bocskei and Keseru, 1994).

2.3.4 Types D and E

Riccardins are another series of this class of compounds. *Riccardia multifida* produces riccardins A (**110a**), B (**109a**) and C (**111**). Although riccardin B (**109a**) has an ether oxygen in the bridge between rings B and D, other riccardins have biphenyl linkages between rings B and D. Thus riccardin B (**109a**) must be classified as another type, namely class D. The structure of riccardin A (**110a**) was finally established by X-ray crystallography (Asakawa *et al.*, 1983b). Riccardin A (**110a**) contains an 18-membered ring, whose individual aromatic rings are connected by an ether oxygen between rings A and C and by a biphenyl bond between rings B and D. In the ^{1}H NMR spectrum a strongly shielded one-proton doublet at δ 5.33 has been assigned to an inner proton (H-3′) on ring C, which is in between rings A and D. This phenomenon is commonly observed in these series of compounds, and is a good marker for identifying the macrocyclic bis(bibenzyl)s.

The IR and ^{1}H NMR spectra of riccardin B (**109a**), its dimethoxy derivative (**109b**) and its diacetate (**109c**) resembled those of **110a**, **110b** and **110c**, indicating that riccardin B might be a desmethoxy derivative of **110a** with an additional ether linkage between C-12′ and C-13 in place of the biphenyl bond. The structures assigned to riccardins A (110a) (Gottsegen *et al.*, 1990) and B (**109a**) (Gottsegen *et al.*, 1990; Iyoda *et al.*, 1985; Kodama *et al.*, 1988; Nógrádi, Vermes and Kajtár-Peredy, 1987; Shiobara *et al.*, 1985) have been confirmed by total synthesis.

Riccardin C (**111**), first isolated from *Reboulia hemisphaerica* (Asakawa and Matsuda, 1982) has also been obtained from South African and Indian *Marchantia polymorpha* (Asakawa, Okada and Perold, 1988; Asakawa *et al.*, 1987) and Indian *M. palmata* (Asakawa *et al.*, 1987) and New Zealand *Monoclea forsteri* (Toyota, Nagashima and Asakawa, 1988) and *Ricciocarpos natans* (Kunz and Becker, 1992). Methylation of **111** gave a trimethyl ether identical to riccardin A dimethyl ether **110b**. The total synthesis of **111** has been achieved (Gottsegen *et al.*, 1988).

Two riccardin-type bis(bibenzyl) derivatives, riccardins D (**112**) and E (**113**) have been isolated from New Zealand thalloid liverwort *Monoclea forsteri,* along with riccardin C (**111**) and perrottetin E (**122**) (Toyota, Nagashima and Asakawa, 1988). Riccardin F (**114**) has been found in

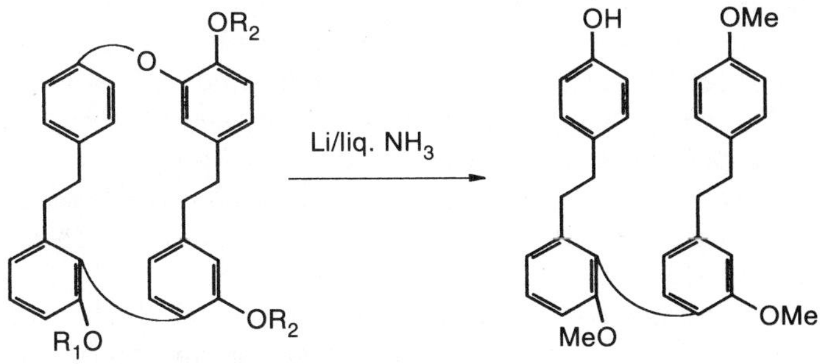

133

Riccardin D (**112**), $R_1=R_2=H$
Riccardin E (**113**), $R_1=CH_3$, $R_2=H$
132 $R_1=R_2=CH_3$

Figure 2.2 Ring opening of cyclic bis(bibenzyl)s.

Marchantia tosana (Asakawa *et al.*, 1984; Asakawa, 1988). Birch reduction of trimethyl ether **132** gave the noncyclic bis(bibenzyl) derivative (**133**), which contained a *p*-methoxybenzyl and a *p*-hydroxyl group (Figure 2.2). The location of these three methoxyls and the single hydroxyl group was established by spin decoupling and NOE experiments (Toyota, Nagashima and Asakawa, 1988). *Blasia pusilla* also produces riccardin F (**114**) (T. Hashimoto and Y. Asakawa, unpublished results). Riccardin G was recently described from *Marchantia chenopoda* (Tori, Aoki and Asakawa, 1994).

Although there was some inconsistency in the names of these compounds (Asakawa, 1988), this is the correct nomenclature and has appeared in the literature (Toyota, Nagashima and Asakawa, 1988).

2.3.5 Type F

Isoriccardin C (**115**) was isolated from the Indian liverworts *Marchantia polymorpha* and *M. palmata* together with isomarchantin C (**105**) (Asakawa *et al.*, 1987). The structure was established by two-dimensional NMR analysis.

2.3.6 Type G

Plagiochila sciophila (= *P. acanthophylla* subsp. *japonica*) biosynthesizes not only the previously known marchantin class bis(benzyl)s marchantins C (**92**), H (**97**) and the perrottetin class, perrottetin E (**122**), but also the new cyclic bis(bibenzyl)s, plagiochins A–D (**116–119**), which possess an ortho biphenyl linkage to the benzyl group (Hashimoto *et al.*, 1987). The structure and conformation of **116** were established by NMR spectroscopy and by an X-ray crystallographic analysis of the methyl ether. This showed that ring A is perpendicular to ring C and parallel with ring D. The proton at C-3′ is strongly shielded by both ring A and ring D, causing a high field shift to δ 4.83 in **116**. Total synthesis of **116**, **117**, **118** and **119** has been achieved (Kametler *et al.*, 1992; Keserú *et al.*, 1992; M. Kodama, personal communication), and high-resolution ^{1}H NMR spectroscopy (Kajtar-Peredy, Keserú and Nógrádi, 1993) of an intermediate in the synthesis of plagiochin C (**118**) confirmed the conformation deduced for plagiochin A (**116**).

2.3.7 Type H

Isoplagiochins A and B (**120**, **121**) have been isolated from the methanol extract of Japanese *Plagiochila fruticosa* (T. Hashimoto and Y. Asakawa, unpublished results). Isoplagiochin A (**120**) showed two more sp^{2} carbons than any other ordinary bis(bibenzyl)s, suggesting the presence of a stilbene moiety in the structure. This is supported by the only two ^{13}C signals owing to benzylic carbons. Hydrogenation of trimethyl ether of **120** gave a dihydroderivative **134**. The structure and conformation have finally been established by X-ray crystallography as its triacetate (T. Hashimoto and Y. Asakawa, unpublished results). Isoplagiochin B (**121**) is a hydroxylated derivative of **120** at the C-10 position.

2.3.8 Type I

Radula species are rich sources not only of bibenzyls but also of bis(bibenzyl)s. From *R. perrottetii* (Toyota *et al.*, 1985) and *R. kojana* (Asakawa *et al.*, 1991b), two novel bis(bibenzyl)s, perrottetins E (**122**) and F (**123**), have been isolated. Perrottetins A–D (**46–48**, **56**) are bibenzyl compounds discussed in section 2.2. The former species elaborates an additional bis(bibenzyl), perrottetin G (**124**) (Toyota *et al.*, 1985). Perrottetin F (**123**) has also been isolated from *Lunularia cruciata* (Hashimoto *et al.*, 1991b). Structure **122** was deduced by extensive NMR studies and was confirmed by total synthesis (Toyota *et al.*, 1985). The presence of one additional hydroxyl group at C-6′ in perrottetin F (**123**) was deduced by similar NMR studies of its tetramethyl ether and confirmed by total synthesis (Mezey-Vándor *et al.*, 1989). Methylation of perrottetin G (**124**) gave a tetramethyl ether which was identical to perrottetin F tetramethyl ether, indicating that **124** was a monomethyl ether of **123** (Toyota *et al.*, 1985). The position of the methoxyl group was suggested to be C-1′ based on the absence of an NOE on irradiation of the methoxyl group. Conclusive evidence for the structure of **124** was also obtained by its total synthesis (Mezey-Vándor *et al.*, 1989).

Perrottetin E (**122**) has also been isolated from *Jungermannia comata* (Nagashima, Toyota and Asakawa, 1990), *Marchantia polymorpha* (Asakawa *et al.*, 1987), *Monocolea forsteri* (Toyota, Nagashima and Asakawa, 1988), *M. gottschei* subsp. *neotropica* (Spörle *et al.*, 1989), *Nardia subclavata* (M. Toyota and Y. Asakawa, unpublished results) *Pellia endiviifolia* (Hashimoto *et al.*, 1991a), *Pellia epiphylla* (Cullmann, Adam and Becker, 1993) and *Plagiochila sciophila* (= *P. acanthophylla* subsp. *japonica*) (Hashimoto *et al.*, 1987). Two bis(bibenzyl) ethers, perrottetin E 11′-methyl ether (**125**) and 14-hydroxyperrottetin E 11′-methyl ether (**126**) have been isolated from *Pellia endiviifolia* (T. Hashimoto and Y. Asakawa, unpublished results). A bis(bibenzyl), 7′,8′-dehydroperrottetin F (**127**) has been isolated from *Lunularia cruciata* (T. Hashimoto and Y. Asakawa, unpublished results). The structure assigned to **127** is based on spectral data and correlation with perrottetin F (**123**). The position of the double bond at C7′-C8′ was confirmed by NOE spectrometry of the tetramethyl ether. Cullmann, Adam and Becker (1993) have identified 14-hydroxyperrottetin E, 14′-hydroxyperrottetin E and 14,14′-dihydroxyperrottetin E in extracts of *Pellia epiphylla*. Another perrottetin was found in *Jubula japonica* (Toyota and Asakawa, 1993).

These phenolic compounds are of interest because they are the linear analogues of the macrocyclic bis(bibenzyl) ethers which have been found in Jungermanniales, Marchantiales, Metzgeriales and Monocleales and are possible biogenetic precursors of the marchantins, plagiochins and riccardins.

2.3.9 Other bis(bibenzyl)s

Further fractionation of the methanol extract of *Marchantia paleacea* var. *diptera* resulted in the isolation of two bis(bibenzyl)s (**128a** and **129**) (Hashimoto *et al.*, 1991b; T. Hashimoto and Y. Asakawa, unpublished results). The structure of **128a** was established by ^{13}C-^{1}H and long-range ^{13}C-^{1}H COSY techniques and NOE spectrometry of its hexamethyl ether (**128b**) and tetraacetate (**128c**). These spectral data were compared with those of marchantin E pentamethyl ether.

In addition to **123** and **127**, Japanese *L. cruciata* also yielded a highly unusual bis(bibenzyl) dimer, cruciatin (**130a**) (T. Hashimoto and Y. Asakawa, unpublished results). The structure of **130a** was established by analysis of the 600-MHz ^{1}H and ^{13}C NMR spectra as well as by chemical degradation. Methylation and acetylation gave a hexamethyl ether (**130b**) and a hexaacetate (**130c**), respectively, indicating that **130a** possessed six phenolic hydroxyl groups. Cruciatin (**130a**) is a dimer of perrottetins E (**122**) and F (**123**). Further bis(bibenzyl) dimers, pusilatins A, B, C and D, have been found in *Blasia pusilla* (Hashimoto *et al.*, 1994).

The minimum strain energies of all possible cyclic bis(bibenzyl) skeletons derivable from a hypothetical acyclic precursor perrottetin E (**122**) by oxidative cyclization have been calculated using the DTMM and MM2 programs (Keserú and Nógrádi, 1992). It can be seen that riccardin B (**109a**) is of extreme low energy (total energy, 1.2 kcal mol^{-1}; free enthalpy change (δG) -26.3 kcal mol^{-1}), whereas the marchantin and plagiochin series have energies of 10.2 and 35.1 kcal mol^{-1} and δG -17.3 and 5.2 kcal mol^{-1}, respectively. However, as pointed out by the authors, the final conformations produced are unrealistic, due to bent aromatic rings. This is presumably a feature of the programs, although the details of the calculations are not known.

2.4 OTHER STILBENOIDS IN BRYOPHYTES

Phenanthrenes with oxygen functions have been isolated and characterized. The substitution patterns have mainly been determined by NOE experiments on their methyl ethers. *Marchantia tosana* produces compound **135** (Asakawa *et al.*, 1984), while *M. paleaceae* var. *diptera* elaborates **136** (Asakawa *et al.*, 1984). Another phenanthrene (**137**) was isolated from *M. polymorpha* by Asakawa *et al.* (1987). 9,10-Dihydrophenanthrenes have also been isolated from liverworts. Three examples (**138–140**) have been recorded so far (Asakawa, 1982; Connally, 1985). These 9,10-dihydrophenanthrenes are presumably biosynthesized from bibenzyls and further oxidized to phenanthrenes.

In an analysis of *in vitro* cultures of *Marchantia polymorpha*, Adam and Becker (1994) identified 2,3-dimethoxy-7-hydroxyphenanthrene, 2,7-dihydroxy-3-methoxyphenanthrene, 3.3′-dimethoxy-2,2′,7,7′-tetrahydroxy-1,1′-biphenanthrene, 3-methoxy-2,2′,3′,7,7′-pentahydroxy-1, 1′-biphenanthrene, 2,2′,3,3′,7,7′-hexahydroxy-1,1′-biphenanthrene and 3-(3,4-dihydroxyphenyl)-8-hydroxyisocoumarin.

The distribution of different stilbenoids in the Hepaticae is summarized in Table 2.2.

2.5 STILBENOIDS IN FERNS

Two bibenzyl-2-carboxylic acids, notholaenic acid (**141**) and isonotholaenic acid (**142**), have been found in the farinose exudates on the lower surface of fronds of species of *Notholaena*, some of which grow in very arid habitats. Notholaenic acid was first identified in *N. dealbata* and *N. limitanea* (Wollenweber and Favre-Bonvin, 1979), and later in *N. chilensis* and *N. nivea* (Wollenweber, 1982). A subsequent investigation (Wollenweber *et al.*, 1993) showed that the compound in *N. chilensis*, *N. dealbata*, *N. nivea* and *N. limitanea* var. *limitanea* was isonotholaenic

Table 2.3 Summary of stilbenoid distribution in the Bryophyta

Taxa	*Bibenzyls*			*Bisbybenzyls*									*Phenanthrenes*	*Phenyldihydroisocoumarins*
	A + B	*C*	*D*	*A*	*B*	*C*	*D*	*E*	*F*	*G*	*H*	*I*		
HEPATICAE														
Jungermanniidae														
Calobryles	+													
Jungermanniales	+	+	+	+		+				+	+	+	+	
Metzgeriales	+						+	+				+	+	
Marchantiidae														
Monocleales	+			+		+		+				+		
Marchantiales	+			+	+	+		+	+			+	+	+
Sphaerocarpales	+													
Anthocerotae	–													
Musci	–													
Takakiophyta	?													

acid, while notholaenic acid was found in *N. limitanea* var. *mexicana*. One of five accessions of the latter variety also contained isonotholaenic acid. Neither stilbenoid was detected in several species of *Notholaena*, which are now classified as *Argyrochosma* species (*A. delicatula*, *A. incana*, *A. pallens*, *A. palmeri*, *A. peninsularis* and *A. pilifera*).

3

Stilbenoids in Gymnosperms

3.1 INTRODUCTION

Two economically important genera of gymnosperms, *Pinus* and *Picea*, contain a number of simple stilbenes and their glucosides. These compounds are of practical interest because of their effects on the process of producing pulp for the paper industry. Significant amounts of these stilbenes are present in heartwood and bark where they may inhibit the growth of fungi, and their accumulation may be stimulated by fungal or insect attack and by a variety of other perturbations. In pines the occurrence of dihydropinosylvin and its monomethyl ether is of taxonomic significance. Resveratrol and its glucoside, piceid, have also been found in leaves of *Juniperus macropoda* (Sethi *et al.*, 1980), but not, as yet, in other Cupressaceae. A number of stilbenes and stilbene oligomers have been isolated from *Gnetum* and *Welwitschia* species.

3.2 STILBENES AND WOOD PULPING

An essential part of the pulping process is the removal of lignin in the sulphite cook. This process is inhibited by a number of constituents of heartwood including resins and polyhydroxyphenols (Gardner and Hillis, 1962). Resin prevents penetration of the sulphite liquor into the wood and residual resin subsequently affects the production of paper. Polyphenols react with lignin and reduce its solubility (Gardner and Hillis, 1962). This is specifically a problem in the pulping of pine heartwood, though not of the sapwood (Erdtman, 1949a; Hachihama *et al.*, 1952; Hata, 1952a, b; Hata and Sogo, 1952; Kondo and Kitamura, 1956). Although this was originally attributed to the high resin content of the heartwood, it was subsequently found that removal of the resin by extraction with nonpolar solvents (ether or benzene) did not solve the problem. Furthermore, other woods with high resin contents do not present problems in the sulphite cook. Hägglund (1927) and Schwalbe and Ekenstam (1929) found that a further extraction of the pine heartwood with a polar solvent (alcohol or acetone) removed the inhibiting polyphenols. The principal inhibitors were identified by Erdtman (1939a, b) as pinosylvin (3,5-dihydroxystilbene, **143a**) and its monomethyl ether (3-hydroxy-5-methoxystilbene, **143b**), and these compounds have been found in the sulphite pulp from *Pinus densiflora* (Kondo and Tanaka, 1955). Spruce wood is normally digested in the sulphite cook, but digestion was inhibited if the spruce wood was impregnated with the acetone extract of pine heartwood. Impregnation with the ether extract did not inhibit digestion (Erdtman, 1949a). The ether extract contained most of the extractives from the pine heartwood (5 to 20% by weight of the wood) whereas the acetone extract contained only 1 to 2% by weight. Some of the acetone extract was soluble in ether, and it was this fraction that was responsible for the inhibition. Impregnation of spruce wood with the ether-insoluble fraction of the acetone extract did not affect delignification. Thus part of the extractives that were soluble in ether appeared to be occluded by ether-insoluble components and hence were not removed by the original ether extraction. Removal of pinosylvin by solvent extraction is not, however, an economically viable process for pulping pine heartwood. Treatment of the wood with formaldehyde (which would more readily form condensation products with the pinosylvins) has also been suggested (Yahama and Kamishiro, 1951).

In the acid conditions of the sulphite cook the lignin reacts with phenols such as resorcinol, phloroglucinol and pinosylvin to form insoluble polymeric condensation products (Erdtman, 1949a). A two-stage process, in which the first sulphonation is performed at a pH of 4 or above, overcomes this problem. It is based on the presence of two types of reactive groups in lignin, one of which (A) reacts with phenols in acid solution and with sulphite at any pH. The other group (B) does not react with phenols, and reacts with sulphite only in acid solution. In the first stage of the Graham two-stage sulphite process the A groups are sulphonated, but no lignin–phenol condensation products are formed. In the second, acidic, stage the B groups are sulphonated and the lignin hydrolysed. Although spruce wood is normally digested in the acid sulphite cook, it was found that unbarked logs which had been immersed in water for several months (e.g. during transport to the mills) were less digestible. This was attributed to the diffusion of condensed tannins from the bark (which contains a high proportion of extractables) to the wood (Erdtman, 1949a; Adler, 1951). Spruce bark also contains a number of polyhydroxy stilbenes (Table 3.2, below). Richtzenhain and von Hofe (1939) reported the isolation of 3,3′-dimethoxy-4,4′-dihydroxystilbene from the spent liquor from a sulphite cook of sprucewood. Hillis, Nelson and Zadow (1966) suggested that this compound was associated with the discoloration of pulp from *Pinus radiata*. The chemistry of lignin–stilbene interactions has been the subject of several investigations (e.g. by Gellerstedt and Agnemo, 1980; Gellerstedt *et al.*, 1976; Gierer and Pettersson, 1977; Gierer, Pettersson and Smedman, 1972; Gierer, Pettersson and Szabo-Lin, 1974; Gierer *et al.*, 1974).

3.3 STILBENES IN *PINUS* SPECIES

A large number of *Pinus* species have been found to contain pinosylvin and pinosylvin monomethyl ether (Table 3.1). The distribution of dihydropinosylvins is more limited. The genus *Pinus* can be divided into two subgenera, *Haploxylon* (white or soft pines usually having bundles of five needles) and *Diploxylon* (hard pines having two to three needles per bundle). *Haploxylon* pines contain bibenzyls (dihydropinosylvin and its monomethyl ether), flavones and flavanones (chrysin, pinostrobin, strobobanksin, strobopinin and cryptostrobin) and the cyclitol pinitol. Within the *Haploxylon* pines, flavones and flavanones with a methylphloroglucinol nucleus may be characteristic of the *Strobi* group (*P. lambertiana, P. monticola, P. strobus, P. parviflora* and *P. peuce*). *Diploxylon* pines generally appear to lack the dehydrogenating enzymes responsible for the formation of bibenzyls from stilbenes and flavones from flavanones (Lindstedt, 1951).

The average pinosylvin phenol (pinosylvin plus pinosylvin monomethyl ether) content of pine heartwood is about 0.9% with the highest concentrations being found in the peripheral heartwood. Higher concentrations may be found in the branches than in the trunk (Erdtman and Rennerfelt, 1944). The ratio of pinosylvin to its monomethyl ether is in the range 1 : 3 to 1 : 4. Little pinosylvin is found in the sapwood or needles, unless its synthesis has been induced by fungal attack or other perturbations (see above). Some 'chemically elite' trees have been found with a pinosylvin phenol content of >1% and a more even distribution between central and peripheral heartwood, although it has not been established whether this difference is genetically or environmentally determined.

A number of other stilbenes have been found in *Pinus* species. Mahesh and Seshadri (1954) reported finding 4-hydroxystilbene and its methylated derivative in *P. griffithii* (= *P. excelsa*), together with pinosylvin mono- and dimethyl ethers. The occurrence of 4-methoxystilbene in this species was confirmed by Billek and Schimpl (1966). The barks of *P. sibirica* and other pines contain several stilbenes, including resveratrol and pinostilbene and their glucosides and trimethoxyresveratrol (Tyukavkina *et al.*, 1972; Lutskii *et al.*, 1973). *Pinus radiata* bark and roots contain piceatannol and astringin (Hillis and Ishikura, 1969b, Yazaki and Hillis, 1977; Markham and Porter, 1973). Armand pine (*P. armandii* Fr. var. *mastersiana* Hay) and *P. morrisonicola* were found to contain, in addition to *cis*- and *trans*-pinosylvin and dihydropinosylvin derivatives, 3,5-dimethoxystilbene oxide (Fang, Su and Cheng, 1988), although whether this is a natural product or an artifact of extraction is not certain.

Table 3.1 Distribution of stilbenes in *Pinus* species. PMME, pinosylvin monomethyl ether; PDME, pinosylvin dimethyl ether; D, dihydropinosylvin; DMME, dihydropinosylvin monomethyl ether; DDME, dihydropinosylvin dimethyl ether

Species	*Compounds*						*Authors*
	Pinosylvin	*PMME*	*PDME*	*D*	*DMME*	*DDME*	
HAPLOXYLON							
Pinus koraiensis	+	+		+	+		Lindstedt and Misiorny (1951)
P. cembra	+	+		+	+		Lindstedt and Misiorny (1951); Erdtman (1944)
P. albicaulis	+	+		+	+		Lindstedt and Misiorny (1951); Lindstrom and Westfelt (1967)
P. flexilis	?	+		+	+		Lindstedt and Misiorny (1951)
P. ayacahuite	+	+		+	+		Lindstedt and Misiorny (1951)
P. lambertiana	–	–		–	–		Lindstedt and Misiorny (1951)
P. parviflora	+	+		+	+		Lindstedt and Misiorny (1951); Fang, Su and Cheng (1988)
P. peuce	–	–		–	–		Lindstedt and Misiorny (1951)
P. griffithii (excelsa)[a]	+	+	+	–	+		Lindstedt and Misiorny (1951); Mahesh and Seshadri (1954)
P. monticola	+	+		–	+		Lindstedt and Misiorny (1951); Lindstedt (1949f)
P. strobus	+	+		–	+		Lindstedt and Misiorny (1951)
P. cembroides	–	+		–	+		Lindstedt and Misiorny (1951)
P. balfouriana	+	+		+	+		Lindstedt and Misiorny (1951)
P. aristata	?	+		–	+		Lindstedt and Misiorny (1951)

DIPLOXYLON					
P. leiophylla	+	+	–	–	Lindstedt and Misiorny (1951)
P. lumholtzii	+	+	?	+	Lindstedt and Misiorny (1951)
P. canariensis	+	+	–	–	Lindstedt and Misiorny (1951)
P. pinea	+	+	–	–	Lindstedt and Misiorny (1951); Alvarez-Novoa, Erdtman and Lindstedt (1950)
P. resinosa (callus)	+	+	–	–	Lindstedt and Misiorny (1951); Jorgensen (1961); Rudolff and Jorgensen (1963); Jorgensen and Balsillie (1969)
P. massoniana	+	+	–	–	Lindstedt and Misiorny (1951)
P. densiflora	+	+	–	–	Lindstedt and Misiorny (1951); Hachihama and Jodai (1951); Hachihama *et al.* (1952); Hata (1952a,b); Hata and Sogo (1952, 1954); Higuchi, Shimada and Watanabe (1967)
(pulp)	+	+		+	Kondo and Tanaka (1955); Kondo and Kitamura (1955)
P. sylvestris[b]	+	+			Erdtman and Rennerfelt (1944); Lindstedt (1951); Lutskii, Gromova and Tyukavkina (1970); Lutskii *et al.* (1973); Rosemann, Heller and Sandermann (1991); Shostakovskii *et al.* (1969); Westfeld (1966)
P. mugo (montana)	+	+	–	–	Lindstedt and Misiorny (1951); Lindstedt (1949a)
P. nigra var. *poiretiana*	+	+	–	–	Lindstedt and Misiorny (1951); Alvarez-Novoa, Erdtman and Lindstedt (1950)
P. nigra var. *austriaca*	+	+	–	–	Lindstedt and Misiorny (1951)
P. nigra var. *maritima*	+	+			Prior (1976)

Table 3.1 (*Contd*)

Species	*Compounds*						*Authors*
	Pinosylvin	*PMME*	*PDME*	*D*	*DMME*	*DDME*	
P. montezumae	+	+		–	–		Lindstedt and Misiorny (1951)
P. ponderosa	+	+		–	–		Lindstedt and Misiorny (1951); Lindstedt (1949e)
P. jeffreyi	+	+		–	–		Lindstedt and Misiorny (1951); Lindstedt (1949d); Anderson (1956)
P. occidentalis	+	+		–	–		Lindstedt and Misiorny (1951)
P. palustris	?	+		–	–		Lindstedt and Misiorny (1951)
P. caribaea	–	+		–	–		Lindstedt and Misiorny (1951)
P. taeda	+	+		–	–		Lindstedt and Misiorny (1951)
			resveratrol?				Hemingway, McGraw and Barras (1977)
P. glabra	+	+		–	–		Lindstedt and Misiorny (1951)
P. echinata	+	+		–	–		Lindstedt and Misiorny (1951)
P. halepensis	+	+		–	–		Lindstedt and Misiorny (1951); Alvarez-Novoa, Erdtman and Lindstedt (1950)
P. pinaster	+	+		–	–		Lindstedt and Misiorny (1951); Alvarez-Novoa, Erdtman and Lindstedt (1950); Hata (1955)
P. virginiana	?	+		–	–		Lindstedt and Misiorny (1951); Lindstedt (1949h)
P. clausa	+	+		–	–		Lindstedt and Misiorny (1951); Lindstedt (1950)
P. rigida	+	+		–	–		Lindstedt and Misiorny (1951)

P. pungens	+	+		–	–		Lindstedt and Misiorny (1951)
P. banksiana	+	+		–	–		Lindstedt and Misiorny (1951); Rowe, Bower and Wagner (1969); Rudolff and Sato (1963)
P. contorta	+	+					Shrimpton and Watson (1971); Shain (1967)
P. contorta var. *latifolia*	+	+		–	–		Lindstedt and Misiorny (1951); Lindstedt (1949c); Loman (1970a,b)
P. muricata	+	+		–	–		Lindstedt and Misiorny (1951)
P. attenuata	?	+		–	–		Lindstedt and Misiorny (1951)
P. radiata	+	+		–	–		Lindstedt (1949b); Lindstedt and Misiorny (1951); Coutts (1970); Hillis and Inoue (1968)
(roots)[c]							Hillis and Ishikura (1969a)
(pulp)[b]							Hillis, Nelson and Zadow (1966)
(bark)[c]							Shain and Hillis (1973)
P. radiata var. *insignis*	+	+		–	–		Lindstedt and Misiorny (1951)
P. sabiniana	+	+		–	–		Lindstedt and Misiorny (1951)
P. coulteri	+	+		–	–		Lindstedt and Misiorny (1951)
P. montana	+	+					Lindstedt (1949a)
OTHER SPECIES							
P. armandii[d]	–	+	+	+	+	+	Fang, Su and Cheng (1988); Hata and Sogo (1954)
P. morrisonicola[d]	+	+	+	–	–	+	Fang, Su and Cheng (1988)
P. longifolia		+					Ahluwalia, Dass and Mehta (1956)
P. gerardiana		+					Ahluwalia, Dass and Mehta (1956)
P. khasya		+					Ahluwalia, Dass and Mehta (1956)
P. krempfii	+	+					Erdtman, Kimland and Norin (1966)

Table 3.1 (*Contd*)

Species	*Compounds*						*Authors*
	Pinosylvin	*PMME*	*PDME*	*D*	*DMME*	*DDME*	
P. sibirica[b,e]	+	+	+			+	Gromova, Lutskii and Tyukavkina (1977); Gromova, Tyukavkina and Lutskii (1975); Lisina *et al.* (1967a,b); Lutskii *et al.* (1970, 1973); Shostakovskii *et al.* (1969); Tyukavkina *et al.* (1968, 1972)
P. thunbergii							Hata and Sogo (1954)
P. pentaphylla	+	+			+		Sogo and Hata (1953); Hata and Sogo (1954)
P. pumila	+	+			+		Kondo, Ito and Miyashi (1955)
P. dansa		+					Lindstedt (1949g)

[a] Also 4-hydroxystilbene and 4-methoxystilbene (Mahesh and Seshadri, 1954)
[b] Also resveratrol and pinostilbene in the bark (Lutskii *et al.*, 1973)
[c] Also piceatannol (3,3′,4′,5-tetrahydroxystilbene and its glucoside (astringin) (Hillis and Ishikura, 1969a; Markham and Porter, 1973; Yazaki and Hillis, 1977
[d] Also 3,5-dimethoxystilbene oxide
[e] Also resveratrol, pinostilbene, trimethoxyresveratrol, pinostilbenoside and resveratroloside (Tyukavkina *et al.*, 1972).

3.4 STILBENES IN *PICEA* SPECIES

Spruce bark, rather than the heartwood, contains three stilbenes (resveratrol (3,4′,5-trihydroxystilbene), piceatannol (3,3′,4′,5-tetrahydroxystilbene) and isorhapontigenin (3,4′,5-trihydroxy-3′-methoxystilbene)) and their glucosides (piceid, astringin and isorhapontin (Toscano Underwood and Pearce, 1991); Table 3.2). The structure for piceatannol proposed by Grassmann and colleagues (Endres, 1957; Grassman and Endres, 1959; Grassmann *et al.*, 1956, 1957; Grassmann, Endres and Pauckner, 1958) was shown by Cunningham, Haslam and Haworth (1963) to be incorrect. Cunningham, Haslam and Haworth (1963) synthesized piceatannol (3,3′,4,5′-tetrahydroxystilbene = 3,3′,4′,5-tetrahydroxystilbene) from 3,4-dihydroxybenzaldehyde and 3,5-dihydroxyphenylacetic acid. Confirmation of the structure of isorhapontigenin was provided by Andrews *et al.* (1968) who synthesized 3,4′,5-trihydroxy-3′-methoxystilbene from 3,5-dihydroxyphenylacetic acid and vanillin. Resveratrol and piceid have also been found in the leaves of *P. glehnii* (Kariyone *et al.*, 1959a, b, c).

The extent of polyphenol (including stilbene) variation within and between spruce trees (*Picea sitchensis*) from different provenances was examined by Forrest (1975a, b, c). Correlations of stilbene content with provenance and vigour were obtained. Toscano Underwood and Pearce (1991) found that astringin and isorhapontin were present in high concentrations in certain cells, and both the ratio and absolute amounts of these stilbene glucosides varied from cambium to periderm.

3.5 STILBENES IN THE GNETACEAE

The order Gnetales comprises three genera of unusual plants: *Ephedra*, *Gnetum* and *Welwitschia*. The genus *Gnetum* contains more than 30 species of lanky trees or climbers distributed in the tropics from Amazonia to southern China. Superficially they resemble dicots, having broad leaves with net venation. Three stilbenes were isolated from *G. ula* by Prakash and associates (Prakash, Ahmed and Zaman, 1981; Prakash *et al.*, 1983, 1985; Zaman *et al.*, 1983). Neither the structure nor the name of the stilbene reported by Prakash, Ahmed and Zaman (1981) are consistent with structures reported later (Table 3.3). Prakash *et al.* (1985) reported a revised structure for the compound previously identified as 3,3′,4-trihydroxy-2-methoxystilbene. These authors also used the trivial name 'gnetin' for 3,4-methylenedioxy-4′-methoxystilbene from *G. ula*. This name had, however, previously been used for a range of stilbene dimers and trimers (gnetins A, B, C, D and E) isolated from *G. leyboldii* and *G. schwackeanum* (Lins *et al.*, 1982). Two further gnetins, J and K, were found in seeds of *Gnetum venosum* by Boralle *et al.* (1993), together with resveratrol, isorhapontigenin and gnetins C and E. Gnetins J and K are trimers in

Table 3.2 Distribution of stilbenes in *Picea* species. R, resveratrol; PD, piceid (resveratrol glucoside); PT, piceatannol (3,3′,4,5′-tetrahydroxystilbene); A, astringin (piceatannol glucoside); IN, isorhapontigenin (3-methoxy-3′,4,5′-trihydroxystilbene); I, isorhapontin (isorhapontigenin glucoside)

Species	*Compounds*						*Authors*
	R	*PD*	*PT*	*A*	*IN*	*I*	
Picea sp. (bark)		+		+		+	Alcubilla-Martin (1970)
P. sp.				+			Endres and Leppmeier (1961)
P. glauca (bark)			+	+	+	+	Andrews *et al.* (1968); Manners and Swan (1971); Grassmann and Endres (1959)
P. sitchensis (bark)		+	+	+	+	+	Aritomi and Donnelly (1976); Hergert (1960); Manners and Swan (1971); Toscano Underwood and Pearce (1991, 1992); Woodward and Pearce (1988)
P. koraiensis	+	+		+	+	+	Gromova, Lutskii and Tyukavkina (1974, 1977)
P. ojanensis (bark)	+		+	+	+		Gromova, Lutskii and Tyukavkina (1975)
P. bicolor	+	?					Ito (1961)

P. glehnii	+	+	+	+	+	+	Kariyone *et al.* (1958, 1959a,b,c); Sano and Sakakibara (1977)
P. koyamai	+						Kariyone *et al.* (1959c)
P. polita	+						Kariyone *et al.* (1959c)
P. engelmannii (bark)			+	+	+	+	Manners and Swan (1971); Pearson, Kriz and Taylor (1977); Taylor and Shaw (1983)
P. mariana			+	+	+	+	Manners and Swan (1971); Manson (1954)
P. rubens			+	+	+	+	Manners and Swan (1971)
P. abies (bark)	+		+	+	+	+	Mannila and Talvitie (1992); Talvitie, Mannila and Kolehmainen (1992); Weissmann (1981)
P. excelsa	+		+				Rolfs and Kindl (1984); Grassmann and Endres (1959)
P. obovata	+	+	+	+	+	+	Shibanova *et al.* (1977); Shibanova and Pentagava (1970)

Table 3.3 Stilbenes and stilbene oligomers in *Gnetum* and *Welwitschia* species

Species	*Compound*	*Authors*
Gnetum ula	**gnetol**, 3,4,5′-trihydroxy-3′-methoxystilbene	Prakash, Ahmed and Zaman (1981) Prakash *et al.* (1983, 1985)
	3,4-methylenedioxy-4′-methoxystilbene	Zaman *et al.* (1983)
	gnetulin	Siddiqui *et al.* (1993)
G. parvifolium	**gnetifolins A, C, D, E, F, resveratrol**	Lin *et al.* (1991, 1992)
	isorhapontigenin, pinosylvin, gnetol, ϵ-viniferin, isorhapontin	Li *et al.* (1991)
G. leyboldii	**gnetins A, B, C, D, E**	Lins *et al.* (1982)
G. schwackeanum	**gnetins C, E**	Lins *et al.* (1982)
G. venosum	**gnetins C, E, J, K, resveratrol, isorhapontigenin**	Boralle *et al.* (1993)
Welwitschia mirabilis	**gnetins F, G, H, I**	Lins *et al.* (1986)

which one of the units has the 3,3′,4′,5-substitution pattern, the other two being resveratrol units. An isorhapontigenin dimer, gnetulin, has been isolated from *Gnetum ula* (Siddiqui *et al.*, 1993). The gnetifolins, a different set of stilbene oligomers, together with some monomeric stilbenes, were described as constituents of *G. parvifolium* by Lin *et al.* (1991, 1992) and Li *et al.* (1991).

Welwitschia mirabilis is a botanical curiosity. It is a long-lived plant (>1000 years) which grows in the Namib desert of Namibia, but only ever consists of a caudex and long tap root, and two leaves of indeterminate growth. These leaves become worn at the edges, but are renewed by growth at the basal end of the leaf. Two resveratrol dimers (gnetins F and G) and two resveratrol trimers (gnetins H and I) were found in this species by Lins *et al.* (1986), but not the gnetins previously found in *Gnetum* species.

The distribution of resveratrol oligomers in plants has been reviewed by Sotheeswaran and Pasupathy (1993). Although some other secondary products characteristic of angiosperms have been found in the Gnetaceae, Lins *et al.* (1986) suggest that the occurrence of resveratrol oligomers and angiosperm secondary compounds in these gymnosperms is a case of convergent chemical evolution, rather than evidence of a close evolutionary relationship. Boralle *et al.* (1993) have indicated that further, and higher, stilbene oligomers also occur in *Gnetum* species.

4

Stilbenoids in angiosperms

4.1 INTRODUCTION

Angiosperms produce a wider variety of stilbenoids than any other group of higher plants. These range from simple bibenzyls such as batatasin III (**167**) from yams to stilbene oligomers such as hopeaphenol in dipterocarps, biphenanthrenes in orchids and the mixed stilbene-flavan-3,4-diols in *Guibourtia coleosperma*. In most cases the biological roles of these compounds in plants are not clear, although *in vitro* growth regulatory and antibiotic activities have been reported (Chapter 7). The wide distribution of stilbenoids in angiosperm taxa (Figure 1.3) suggests that the ability to synthesize stilbenoids has evolved in parallel in a number of families of seed plants. This is not surprising given the widespread distribution of the flavonoid pathway and the similarity of chalcone synthase and stilbene synthase (Chapter 5).

Three stilbenoids have been reported in grasses, resveratrol phenanthrene in wheat and piceatannol in sugarcane (Table 4.1). The recent finding of stilbenes and stilbene oligomers in *Carex* and *Scirpus* species suggests that more such compounds are waiting to be discovered in monocots. *Juncus* species produce a variety of biologically active phenanthrene and 9,10-dihydrophenanthrene derivatives with methyl substituents which may be synthesized by a different pathway, but which are listed in Table 4.1 pending biosynthetic studies. The earliest reports of stilbenoids in monocots were those describing resveratrol (**144a**) and oxyresveratrol (**204**) from *Veratrum grandiflorum*. Other members of the Liliaceae contain lunularic acid (*Allium* spp.) and 2,2′,4,4′-tetrahydroxystilbene-2-*O*-xylopyranoside (*Chlorophytum arundinaceum*). The former compound is common in liverworts and has been reported from *Hydrangea* (Pryce, 1971a, b; Gorham, 1977a), but its occurrence in *Allium* merits further investigation.

Combretum species synthesize a large number of bibenzyls, stilbenes, phenanthrenes and 9,10-dihydrophenanthrenes (Table 4.2), many of which have interesting biological properties (Chapter 7). A systematic search of other Combretaceae could lead to the discovery of further stilbenoids in this family.

Yams (*Dioscorea* species) also contain several bibenzyls, phenanthrenes and 9,10-dihydrophenanthrenes, notably the plant growth regulatory batatasins (Table 4.3). Stilbenes have not been found in the Dioscoreaceae, and are less common than other stilbenoids in the Combretaceae.

The Dipterocarpaceae (Table 4.4) produce a number of resveratrol oligomers ranging from the dimer balanocarpol (**229**) to the tetramer hopeaphenol (**245**), although resveratrol (**144a**) and other monomers have not been reported from this family.

Bibenzyls and, particularly, stilbenes are characteristic of the Fabaceae (Table 4.5), a family which apparently does not contain phenanthrenes (with the exception of the phenanthraquinone latinone (**291**) of uncertain biosynthetic origin). Prenylated stilbenoids are found in *Amorpha, Arachis, Cajanus, Derris* and *Glycyrrhiza* species. Resveratrol oligomers have been found in *Cassia garrettiana*, and stilbene-flavan-3,4-diol oligomers were found in *Guibourtia coleosperma*. In groundnut (*Arachis hypogaea*) stilbenes function as phytoalexins (Chapter 7).

Another family which contains only bibenzyl and stilbene derivatives, particularly resveratrol (**144a**) and oxyresveratrol (**204**), is the Moraceae (Table 4.6). Lunularin (**6**), a liverwort constituent, has been found in *Morus laevigata*, and a geranyl-tetrahydroxystilbene (chlorophorin, **218b**) and its dimer (excelsaoctaphenol, **257**) occur in *Chlorophora excelsa*.

Resveratrol (**144a**) and its glucoside piceid (**144b**) are common in *Eucalyptus* species (Table 4.7). Variation for polyphenol composition of *Eucalyptus* species was documented by Hillis (1966). Piceid and astringin (piceatannol glucoside, **147b**) have also been found in another member of the Myrtaceae, *Angophora cordifolia*.

Members of the large family Orchidaceae contain bibenzyls (Table 4.8), 9,10-dihydrophenanthrenes (Table 4.9), phenanthrenes (Table 4.10), phenanthropyrans and phenanthraquinones (Table 4.11) and biphenanthrenes (Table 4.12). At present there are no reports of stilbenes in orchids. The 9,10-dihydrophenanthrenes orchinol (**273**), hircinol (**274b**) and loroglossol (**274a**) were originally isolated as phytoalexins in European orchids (Chapter 7). Most of the later reports describe stilbenoids from Asian species, but only a small sample of the estimated 20 000 species, distributed across all continents except Antarctica, have been examined.

The occurrence of rhapontin (**205b**) in rhubarb (*Rheum rhaponticum*) was first reported by Hornemann in 1822 (as rhaponticin), Hesse in 1899 (as rhapontin) and Gilson in 1903 (as ponticin). The identity of the compound was finally established by Tschirch and Christofoletti (1905). Other members of the Polygonaceae contain a variety of stilbenes (Table 4.13), but not other stilbenoids.

Another family in which stilbenoids are restricted to stilbenes, and

Table 4.1 Stilbenoids in monocotyledonous plants, except orchids

Species	*Compounds*	*Authors*
POACEAE		
Triticum aestivum	phenanthrene	Morgante (1976)
Festuca argentina	resveratrol	Casabuono and Pomilio (1994)
Festuca versuta	resveratrol	Powell *et al.* (1994)
Festuca arundinacea	resveratrol	Powell *et al.* (1994)
Stipa robusta	resveratrol	Powell *et al.* (1994)
Poa alsodes	resveratrol	Powell *et al.* (1994)
Lolium spp.	resveratrol	Powell *et al.* (1994)
Hordeum bogdanii	resveratrol	Powell *et al.* (1994)
Hordeum brevisubulatum	resveratrol	Powell *et al.* (1994)
Saccharum sp.	3,3′,4,5′-tetrahydroxystilbene (piceatannol)	Brinker and Seigler (1991, 1993)
CYPERACEAE		
Carex fedia var. *miyabenei*	miyabenol A	Suzuki *et al.* (1987)
Carex kobomugi	ϵ-viniferin	Kurihara *et al.* (1991)
		Kawabata *et al.* (1989)
	miyabenol C	Kurihara *et al.* (1991)
		Kawabata *et al.* (1989)
	kobophenol A	Kurihara *et al.* (1991)
		Kawabata *et al.* (1989)

Carex pumila	kobophenol B	Kawabata *et al.* (1991)
Scirpus fluviatilis	3,4′,5-trihydroxystilbene (resveratrol)	Nakajima *et al.* (1978)
	3,3′,4,5′-tetrahydroxystilbene (piceatannol)	Nakajima *et al.* (1978)
	scirpusin A resveratrol/piceatannol dimer	Nakajima *et al.* (1978)
	scirpusin B piceatannol dimer	Nakajima *et al.* (1978)
Scirpus maritimus	3,4′,5-trihydroxystilbene (resveratrol)	Powell, Bajaj and McLauchlin (1987)
	3,3′,4,5′-tetrahydroxystilbene (piceatannol)	Powell, Bajaj and McLauchlin (1987)
	ϵ-viniferin	Powell, Bajaj and McLauchlin (1987)
	scirpusin A	Powell, Bajaj and McLauchlin (1987)
	scirpusin B	Powell, Bajaj and McLaughlin (1987)
JUNCACEAE		
Juncus roemerianus	5-acetyl-2,6-dihydroxy-1,7-dimethyl-9,10-dihydrophenanthrene (juncunone)	Miles *et al.* (1981)
	1,7-dimethyl-2-hydroxy-5-vinyl-9,10-dihydrophenanthrene (juncunol)	Bhattacharyya and Miles (1977); Sarkar, Zerezghi and Bhattacharyya (1988); Cossey, Gunter and Mander (1980)[a]
	2,7-dihydroxy-1,6-dimethyl-5-vinyl-9,10-dihydrophenanthrene (juncusol)	Kende and Curran (1979); Miles and De al Cruz (1976); Miles *et al.* (1977); Sarkar, Zerezghi and Bhattacharyya (1988)

Table 4.1 *(Contd)*

Species	*Compounds*	*Authors*
Juncus roemerianus	2,7-dihydroxy-1,6-dimethyl-5-vinyl-phenanthrene (dehydrojuncusol)	Sarkar, Zerezghi and Bhattacharyya (1988)
Juncus effusus	2,7-dihydroxy-1,6-dimethyl-5-vinyl-9,10-dihydrophenanthrene (juncusol)	Bhattacharyya (1980); Shima, Toyota and Asakawa (1991)
	2,7-dihydroxy-1-methyl-5-vinyl-9,10-dihydrophenanthrene (effusol)	Bhattacharyya (1980); Shima, Toyota and Asakawa (1991); Mody *et al.* (1982)
	2,7-dihydroxy-1,6-dimethyl-5-vinyl-phenanthrene (dehydrojuncusol)	Shima, Toyota and Asakawa (1991)
	2,7-dihydroxy-1-methyl-5-vinyl-phenanthrene (dehydroeffusol)	Shima, Toyota and Asakawa (1991)
	2,7-dihydroxy-1-methyl-phenanthren-5-al (dehydroeffusal)	Shima, Toyota and Asakawa (1991)
	2,8-dihydroxy-1,7-dimethyl-6-vinyl-10,11-dihydrodibenz [b,f] oxepin	Della Greca *et al.* (1993)

LILIACEAE		
Veratrum grandiflorum	3,4′,5-trihydroxystilbene (resveratrol)	Takaoka (1940a,b); Hanawa, Tahara and Mizutani (1992)
	2,3′,4,5′-tetrahydroxystilbene (oxyresveratrol)	Takaoka (1939, 1940a,b,c,d); Hanawa, Tahara and Mizutani (1992)
	3,4′,5-trihydroxystilbene-3-*O*-D-glucopyranoside (piceid)	Hanawa, Tahara and Mizutani (1992)
	2,3′,4,5′-tetrahydroxystilbene-3′-*O*-D-glucopyranoside	Hanawa, Tahara and Mizutani (1992)
Allium spp.	3,4′-dihydroxybibenzyl-2-carboxylic acid (lunularic acid)	Goda and Sankawa (1985)
Chlorophytum arundinaceum	2,2′,4,4′-tetrahydroxybibenzyl-2-*O*-xylopyranoside	Tandon and Shukla (1993)

[a] Revised structure; originally thought to be the 1,6-dimethyl-7-vinyl isomer.

Table 4.2 Stilbenoids in the Combretaceae

Species	*Compounds*	*Authors*
Combretum apiculatum	3,4′-dihydroxy-5-methoxybibenzyl	Malan and Swinny (1993)
	3,3′-dihydroxy-4′,5-dimethoxybibenzyl	Malan and Swinny (1993)
	3,4′-dihydroxy-3′,4,5-trimethoxybibenzyl	Malan and Swinny (1993)
	3,4-dihydroxy-3,4′,5-trimethoxybibenzyl	Malan and Swinny (1993)
	4,5-dimethoxy-3,3′,4′-trihydroxybibenzyl	Malan and Swinny (1993)
	3,4-dihydroxy-4′,5-dimethoxybibenzyl	Malan and Swinny (1993)
	3,4′-dihydroxy-4,5-dimethoxybibenzyl	Malan and Swinny (1993)
	4,4′-dihydroxy-3,5-dimethoxybibenzyl	Malan and Swinny (1993)
	2,3-dimethoxy-4,6,7-trihydroxy-9,10-dihydrophenanthrene	Malan and Swinny (1993)
	2,7-dihydroxy-3,4,6-trimethoxy-9,10-dihydrophenanthrene	Malan and Swinny (1993)
	7-hydroxy-2,3,4,6-tetramethoxy-9,10-dihydrophenanthrene	Malan and Swinny (1993)
	6,7-dihydroxy-2,3,4-trimethoxy-9,10-dihydrophenanthrene	Malan and Swinny (1993)
	3-hydroxy-2,4,6,7-tetramethoxy-9,10-dihydrophenanthrene	Malan and Swinny (1993)
	4,7-dihydroxy-2,6-dimethoxy-9,10-dihydrophenanthrene	Letcher and Nhamo (1971)
	4,7-dihydroxy-2,3,6-trimethoxy-9,10-dihydrophenanthrene	Letcher and Nhamo (1971)
	2,6-dihydroxy-3,4,7-trimethoxy-9,10-dihydrophenanthrene	Letcher and Nhamo (1971)
	2,6-dihydroxy-3,4,7-trimethoxy-phenanthrene	Letcher and Nhamo (1971)

	2,3-dimethoxy-4,6,7-trihydroxy-phenanthrene	Letcher and Nhamo (1971)
	4,6-dimethoxy-2,3,7-trihydroxy-phenanthrene	Malan and Swinny (1993)
	2,4-dimethoxy-3,6,7-trihydroxy-phenanthrene	Malan and Swinny (1993)
	3,7-dihydroxy-2,4,6-trimethoxy-phenanthrene	Malan and Swinny (1993)
	4,7-dihydroxy-2,3,6-trimethoxy-phenanthrene	Malan and Swinny (1993)
Combretum caffrum	3,4-dihydroxy-3′,5′-dimethoxybibenzyl (combretastatin B-4)	Pettit, Cragg and Singh (1987); Pettit, Smith and Singh (1987); Pettit *et al.* (1987)
	3′,4′-dihydroxy-3,4,5-trimethoxybibenzyl	Pettit, Cragg and Singh (1987); Pettit, Smith and Singh (1987); Pettit *et al.* (1987)
	3′,7-dihydroxy-3,4,4′,5-tetramethoxy-bibenzyl (combretastatin)	Pettit *et al.* (1982); Pettit, Singh and Cragg (1985); Pettit, Cragg and Singh (1987); Pettit, Smith and Singh (1987); Pettit *et al.* (1987)
	2,3-dihydroxy-3′,4,4′,5-tetramethoxy-bibenzyl (combretastatin B-1)	Pettit *et al.* (1982, 1987b); Pettit, Singh and Cragg (1985)
	2,3-dihydroxy-4,5′-dimethoxy-3′,4′-methylenedioxybibenzyl (combretastatin B-2)	Pettit *et al.* (1982); Pettit, Singh and Cragg (1985) Pettit and Singh (1987)
	3′,5-dihydroxy-3,4,4′-trimethoxystilbene (combretastatin A-3)[a]	Pettit and Singh (1987)
	3′-hydroxy-3,4,4′,5-tetramethoxystilbene (combretastatin A-4)	El-Zayat *et al.* (1993)

Table 4.2 (*Contd*)

Species	*Compounds*	*Authors*
	3,4′-dimethoxy-3′-hydroxy-4,5-methylenedioxystilbene (combretastatin A-2)[a]	Pettit and Singh (1987)
	2,3-dihydroxy-3′,4,4′,5′-tetramethoxystilbene (combretastatin A-1)[a]	Pettit *et al.* (1987)
	6,7-dihydroxy-2,3,4-trimethoxy-9,10-dihydrophenanthrene	Pettit *et al.* (1988b)
	2,7-dihydroxy-3,4,6-trimethoxy-9,10-dihydrophenanthrene	Pettit *et al.* (1988b)
	7-hydroxy-2,3,4,6-tetramethoxy-9,10-dihydrophenanthrene	Pettit *et al.* (1988b)
	2-hydroxy-3,4,6,7-tetramethoxy-9,10-dihydrophenanthrene	Pettit *et al.* (1988b)
	7-hydroxy-2,3,4,6-tetramethoxyphenanthrene	Pettit *et al.* (1988b)
Combretum hereroense	2,7-dihydroxy-3,4,6-trimethoxy-9,10-dihydrophenanthrene	Letcher and Nhamo (1973a,b)
	3,6-dihydroxy-2,4,7-trimethoxy-9,10-dihydrophenanthrene	Letcher and Nhamo (1973a,b)
	6,7-dihydroxy-2,4-dimethoxyphenanthrene	Letcher and Nhamo (1973a,b)
	6,7-dihydroxy-2,3,4-trimethoxyphenanthrene	Letcher and Nhamo (1973a,b)
	2,7-dihydroxy-3,4,6-trimethoxyphenanthrene	Letcher and Nhamo (1973a,b)

	3,6-dihydroxy-2,4,7-trimethoxy-phenanthrene	Letcher and Nhamo (1973a,b)
Combretum molle	3,4′-dihydroxy-4,5-dimethoxybibenzyl	Letcher, Nhamo and Gumiro (1972)
	4,7-dihydroxy-2,6-dimethoxy-9,10-dihydrophenanthrene	Letcher, Nhamo and Gumiro (1970)
	2,3-dimethoxy-4,6,7-trihydroxy-9,10-dihydrophenanthrene	Letcher, Nhamo and Gumiro (1972)
	2,6-dihydroxy-3,4,7-trimethoxy-9,10-dihydrophenanthrene	Letcher, Nhamo and Gumiro (1972)
	6,7-dihydroxy-2,3,4-trimethoxy-9,10-dihydrophenanthrene	Letcher, Nhamo and Gumiro (1972)
	4,7-dihydroxy-2,3,6-trimethoxy-phenanthrene	Letcher, Nhamo and Gumiro (1972)
	2,3-dimethoxy-4,6,7-trihydroxy-phenanthrene	Letcher, Nhamo and Gumiro (1972)
	6,7-dihydroxy-2,3,4-trimethoxy-phenanthrene	Letcher, Nhamo and Gumiro (1972)
Combretum psidioides	4,4′-dihydroxy-3,5-dimethoxybibenzyl	Letcher and Nhamo (1972a)
	3,4′-dihydroxy-4,5-dimethoxybibenzyl	Letcher and Nhamo (1972a)
	4′-hydroxy-3,4,5-trimethoxybibenzyl	Letcher and Nhamo (1972a)
	4,7-dihydroxy-2,6-dimethoxy-9,10-dihydrophenanthrene	Letcher and Nhamo (1972a)
	6,7-dihydroxy-2,3,4-trimethoxy-9,10-dihydrophenanthrene	Letcher and Nhamo (1972a)
	4,7-dihydroxy-2,3,6-trimethoxy-9,10-dihydrophenanthrene	Letcher and Nhamo (1972a)
	7-hydroxy-2,3,4,6-tetramethoxy-9,10-dihydrophenanthrene	Letcher and Nhamo (1972a)

Table 4.2 (*Contd*)

Species	*Compounds*	*Authors*
	7-hydroxy-2,4,6-trimethoxy-9,10-dihydrophenanthrene	Letcher and Nhamo (1972a)
	2,3-dimethoxy-4,6,7-trihydroxy-9,10-dihydrophenanthrene	Letcher and Nhamo (1972a)
	2,7-dihydroxy-3,4,6-trimethoxy-9,10-dihydrophenanthrene	Letcher and Nhamo (1972a)
	3,4-dimethoxy-2,6,7-trihydroxy-9,10-dihydrophenanthrene	Letcher and Nhamo (1972a)
	2,3-dimethoxy-4,6,7-trihydroxy-phenanthrene	Letcher and Nhamo (1972a)
	6,7-dihydroxy-2,3,4-trimethoxy-phenanthrene	Letcher and Nhamo (1972a)
	7-hydroxy-2,4,6-trimethoxyphenanthrene (**Ta VI**)	Letcher and Nhamo (1972a)
	7-hydroxy-2,3,4,6-tetramethoxy-phenanthrene	Letcher and Nhamo (1972a)
	2,7-dihydroxy-3,4,6-trimethoxy-phenanthrene	Letcher and Nhamo (1972a)
	3,4-dimethoxy-2,6,7-trihydroxy-phenanthrene	Letcher and Nhamo (1972a)
	2,3-dimethoxy-4,6,7-trihydroxy-phenanthrene	Letcher and Nhamo (1972a)
Terminalia sericea	3,4′,5-trihydroxystilbene-*O*-rutinoside	Bombardelli, Martinelli and Mustich (1975)

[a] *cis*-configuration

Table 4.3 Bibenzyls and phenanthrenes in the Dioscoreaceae

Species	*Compounds*	*Authors*
Dioscorea		
D. alata	6-hydroxy-2,4,7-trimethoxyphenanthrene (batatasin I)	Ireland, Schwabe and Coursey (1981)
	batatasin II[a]	Ireland, Schwabe and Coursey (1981)
	3,5-dihydroxybibenzyl (dihydropinosylvin)	Cline *et al.* (1989)
	2′,3,5-trihydroxybibenzyl (demethyl batatasin IV)	Cline *et al.* (1989)
	2′,3-dihydroxy-5-methoxybibenzyl (batatasin IV)	Cline *et al.* (1989)
	3,3′-dihydroxy-5-methoxybibenzyl (batatasin III)	Cline *et al.* (1989)
D. bulbifera	2,4,6,7-tetrahydroxy-9,10-dihydrophenanthrene	Wij and Rangaswami (1978)
	2,4,5,6-tetrahydroxyphenanthrene	Wij and Rangaswami (1978)
	6-hydroxy-2,4,7-trimethoxyphenanthrene (batatasin I)	Ireland, Schwabe and Coursey (1981)
	2′,3,5-trihydroxybibenzyl (demethyl batatasin IV)	Adesanya, Ogundana and Roberts (1989)
D. cayenensis	6-hydroxy-2,4,7-trimethoxyphenanthrene (batatasin I)	Ireland, Schwabe and Coursey (1981)
D. composita	6-hydroxy-2,4,7-trimethoxyphenanthrene (batatasin I)	Ireland, Schwabe and Coursey (1981)
	batatasin II	Ireland, Schwabe and Coursey (1981)
D. decipiens	2,7-dihydroxy-1,3,5-trimethoxy-9,10-dihydrophenanthrene	Sunder, Rangaswami and Reddy (1978)

Table 4.3 (*Contd*)

Species	*Compounds*	*Authors*
D. dumetorum	3,4′,5-trihydroxybibenzyl (dihydroresveratrol)	Adesanya, Ogundana and Roberts (1989)
	6-hydroxy-2,4,7-trimethoxyphenanthrene (batatasin I)	El-Olemy and Reisch (1979a, b); Ireland, Schwabe and Coursey (1981)
	3,3′-dihydroxy-5-methoxybibenzyl (batatasin III)	Ireland, Schwabe and Coursey (1981)
D. floribunda	batatasin II	Ireland, Schwabe and Coursey (1981)
	2′-hydroxy-3,4,5-trimethoxybibenzyl (batatasin V)	Ireland, Schwabe and Coursey (1981)
D. opposita[b]	3,5-dihydroxybibenzyl (dihydropinosylvin)	Takasugi *et al.* (1987)
	3,3′-dihydroxy-5-methoxybibenzyl (batatasin III)	Hashimoto, Hasegawa and Kawarada (1972); Hashimoto *et al.* (1974); Hasegawa and Hashimoto (1975); Ireland, Schwabe and Coursey (1981)
	2′,3-dihydroxy-5-methoxybibenzyl (batatasin IV)	Hashimoto and Tajima (1978); Ireland, Schwabe and Coursey (1981); Takasugi *et al.* (1987)
	3,5-dimethoxy-2′-hydroxybibenzyl	Takasugi *et al.* (1987)
	2,2′-dihydroxy-3,5-dimethoxybibenzyl	
	2′-hydroxy-3,4,5-trimethoxybibenzyl (batatasin V)	Ireland, Schwabe and Coursey (1981)
	6-hydroxy-2,4,7-trimethoxyphenanthrene (batatasin I)	Hashimoto, Hasegawa and Kawarada (1972); Hashimoto *et al.* (1974); Ireland, Schwabe and Coursey (1981); Letcher (1973); Hasegawa and Hashimoto (1975); Hashimoto and Hasegawa (1974); Takasugi *et al.* (1987)
	6,7-dihydroxy-2,4-dimethoxyphenanthrene	Takasugi *et al.* (1987)
	2,7-dihydroxy-4,6-dimethoxyphenanthrene	Takasugi *et al.* (1987)
	batatasin II	Ireland, Schwabe and Coursey (1981)

D. prazeri	5,6-dihydroxy-2,4-dimethoxy-9,10-dihydro-penanthrene	Rajaraman and Rangaswami (1975)
	5,6-dihydroxy-1,3,4-trimethoxy-9,10-dihydro-phenanthrene	Rajaraman and Rangaswami (1975)
D. preusii	2′-hydroxy-3,4,5-trimethoxybibenzyl (batatasin V)	Ireland, Schwabe and Coursey (1981)
D. rotundata	3,5-dihydroxybibenzyl (dihydropinosylvin)	Fagboun *et al.* (1987)
	2′,3,5-trihydroxybibenzyl (demethyl batatasin IV)	Fagboun *et al.* (1987)
	2′,3-dihydroxy-5-methoxybibenzyl (batatasin IV)	Fagboun *et al.* (1987)
	7-hydroxy-2,4,6-trimethoxyphenanthrene (Ta VI)	Coxon, Ogundana and Dennis (1982)
	2,5-dihydroxy-4-methoxy-9,10-dihydro-phenanthrene (hircinol)	Coxon, Ogundana and Dennis (1982)
	6-hydroxy-2,4,7-trimethoxyphenanthrene (batatasin I)	Coxon, Ogundana and Dennis (1982)
D. sansibarensis	6-hydroxy-2,4,7-trimethoxyphenanthrene (batatasin I)	Ireland, Schwabe and Coursey (1981)
	batatasin II	Ireland, Schwabe and Coursey (1981)
D. trifida	6-hydroxy-2,4,7-trimethoxyphenanthrene (batatasin I)	Ireland, Schwabe and Coursey (1981)

Table 4.3 *(Contd)*

Species	*Compounds*	*Authors*
Tamus		
T. communis	4-hydroxy-2,6,7-trimethoxy-9,10-dihydrophenanthrene	Aquino *et al.* (1985a,b)
	4,7-dihydroxy-2,6-dimethoxy-9,10-dihydrophenanthrene	Aquino *et al.* (1985a,b)
	4,7-dihydroxy-2,3-dimethoxyphenanthrene	Reisch *et al.* (1973)
	3,7-dihydroxy-2,4-dimethoxyphenanthrene	Letcher and Wong (1979); Aquino *et al.* (1985a,b)
	7-hydroxy-2,4,6-trimethoxyphenanthrene (Ta VI)	Reisch *et al.* (1970)[c]; Letcher and Nhamo (1972b)
	6-hydroxy-2,4,7-trimethoxyphenanthrene (batatasin I)	Ireland, Schwabe and Coursey (1981)
	8-hydroxy-2,3,4,7-tetramethoxyphenanthrene (Ta V)	Reisch *et al.* (1969); Letcher and Wong (1978)
	4,8-dihydroxy-2,3,7-trimethoxyphenanthrene (Ta IX)	Reisch *et al.* (1973); Aquino *et al.* (1985a,b)
	1,2-methylenedioxy-5,6,7-trimethoxyphenanthrene (Ta I)	Reisch *et al.* (1970); Letcher and Wong (1978)
	2,8-dimethoxy-7-hydroxy-3,4-methylenedioxyphenanthrene (Ta IV)	Reisch *et al.* (1970)
	5,7-dimethoxy-6-hydroxy-1,2-methylenedioxyphenanthrene (Ta IV)	Letcher and Wong (1979)
	1,3,5-dimethoxyphenanthren-2,7-dione (Ta II)	Reisch *et al.* (1970)
	2,3,4,7,8-pentamethoxyphenanthrene (Ta DC)	Reisch *et al.* (1973)

[a] Structure unknown
[b] *D. opposita* Thunb. = *D. batatas* Decne
[c] Structures reported were incorrect.

Table 4.4 Stilbenes in the Dipterocarpaceae

Species	*Compounds*	*Authors*
Balanocarpus heimii	hopeaphenol (resveratrol tetramer)	Coggan *et al.* (1965); Coggan, King and Wallwork (1966)
Balanocarpus zeylandicus	balanocarpol (resveratrol dimer)	Diyasena *et al.* (1985)
Balanocarpus zeylandicus	copalliferol-A (resveratrol trimer)	Sotheeswaran *et al.* (1987)
Hopea odorata	hopeaphenol (resveratrol tetramer)	Coggan *et al.* (1965); Coggan, King and Wallwork (1966)
Hopea jucunda	balanocarpol (resveratrol dimer)	Diyasena *et al.* (1985)
Hopea cordifolia	copalliferol-A (resveratrol trimer)	Sotheeswaran *et al.* (1987)
Shorea disticha	ϵ-viniferin	Sultanbawa, Surendrakumar and Bladon (1987)
Shorea disticha	distichol (resveratrol trimer)	Sultanbawa, Surendrakumar and Bladon (1987)
Shorea stipularis	stemonoporol (resveratrol trimer)	Sotheeswaran *et al.* (1983)
Shorea stipularis	copalliferol-A (resveratrol trimer)	Sotheeswaran *et al.* (1987)
Shorea robusta	shoreaphenol	Saraswathy *et al.* (1992)
Shorea robusta	hopeaphenol (resveratrol tetramer)	Madhav, Seshadri and Subramanian (1967)
Shorea talura	hopeaphenol (resveratrol tetramer)	Madhav, Seshadri and Subramanian (1967)
Stemonoporus canaliculatus	canaliculatol (resveratrol trimer)	Bokel *et al.* (1988)
Stemonoporus canaliculatus	vaticaffinol (resveratrol tetramer)	Sotheeswaran *et al.* (1987); Bokel *et al.* (1988)
Stemonoporus affinis	stemonoporol (resveratrol trimer)	Samaraweera, Sotheeswaran and Sultanbawa (1982)

Table 4.4 (*Contd*)

Species	*Compounds*	*Authors*
Stemonoporus affinis	copalliferol-A (resveratrol trimer)	Samaraweera, Sotheeswaran and Sultanbawa (1982)
Stemonoporus cordifolius	vaticaffinol (resveratrol tetramer)	Samaraweera, Sotheeswaran and Sultanbawa (1982)
Stemonoporus elegans	stemonoporol (resveratrol trimer)	Samaraweera, Sotheeswaran and Sultanbawa (1982)
Stemonoporus elegans	copalliferol-A (resveratrol trimer)	Samaraweera, Sotheeswaran and Sultanbawa (1982)
Stemonoporus kanneliensis	stemonoporol (resveratrol trimer)	Samaraweera, Sotheeswaran and Sultanbawa (1982)
Stemonoporus kanneliensis	copalliferol-A (resveratrol trimer)	Samaraweera, Sotheeswaran and Sultanbawa (1982)
Stemonoporus lancifolius	vaticaffinol (resveratrol tetramer)	Samaraweera, Sotheeswaran and Sultanbawa (1982)
Stemonoporus oblongifolius	stemonoporol (resveratrol trimer)	Samaraweera, Sotheeswaran and Sultanbawa (1982)
Stemonoporus oblongifolius	copalliferol-A (resveratrol trimer)	Samaraweera, Sotheeswaran and Sultanbawa (1982)
Vatica affinis	vaticaffinol (resveratrol tetramer)	Sultanbawa *et al.* (1981); Sotheeswaran *et al.* (1985)
Vatica affinis	ϵ-viniferin	Sultanbawa *et al.* (1981); Sotheeswaran *et al.* (1985)
Vateria copallifera	stemonoporol (resveratrol trimer)	Sotheeswaran *et al.* (1983)
Vateria copallifera	copalliferol-A (resveratrol trimer)	Sultanbawa and Surendrakumar (1980); Sotheeswaran *et al.* (1987)
Vateria copallifera	copalliferol-B (resveratrol trimer)	Geewanada *et al.* (1986)

Table 4.5 Stilbenoids in the Fabaceae

Species	*Compounds*	*Authors*
Amorpha canescens	3,5-dihydroxy-4-geranylstilbene (amorphastilbol)	Kemal *et al.* (1979)
Amorpha fruticosa	3,5-dihydroxy-4-geranylstilbene (amorphastilbol)	Kemal *et al.* (1979)
	3-hydroxy-5-methoxy-4-isopentenylbibenzyl-2-carboxylic acid (amorfrutin A)	Mitscher *et al.* (1981)
	3-hydroxy-5-methoxy-4-geranylbibenzyl-2-carboxylic acid (amorfrutin B)	Mitscher *et al.* (1981)
Amorpha nana	3,5-dihydroxy-4-geranylstilbene (amorphastilbol)	Kemal *et al.* (1979); Mitscher *et al.* (1985)
Arachis hypogaea	3,4′,5-trihydroxystilbene (resveratrol)[a]	Ingham (1976); Keen and Ingham (1976); Cooksey *et al.* (1988); Schöppner and Kindl (1984); Rolfs, Fritzemeir and Kindl (1981); Fritzemeir *et al.* (1983)
	4-isopentenyl-3,4′,5-trihydroxystilbene[a] (arachidin II)[b]	Keen and Ingham (1976); Aguamah *et al.* (1981); Fritzemeier *et al.* (1983); Wotton and Strange (1985)
	4-(3-methyl-but-1-enyl)-3,4′,5-trihydroxystilbene (arachidin III)[b]	Aguamah *et al.* (1981); Wotton and Strange (1985)
	3-isopentadienyl-3′,4,5′-trihydroxystilbene	Cooksey *et al.* (1988)
	4-(3-methyl-but-1-enyl)-3,3′,4′,5-tetrahydroxystilbene (arachidin I)[b]	Aguamah *et al.* (1981); Wotton and Strange (1985)
Bauhinia racemosa	3,4′,5-trihydroxystilbene (resveratrol)	Anjaneyulu *et al.* (1984)
	pacharin	Anjaneyulu *et al.* (1984)

Table 4.5 (*Contd*)

Species	*Compounds*	*Authors*
Cajanus cajan	3-hydroxy-5-methoxystilbene-2-carboxylic acid	Ohwaki, Ogino and Shibano (1993)
	3-hydroxy-5-methoxy-6-isopentenylstilbene-2-carboxylic acid	Cooksey *et al.* (1982)
	3-hydroxy-5-methoxy-4-isopentenylstilbene-2-carboxylic acid	Cooksey *et al.* (1982)
Cassia dentata	3,4′,5-trihydroxystilbene (resveratrol)	De Oliveira *et al.* (1977)
	3,3′,4,5′-tetrahydroxystilbene (piceatannol)	De Oliveira *et al.* (1977)
Cassia marginata	3,3′,4,5′-tetrahydroxystilbene (piceatannol)	Rao and Rajadurai (1968)
Cassia garrettiana	3,3′,4,5′-tetrahydroxystilbene (piceatannol)	Inamori *et al.* (1984a); Hata, Baba and Kozawa (1979)
	3,3′-dihydroxybibenzyl	Hata, Baba and Kozawa (1979)
	3,3′,4-trihydroxybibenzyl	Hata, Baba and Kozawa (1979)
	3,3′,4,5′-tetrahydroxybibenzyl	Hata, Baba and Kozawa (1979)
	2,3,5,7-tetrahydroxy-9,10-dihydrophenanthrene	Hata, Baba and Kozawa (1979)
	cassigerol A, cassigerol B	Hata, Baba and Kozawa (1979); Baba *et al.* (1988)
	cassigerol C, cassigerol D	Baba *et al.* (1992)
Cassia siamea	2,3,2′,3′-tetrahydroxystilbene	Updhyay and Dutta (1968)
Cassia didymobotrya[c]	3′-hydroxy-3,4,5′-trimethoxystilbene[a]	Delle Monache *et al.* (1991)
Cassia didymobotrya[c]	3′,4-dihydroxy-3,5′-dimethoxystilbene	Delle Monache *et al.* (1991)
Centrolobium robustum	3,3′,4,5′-tetrahydroxystilbene (piceatannol)	Craveiro and Gottlieb (1968); Craveiro *et al.* (1970)
Dalbergia sissoo	3,5-dihydroxystilbene (pinosylvin)	Soni (1975)
Dalbergia latifolia	latinone	Criodain *et al.* (1981)
Derris rariflora	3,5-dimethoxy-4-isopentenylstilbene	Braz Filho, Gottlieb and Mourao (1975a); Braz Filho *et al.* (1975)
Derris floribunda	3,5-dimethoxy-4-isopentenylstilbene	Braz Filho, Gottlieb and Mourao (1975b)
	4-isopentenyl-3,4′,5-trimethoxystilbene	Braz Filho, Gottlieb and Mourao (1975b)

Glycyrrhiza acanthocarpa	3-hydroxy-4-isopentenyl-5-methoxybibenzyl	Ghisalberti, Jefferies and McAdam (1981)
	3-hydroxy-4-isopentenyl-5-methoxybibenzyl-2-carboxylic acid (amorfrutin A)	Ghisalberti, Jefferies and McAdam (1981)
Glycyrrhiza lepidota	3,5-dihydroxy-4-(2-hydroxy-3-methylbut-3-enyl)bibenzyl (glepidotin C)	Gollapudi *et al.* (1989)
	3,5-dihydroxy-4-isopentenylbibenzyl	Mitscher *et al* (1983)
Glycyrrhiza uralensis	2,6-diisopentenyl-3,3′,4′,5-tetrahydroxybibenzyl (gancaonin R)	Fukai, Wang and Nomura (1981)
	2,4-diisopentenyl-3,3′,4′,5-tetrahydroxybibenzyl (gancaonin S)	Fukai, Wang and Nomura (1991)
	gancaonin T	Fukai, Wang and Nomura (1991)
	1-isopentenyl-2,4,6,7-tetrahydroxy-9,10-dihydrophenanthrene (gancaonin V)	Fukai, Wang and Nomura (1991)
	1,3-diisopentenyl-2,4,6,7-tetrahydroxy-9,10-dihydrophenanthrene (gancaonin U)	Fukai, Wang and Nomura (1991)
Guibourtia coleosperma	4′-methoxy-3,3′,5-trihydroxystilbene-3-*O*-β-D-glucopyranoside (rhapontin, rhaponticin)	Steynberg *et al.* (1988)
	3,3′,4,5′-tetrahydroxystilbene-3-*O*-β-D-glucopyranoside (astringin, piceatannol glucoside)	Steynberg, Ferriera and Roux (1983); Steynberg (1988)
	4′,5-dimethoxy-3-hydroxystilbene-3-[*O*-β-L-rhamnopyranosyl-(1–6)-β-D-glucopyranoside]	Steynberg *et al.* (1988)
	3,3′,5-trihydroxy-4′-methoxy-3-[*O*-β-L-rhamnopyranosyl-(1–6)-β-D-glucopyranoside]	Steynberg *et al.* (1988)

Table 4.5 (*Contd*)

Species	*Compounds*	*Authors*
	condensed tannins, stilbene/flavan-3,4-diol dimers (5 structures) and trimers	Steynberg, Ferriera and Roux (1983, 1987)
Haplormosia monophylla	3,4′,5-trihydroxystilbene (resveratrol)	Hayashi *et al.* (1974)
Intsia bijunga	3,4′,5-trihydroxystilbene (resveratrol)	Hillis and Yazaki (1973)
Intsia bijunga	3,3′,4,5′-tetrahydroxystilbene (piceatannol)	Hillis and Yazaki (1973)
Laburnum alpinum	3,3′,4,5′-tetrahydroxystilbene (piceatannol)	Erdtman and Norin (1963); Erdtman (1939a,b); Billek (1964)
Laburnum anagyroides	3,3′,4,5′-tetrahydroxystilbene (piceatannol)	Erdtman (1939a)
Lonchocarpus longistylus	3,5-dihydroxy-2,4-diisopentenylstilbene (longistyline B)	Delle Monarche *et al.* (1977), Marta *et al.* (1979)
	3,5-dihydroxy-2,6-diisopentenylstilbene (longistyline D)	Delle Monarche *et al.* (1977), Marta *et al.* (1979)
	3-hydroxy-4-isopentenyl-5-methoxystilbene (longistyline A)	Delle Monarche *et al.* (1977); Marta *et al.* (1979)
	5-hydroxy-2-isopentenyl-3-methoxystilbene (longistyline C)	Delle Monarche *et al.* (1977); Marta *et al.* (1979)
Lonchocarpus nicou	lonchocarpene	Kaouadji *et al.* (1986)
Maackia amurensis	3,4′,5-trihydroxystilbene (resveratrol)	Maksimov *et al.* (1985)
	3,3′,4,5′-tetrahydroxystilbene (piceatannol)	Maksimov *et al.* (1985)
Pericopsis angolensis	3,3′,4,5′-tetrahydroxystilbene (piceatannol)	Morgan and Orsler (1968)
Pericopsis elata	3,3′,4,5′-tetrahydroxystilbene (piceatannol)	Fitzgerald, Gunning and Donnelly (1976)
Pericopsis schliebenii	3,3′,4,5′-tetrahydroxystilbene (piceatannol)	Fitzgerald, Gunning and Donnelly (1976)
Pterocarpus santalinus	3,5-dimethoxy-4′-hydroxystilbene (pterostilbene)	Akisanya, Bevan and Hirst (1959); Späth and Schläger (1940); Kumar, Ravindranath and Seshadri (1974); Sawhney and Seshadri (1956)
Pterocarpus marsupium	3,5-dimethoxy-4′-hydroxystilbene (pterostilbene)	Mathew, Subba Rao and Subba Rao (1977)
Pterocarpus dalbergoides	3,5-dimethoxy-4′-hydroxystilbene (pterostilbene)	King *et al.* (1953); Sawhney and Seshadri (1956)

Pterocarpus macrocarpus	3,5-dimethoxy-4′-hydroxystilbene (pterostilbene)	King *et al.* (1953); Sawhney and Seshadri (1956)
Pterocarpus soyauxii	3,5-dimethoxy-4′-hydroxystilbene (pterostilbene)	King *et al.* (1953)
Pterocarpus tinctorius	3,5-dimethoxy-4′-hydroxystilbene (pterostilbene)	King *et al.* (1953)
Pterolobium hexapetallum	phenanthrene	Kumar *et al.* (1988)
	3,4′,5-trihydroxystilbene (resveratrol)	Kumar *et al.* (1988)
	3,5-dimethoxy-4′-hydroxystilbene (pterostilbene)	Kumar *et al.* (1988)
	3,4′,5-trimethoxystilbene	Kumar *et al.* (1988)
Schotia brachypetala	3,3′,4,5′-tetrahydroxystilbene (piceatannol)	Drewes and Fletcher (1974)
	3,3′,4,5,5′-pentahydroxystilbene[a]	Drewes (1971); Drewes and Flecther (1974)
Sophora leachiana	leachianone B	Iinuma *et al.* (1991)
	leachianone C (= sophoraflavanone I)	Iinuma *et al.* (1991)
Sophora moorcroftiana	sophoraflavanone H	Shirataki *et al.* (1991)
	sophoraflavanone I	Shirataki *et al.* (1991)
	sophoraflavanone J	Shirataki *et al.* (1991)
Trifolium dubium	3,4′,5-trihydroxystilbene (resveratrol)	Ingham (1978)
Trifolium campestre	3,4′,5-trihydroxystilbene (resveratrol)	Ingham (1978)
Vouacapoua macropetala	3,3′,4,5′-tetrahydroxystilbene (piceatannol)	King *et al.* (1956)
Vouacapoua americana	3,3′,4,5′-tetrahydroxystilbene (piceatannol)	King *et al.* (1956)
	3,3′,4 ,4′,5-pentahydroxystilbene	King *et al.* (1956)

[a] *cis-* and *trans*-isomers
[b] Trivial name given by Wotton and Strange (1985)
[c] From *in vitro* cell cultures

Table 4.6 Stilbenes and bibenzyls in the Moraceae

Species	*Compounds*	*Authors*
Artocarpus lakoocha	2,3′,4,5′-tetrahydroxystilbene (oxyresveratrol)	Mongulsuk, Robertson and Towers (1957); Ratanabanangkoon, Poopyruchpong and Charoensiri (1976); Sambhandkaraksa, Thantivantana and Ratanachai (1962)
Artocarpus chaplasha	2,3′,4,5′-tetrahydroxystilbene (oxyresveratrol)	Rathi (1970)
	3,4′,5-trihydroxystilbene (resveratrol)	Rathi (1970)
Chlorophora excelsa	2,3′,4,5′-tetrahydroxystilbene (oxyresveratrol)	Malan *et al.* (1988); Morgan and Orsler (1968)
	4-geranyl-3,4′,5-trihydroxystilbene	Christensen, Lam and Sigsgaard (1988); Malan *et al.* (1988)
	4-geranyl-2′,3,4′,5-tetrahydroxystilbene. (chlorophorin)	King and Grundon (1949, 1950); Morgan and Orsler (1968); Christensen, Lam and Sigsgaard (1988); Malan *et al.* (1988); Grundon and King (1949); Nunn and Rapson (1949)
	excelsaoctaphenol, (chlorophorin dimer)	Christensen and Lam (1989)
Chlorophora regia	2,3′,4,5′-tetrahydroxystilbene (oxyresveratrol)	Morgan and Orsler (1968)
	4-geranyl-2′,3,4′,5-tetrahydroxystilbene (chlorophorin)	Morgan and Orsler (1968)
Clarisia racemosa	3,5-dihydroxy-4′-methoxystilbene	Gottllieb *et al.* (1975)
Cudrania javanensis	2,3′,4,5′-tetrahydroxystilbene (oxyresveratrol)	Venkatamaran (1972)
Morus laevigata	3,4′-dihydroxybibenzyl (lunularin)	Deshpande, Srinivasan and Rao (1975); Venkatamaran (1972)
	2,3′,4,5′-tetrahydroxystilbene (oxyresveratrol)	Venkatamaran (1972); Deshpande, Srinivasan and Rao (1975)
	2,3′,4,5′-tetrahydroxybibenzyl (dihydrooxyresveratrol)	Venkatamaran (1972); Deshpande, Srinivasan and Rao (1975)

Morus spp.	2,3′,4-trihydroxybibenzyl	Deshpande, Srinivasan and Rao (1975); Venkatamaran (1972)
Morus spp.	3,4′,5-trihydroxystilbene (resveratrol)	*
Morus spp.	3,3′,4,5′-tetrahydroxystilbene (piceatannol)	*
	2,3′,4,5′-tetrahydroxystilbene (oxyresveratrol)	Takasugi *et al.* (1978)*
Morus alba	2,3′,4,5′-tetrahydroxystilbene (oxyresveratrol)	Venkatamaran (1972); Deshpande, Srinivasan and Rao (1975); Laidlaw and Smith (1959); Suzushino (1954a,b)
	2,3′,4,5′-tetrahydroxybibenzyl (dihydrooxyresveratrol)	Venkatamaran (1972); Deshpande, Srinivasan and Rao (1975)
Morus bombycis	2,3′,4,5′-tetrahydroxystilbene (oxyresveratrol)	Kondo, Ito and Suda (1956)
Morus indica	2,3′,4,5′-tetrahydroxystilbene (oxyresveratrol)	Venkatamaran (1972); Deshpande, Srinivasan and Rao (1975)
	2,3′,4,5′-tetrahydroxybibenzyl (dihydrooxyresveratrol)	Venkatamaran (1972); Deshpande, Srinivasan and Rao (1975)
Morus serrata	2,3′,4,5′-tetrahydroxystilbene (oxyresveratrol)	Venkatamaran (1972); Deshpande, Srinivasan and Rao (1975)
	2,3′,4,5′-tetrahydroxybibenzyl (dihydrooxyresveratrol)	Venkatamaran (1972); Deshpande, Srinivasan and Rao (1975)
Morus rubra	2,3′,4,5′-tetrahydroxystilbene (oxyresveratrol)	Venkatamaran (1972); Deshpande, Srinivasan and Rao (1975)
	3,3′,4,5′-tetrahydroxystilbene (piceatannol)	Venkatamaran (1972); Deshpande, Srinivasan and Rao (1975)
Toxylon pomifera	3,4′,5-trihydroxystilbene (resveratrol)	Gerber (1986)
	2,3′,4,5′-tetrahydroxystilbene (oxyresveratrol)	Gerber (1986); Laidlaw and Smith (1959); Barnes and Gerber (1955); Wolfrom and Bhat (1965)

* For early references, see Gorham (1980)

Table 4.7 Stilbenes in the Myrtaceae. R, 3,4′,5-trihydroxystilbene (resveratrol); P, 3,4′,5-trihydroxystilbene-3-*O*-β-D-glucopyranoside (piceid); A, 3,3′,4,5′-tetrahydroxystilbene-3-*O*-β-D-glucopyranoside (astringin, piceatannol glucoside); RH, 4′-methoxy-3,3′,5-trihydroxystilbene-3-*O*-β-D-glucopyranoside (rhapontin, rhaponticin)

Species	*R*	*P*	*A*	*RH*	*Authors*
Angophora cordifolia		+	+		Hillis, Hart and Yazaki (1974)
Eucalyptus					
E. abergiana	+	+			Hathaway (1962)
E. astringens	+	+			Hathaway (1962)
E. astringens	+	+			Hillis and Carle (1962)[a]
E. caesia	+	+			Hathaway (1962)
E. calophylla	+	+			Hathaway (1962)
E. campaspe	+	+			Hathaway (1962)
E. citriodora	+	+			Hathaway (1962)
E. crebra	+				Hillis, Hart and Yazaki (1974)
E. decorticans	+	+			Hillis, Hart and Yazaki (1974)
E. dichromophloia	+	+			Hathaway (1962)
E. eremophila	+	+			Hathaway (1962)
E erythrophloia	+	+			Hathaway (1962)
E. ficifolia	+	+			Hathaway (1962)
E. gardneri	+	+			Hathaway (1962)
E. griffithsii	+	+			Hathaway (1962)
E. grossa	+	+			Hathaway (1962)
E. gummifera	+	+			Hathaway (1962)
E. haematoxylon	+	+			Hathaway (1962)
E. intermedia	+	+			Hathaway (1962)
E. maculata	+	+			Hathaway (1962)
E. melanophloia	+	+			Hillis, Hart and Yazaki (1974)
E. nowraensis	+	+			Hathaway (1962)
E. nutans	+				Hathaway (1962)
E. occidentalis		+			Hathaway (1962)
E. papuana	+	+			Hathaway (1962)
E. platypus	+				Hathaway (1962)
E. pruinosa	+	+			Hillis, Hart and Yazaki (1974)
E. polycarpa	+	+			Hathaway (1962)
E. sargenti	+	+			Hathaway (1962)
E. sideroxylon	+	+			Hillis and Isoi (1965a); Hillis, Hart and Yazaki (1974); Hart and Hillis (1974)
E. sideroxylon		+		+	Hillis and Hasegawa (1962b)[b]
E. sideroxylon			+		Hillis and Isoi (1965b)[c]
E. spathulata		+			Hathaway (1962)
E. stricklandi	+	+			Hathaway (1962)
E. terminalis		+			Hathaway (1962)

Table 4.7 *(Contd)*

Species	*R*	*P*	*A*	*RH*	*Authors*
E. tesselaris		+			Hathaway (1962)
E. trachyphloia	+	+			Hathaway (1962)
E. woodwardi	+	+			Hathaway (1962)
E. wandoo	+	+			Hathaway (1962) Hathaway and Seakins (1959)

[a] Also 3,3′,4,5,5′-pentahydroxystilbene and its *cis*- and *trans*-3-*O*-β-D-glucopyranosides
[b] Also 3,3′,4,5,5′-pentahydroxystilbene-3-*O*-β-D-glucopyranoside
[c] Hillis and Isoi (1965b) report the occurrence of unnamed stilbenes in *E. accendens, E. anceps, E. dundasi, E. foecunda, E. incrassata, E. lesouefii, E. salmonophloia, E. salubris* and *E. decorticans*. Hillis and Isoi (1965a) obtained incorporation of ^{14}C-phenylalanine and ^{14}C-tyrosine into stilbenes in *E. dalrympleana* and *E. guilfoylei*.

stilbene oligomers, is the Vitaceae (Table 4.14). In grapevines, resveratrol and its oligomers were identified as phytoalexins (Chapter 7).

Varieties of *Hydrangea macrophylla* contain phenyldihydroisocoumarins and related bibenzyl- and stilbene-2-carboxylic acids (Table 4.15). Other members of the Saxifragaceae appear not to synthesize these compounds, and other stilbenoids have not been found in this family. The phenyldihydroisocoumarin phyllodulcin (**224a**) is responsible for the sweet taste of tea made from *H. macrophylla* var. *thunbergii*, and is sweeter than sucrose.

In the Cannabinaceae, *Cannabis sativa* (Table 4.16) contains bibenzyls, 9,10-dihydrophenanthrenes and related spirans. Other families which contain stilbenoids are listed in Table 4.17. The micrandrols from *Micrandropsis scleroxylon* and *Sagotia racemosa* (Euphorbiaceae), like the *Juncus* phenanthrenes, are probably products of a different biosynthetic pathway. Stilbene- and bibenzyl-2-carboxylic acids are found in *Gaylussacia* species (Ericaceae) and *Helichrysum umbraculigerum* (Asteraceae).

As mentioned in Chapter 1, our present knowledge of the distribution of stilbenoids in higher plants (summarized in Tables 4.1 to 4.17) can only be regarded as incomplete. The range of structures of the known compounds will now be described in more detail.

4.2 BIBENZYLS

The parent hydrocarbon has not been found in plants, but 54 hydroxylated derivatives occur in several higher plants, often together with stilbenes and/or 9,10-dihydrophenanthrenes. The structures range from dihydropinosylvin (3,5-dihydroxybibenzyl, **165a**) to arundin (2,6-bis(4-hydroxybenzyl)-3-hydroxy-5-methoxybibenzyl, **186**). One notable group are the

Table 4.8 Orchid bibenzyls

Species	*Compounds*	*Authors*
Anthogonium gracile	3,3′-dihydroxy-5-methoxybibenzyl (batatasin III)	Veerraju *et al.* (1989b)
Arundina bambusifolia	3,3′-dihydroxy-5-methoxybibenzyl (batatasin III)	Majumder and Ghosal (1993, 1994)
	2,6-bis(4-hydroxybenzyl)-3-hydroxy-5-methoxybibenzyl (arundin)	Majumder and Ghosal (1993, 1994)
	2,6-bis(4-hydroxybenzyl)-3,3′-dihydroxy-5-methoxybibenzyl	Majumder and Ghosal (1993, 1994)
	4-(4-hydroxybenzyl)-3,3′-dihydroxy-5-methoxybibenzyl (arundinin)	Majumder and Ghosal (1993, 1994)
	3′,5-dihydroxy-2-(4-hydroxybenzyl)-3-methoxybibenzyl (isoarundinin-I)	Majumder and Ghosal (1994)
	3,3′-dihydroxy-2-(4-hydroxybenzyl)-5-methoxybibenzyl (isoarundinin-II)	Majumder and Ghosal (1994)
Bletilla striata	2,6-bis(4-hydroxybenzyl)-3,3′-dihydroxy-5-methoxybibenzyl	Takagi, Yamaki and Inoue (1983)
	2,6-bis(4-hydroxybenzyl)-3′,5-dimethoxy-3-hydroxybibenzyl	Takagi, Yamaki and Inoue (1983)
	3,3′-dihydroxy-5-methoxy-2,5′,6-tris(4-hydroxybenzyl) bibenzyl	Takagi, Yamaki and Inoue (1983)
	3,3′-dihydroxy-4-(4-hydroxybenzyl)-5-methoxybibenzyl	Bai *et al.* (1993)
	3,3′-dihydroxy-2-(4-hydroxybenzyl)-5-methoxybibenzyl	Bai *et al.* (1993)
	3′,5-dihydroxy-2-(4-hydroxybenzyl)-3-methoxybibenzyl	Bai *et al.* (1993)
	3,5-dimethoxybibenzyl (dihydropinosylvin dimethyl ether)	
	3,3′,5-trimethoxybibenzyl	
	3,3′-dihydroxy-5-methoxybibenzyl (batatasin III)	Yamaki *et al.* (1989)
	3′,5-dimethoxy-3-hydroxybibenzyl (3′-*O*-methylbatatasin III)	Yamaki *et al.* (1989)

Bulbophyllum fuscopurpureum	3,3′-dihydroxy-5-methoxybibenzyl (batatasin III)	Veerraju *et al.* (1989b)
Bulbophyllum guttulatum	3,3′-dihydroxy-5-methoxybibenzyl (batatasin III)	Veerraju *et al.* (1989b)
Bulbophyllum triste	3,4′,5-trihydroxybibenzyl (dihydroresveratrol)	Majumder and Pal (1993)
	3,4′-dihydroxy-5-methoxybibenzyl	Majumder and Pal (1993)
	3′-methoxy-3,4′,5-trihydroxybibenzyl (tristin)	Majumder and Pal (1993)
Calanthe masuca	3,3′-dihydroxy-5-methoxybibenzyl (batatasin III)	Veerraju *et al.* (1989b)
Cirrhopetalum andersonii	2′,3-dihydroxy-5′-methoxy-4,5-methylenedioxybibenzyl (cirrhopetalidinin)	Majumder and Basak (1991b)
	3,3′-dihydroxy-5-methoxybibenzyl (batatasin III)	Majumder and Basak (1991a)
	2′,3-dihydroxy-3′-methoxy-4,5-methylenedioxybibenzyl (cirrhopetalidin)	Majumder and Basak (1991a)
	3,3′-dihydroxy-4,5-methylenedioxybibenzyl (cirrhopetalinin)	Majumder and Basak (1991a)
	3,4′-dihydroxy-3′,5-dimethoxybibenzyl (gigantol)	Majumder and Basak (1991b)
Coelogyne barbata	3,3′-dihydroxy-5-methoxybibenzyl (batatasin III)	Veerraju *et al.* (1989b)
Coelogyne corymbosa	3,3′-dihydroxy-5-methoxybibenzyl (batatasin III)	Veerraju *et al.* (1989b)
Coelogyne elata	3,3′-dihydroxy-5-methoxybibenzyl (batatasin III)	Veerraju *et al.* (1989b)
Coelogyne flaccida	3,3′-dihydroxy-5-methoxybibenzyl (batatasin III)	Veerraju *et al.* (1989b)
Coelogyne nitida	3,3′-dihydroxy-5-methoxybibenzyl (batatasin III)	Veerraju *et al.* (1989b)
Coelogyne ochracea	3,3′-dihydroxy-5-methoxybibenzyl (batatasin III)	Veerraju *et al.* (1989b)
Coelogyne odorotissima	3,3′-dihydroxy-5-methoxybibenzyl (batatasin III)	Veerraju *et al.* (1989b)
Coelogyne ovalis	3,3′-dihydroxy-5-methoxybibenzyl (batatasin III)	Sachdev and Kulshreshtha (1986)
	3′,5-dimethoxy-3-hydroxybibenzyl (3′-*O*-methylbatatasin III)	Sachdev and Kulshreshtha (1986)
Cymbidium aloifolium	3,3′-dihydroxy-5-methoxybibenzyl (batatasin III)	Juneja, Sharma and Tandon (1987)
	3,4′-dihydroxy-3′,5-dimethoxybibenzyl (gigantol)	Juneja, Sharma and Tandon (1987)
	3′,4-dihydroxy-3,5-dimethoxybibenzyl (aloifol I)	Juneja, Sharma and Tandon (1987)
	4,4′-dihydroxy-3,3′,5-trimethoxybibenzyl (aloifol II)	Juneja, Sharma and Tandon (1987)

Table 4.8 (*Contd*)

Species	*Compounds*	*Authors*
Cymbidium giganteum	3,4′-dihydroxy-3′,5-dimethoxybibenzyl (gigantol)	Juneja, Sharma and Tandon (1985)
Dendrobium amoenum	3,3′-dihydroxy-5-methoxybibenzyl (batatasin III)	Veerraju *et al.* (1989b)
Dendrobium densiflorum	3,3′-dihydroxy-5-methoxybibenzyl (batatasin III)	Veerraju *et al.* (1989b)
Dendrobium chrysanthum	3,3′-dihydroxy-5-methoxybibenzyl (batatasin III)	Veerraju *et al.* (1989b)
Dendrobium crepidatum	4′-hydroxy-3,3′,4,5-tetramethoxybibenzyl (crepidatin)	
Dendrobium cumulatum	3,3′-dihydroxy-4,4′,5,5′-tetramethoxybibenzyl (cumulatin)	Majumder and Pal (1993)
Dendrobium graminifolium	3,3′-dihydroxy-5-methoxybibenzyl (batatasin III)	Veerraju *et al.* (1989b)
Dendrobium herbaceum	3,3′-dihydroxy-5-methoxybibenzyl (batatasin III)	Veerraju *et al.* (1989b)
Dendrobium lituiflorum	3,3′-dihydroxy-5-methoxybibenzyl (batatasin III)	Veerraju *et al.* (1989b)
Dendrobium moscatum	4,4′-dihydroxy-3,3′,5-trimethoxybibenzyl (aloifol II)[a]	Majumder and Sen (1987)
Dendrobium nobile	3,3′-dihydroxy-5-methoxybibenzyl (batatasin III)	Veerraju *et al.* (1989b)
Dendrobium pierardii	3,3′-dihydroxy-5-methoxybibenzyl (batatasin III)	Veerraju *et al.* (1989b)
Dendrobium spathaceum	3,3′-dihydroxy-5-methoxybibenzyl (batatasin III)	Veerraju *et al.* (1989b)
Dendrobium terminale	3,3′-dihydroxy-5-methoxybibenzyl (batatasin III)	Veerraju *et al.* (1989b)
Dendrobium rotundatum	3,3′-dihydroxy-5-methoxybibenzyl (batatasin III)	Majumder and Pal (1992)
Epipactis palustris	3,3′-dihydroxy-4′,5-dimethoxybibenzyl	Gehlert and Kindl (1991)
Eria carinata	3′-hydroxy-3,4,4′,5-tetramethoxybibenzyl (erianin)	Majumder and Joardar (1984)
Eulophia nuda	3,4′-dihydroxy-3′,5,5′-trimethoxybibenzyl	Tuchinda *et al.* (1989)
Lusia indivisa	3′,4-dihydroxy-3,5′-dimethoxybibenzyl	Majumder and Lahiri (1989)
Pholidota imbricata	3,3′-dihydroxy-5-methoxybibenzyl (batatasin III)	Veerraju *et al.* (1989b)

[a] Also named moscatilin by Majumder and Sen (1987).

Table 4.9 Orchid 9,10-dihydrophenanthrenes

Species	*Compounds*	*Authors*
Anacamptis pyramidalis	2,4-dimethoxy-7-hydroxy- (orchinol)	
Arundina bambusifolia	4,7-dihydroxy-2-methoxy- (lusianthridin)	Majumder and Ghosal (1993a,b)
Bletilla striata	2,7-dihydroxy-4-methoxy-[a] (coelonin)	Takagi, Yamaki and Inoue (1983); Yamaki *et al.* (1990)
	2,7-dihydroxy-1-(4-hydroxybenzyl)-4-methoxy-[b]	Takagi, Yamaki and Inoue (1983); Yamaki *et al.* (1990)
	2,7-dihydroxy-3-(4-hydroxybenzyl)-4-methoxy-	Yamaki *et al.* (1990)
	2,7-dihydroxy-1,6-bis(4-hydroxybenzyl)-4-methoxy-	Yamaki *et al.* (1990)
	2,4,7-trimethoxy-	Yamaki *et al.* (1991)
	2,7-dihydroxy-1,3-bis(4-hydroxybenzyl)-4-methoxy-	Bai *et al.* (1993)
	2,7-dihydroxy-1-(4-hydroxybenzoyl)-4-methoxy-	Bai *et al.* (1993)
	2,7-dihydroxy-1-(4-glucopyranosylbenzyl)-4-methoxy-	Yamaki *et al.* (1993c)
Bulbophyllum gymnopus	2,7-dihydroxy-3,4,6-trimethoxy-	Majumder and Banerjee (1988a,b)
Cirrhopetalum andersonii	4,7-dihydroxy-2,3-methylenedioxy- (cirrhopetalanthridin)	Majumder and Basak (1991b)

Table 4.9 (*Contd*)

Species	*Compounds*	*Authors*
Coeloglossum viride	2,4-dimethoxy-7-hydroxy- (orchinol)	
Coelogyne elata	2,7-dihydroxy-4-methoxy- (coelonin)	Majumder, Laha and Datta (1982)
Coelogyne ochracea	2,7-dihydroxy-4-methoxy- (coelonin)	Majumder, Laha and Datta (1982)
Cymbidium aloifolium	2,7-dihydroxy-4,6-dimethoxy- (6-*O*-methylcoelonin)	Juneja, Sharma and Tandon (1987)
	2,7-dihydroxy-4-methoxy- (coelonin)	Juneja, Sharma and Tandon (1987)
Dendrobium rotundatum	2,7-dihydroxy-3,4,6-trimethoxy-	Majumder and Pal (1992)
	2,5-dihydroxy-4-methoxy- (moscatin) [=hircinol?]	Majumder and Pal (1992)
	4-methoxy-2,5,9-trihydroxy- (rotundatin)	Majumder and Pal (1992)
Epipactis palustris	1,7-dihydroxy-5-methoxy-	Gehlert and Kindl (1991)
	1,7-dihydroxy-2,5-dimethoxy-	Gehlert and Kindl (1991)
Eria flava	2,7-dihydroxy-4-methoxy- (coelonin)	Majumder and Banerjee (1990)
	3,7-dihydroxy-2,4-dimethoxy- (flavanthridin)	Majumder and Banerjee (1990)
Eulophia nuda	1,5-dihydroxy-2,7-dimethoxy- (eulophiol)	Bhandari and Kapadi (1983)
	1,7-dihydroxy-2,5-dimethoxy-	Tuchinda *et al.* (1988)
	2,7-dihydroxy-4-methoxy- (coelonin)	Tuchinda *et al.* (1988)
	2,8-dihydroxy-4,7-dimethoxy-1-(4-hydroxybenzyl)-	Tuchinda *et al.* (1989)
Gymnadenia albida	2,4-dimethoxy-7-hydroxy- (orchinol)	Braun (1963)
Loroglossum hircinum	2,5-dihydroxy-4-methoxy- (hircinol)	Urech *et al.* (1963); Fisch and Arditti (1972)
	5-hydroxy-2,4-dimethoxy- (loroglossol)	Urech *et al.* (1963); Fisch and Arditti (1972); Ward, Unwin and Stoessl (1975a,b)

Lusia indivisa	4,7-dihydroxy-2-methoxy- (lusianthridin)	Majumder and Lahiri (1990a,b)
Nigritella nigra	2,4-dimethoxy-7-hydroxy- (orchinol)	
Orchis mascula	2,4-dimethoxy-7-hydroxy- (orchinol)	Gäumann and Kern (1959a,b); Birch (1966); Hardegger, Biland and Carrodi (1963); Hardegger, Schellenbaum and Carrodi (1963); Hardegger *et al.* (1963); Gäumann (1963); Braun (1963), Arditti (1979); Ingham (1972); Nüesch (1963), Fisch and Arditti (1972); Letcher and Nhamo (1973a,b)
Orchis morio		
Orchis militaris	2,4-dimethoxy-7-hydroxy- (orchinol)	Fisch, Flick and Arditti (1973)
Orchis sambucina	2,4-dimethoxy-7-hydroxy- (orchinol)	
Serapias lingua	2,4-dimethoxy-7-hydroxy- (orchinol)	Braun (1963)
Spiranthes sinensis var. *amoena*	spiranthol-A	Tezuka *et al.* (1989)
	spiranthol-B	Tezuka *et al.* (1989)
	spiranthol-C	Tezuka *et al.* (1990)
	spirasineol-A	Tezuka *et al.* (1989)
	spirasineol-B	Tezuka *et al.* (1990)

[a] Previously 4,7-dihydroxy-2-methoxy-9,10-dihydrophenanthrene (Takagi, Yamaki and Inoue, 1983); also suggested to be coelonin by Majumder and Lahiri (1990b)
[b] Previously 4,7-dihydroxy-1-(4-hydroxybenzyl)-2-methoxy-9,10-dihydrophenanthrene (Takagi, Yamaki and Inoue, 1983)

Table 4.10 Orchid phenanthrenes

Species	*Compounds*	*Authors*
Bletilla striata	2,4,7-trimethoxy-	Yamaki *et al.* (1991)
	2,3,4,7-tetramethoxy-	Yamaki *et al.* (1991)
	2,7-dihydroxy-1-(4-hydroxybenzyl)-4-methoxy-	Yamaki *et al.* (1990)
	1,8-bis(4-hydroxybenzyl)-2,7-dihydroxy-4-methoxy-	Bai *et al.* (1991)
	2-glucopyranosyl-7-hydroxy-4-methoxy-	Yamaki *et al.* (1993c)
	2,7-diglucopyranosyl-4-methoxy-	Yamaki *et al.* (1993c)
	2,4-dimethoxy-3-glucopyranosyl-7-hydroxy-	Yamaki *et al.* (1993c)
Bulbophyllum gymnopus	7,9-dihydroxy-2,3,4-trimethoxy- (gymnopusin)	Majumder and Banerjee (1988a,b)
	2,7-dihydroxy-3,4,6-trimethoxy-	Majumder and Banerjee (1988a,b)
	2,7-dihydroxy-3,4,8-trimethoxy- (confusarin)	Majumder and Banerjee (1988a,b,)
Bulbophyllum leopardium	3,5-dihydroxy-2,4-dimethoxy- (bulbophyllanthrin)	Majumder, Kar and Shooley (1985)
Cirrhopetalum andersonii	7-hydroxy-4-methoxy-2,3-methylenedioxy-(cirrhopetalin)	Majumder and Basak (1990)
Cymbidium pendulum	3-hydroxy-2,4,6,7,8-pentamethoxy- (pendulin)	Majumder and Sen (1991a)
	2,6-dihydroxy-1,5,7-trimethoxy-	Majumder and Sen (1991a)
Dendrobium rotundatum	2,7-dihydroxy-3,4,6-trimethoxy-	Majumder and Pal (1992)
	2,7-dihydroxy-3,4-dimethoxy- (nudol)	Majumder and Pal (1992)
Eria carinata	2,7-dihydroxy-3,4-dimethoxy- (nudol)	Bhandari *et al.* (1985)

Eria confusa	2,6-dihydroxy-3,4,7,8-tetramethoxy- (confusaridin)	Majumder and Kar (1987)
	2,7-dihydroxy-3,4,8-trimethoxy- (confusarin)	Majumder and Kar (1987)
Eria flava	3,7-dihydroxy-2,4-dimethoxy- (TaIII)	Majumder and Banerjee (1990)
	2,7-dihydroxy-4-methoxy- (flavanthrinin)	Majumder and Banerjee (1990)
Eria stricta	2,7-dihydroxy-3,4-dimethoxy- (nudol)	Bhandari *et al.* (1985)
Eulophia nuda	2,7-dihydroxy-3,4-dimethoxy- (nudol)	Bhandari *et al.* (1985)
	2,7-dihydroxy-1,5-dimethoxy-	Tuchinda *et al.* (1988)
	2,6-dihydroxy-5,7-dimethoxy-(=3,7-dihydroxy-2,4-dimethoxy- [TaIII])	Tuchinda *et al.* (1988)
	2,6-dihydroxy-1,5,7-trimethoxy-(=3,7-dihydroxy-2,4,8-trimethoxy-)	Tuchinda *et al.* (1988)
	2,7-dihydroxy-4,8-dimethoxy-1-(4-hydroxybenzyl)-	Tuchinda *et al.* (1989)
Lusia indivisa	4,7-dihydroxy-2-methoxy- (lusianthrin)	Majumder and Lahiri (1990a,b)
Oncidium cebolleta	2,7-dihydroxy-3,4-dimethoxy- (nudol)	Stermitz *et al.* (1983)

Table 4.11 Orchid phenanthro-pyrones and -quinones

Orchidaceae	flavidinin	Majumder and Datta (1981)
Orchidaceae	oxoflavidinin	Majumder and Datta (1981)
Agrostophyllum khasiyanum	agrostophyllin	Majumder and Sabzabadi (1988)
Arundina bambusifolia	flavidin	Majumder and Ghosal (1993a,b)
Bletilla striata	bletilol A	Yamaki *et al.* (1993a)
	bletilol B	Yamaki *et al.* (1993a)
	bletilol C	Yamaki *et al.* (1993a)
	blespirol	Bai *et al.* (1993)
Bulbophyllum fuscopurpureum	flavidin	Veerraju *et al.* (1989b)
	imbricatin	Veerraju *et al.* (1989b)
Bulbophyllum guttulatum	flaccidin	Veerraju *et al.* (1989b)
Bulbophyllum odoratissimum	bulbophyllanthrone	Majumder and Sen (1991b)
Coelogyne barbata	flavidin	Veerraju *et al.* (1989b)
	coelogin	Veerraju *et al.* (1989b)
Coelogyne corymbosa	coelogin	Veerraju *et al.* (1989b)
Coelogyne cristata	coeloginin	Majumder, Bandyopadhyay and Joarder (1982)
	coelogin	Majumder, Bandyopadhyay and Joarder (1982)
Coelogyne elata	flavidin	Veerraju *et al.* (1989b)
Coelogyne flavida	flavidin	Majumder *et al.* (1982)
Coelogyne flaccida	imbricatin	Majumder and Maiti (1988, 1989); Veerraju *et al.* (1989b)
	flaccidin	Majumder and Maiti (1988, 1989); Veerraju *et al.* (1989b)
	isoflaccidinin	Majumder and Maiti (1991)
	isooxoflaccidin	Majumder and Maiti (1991)
	flaccidinin	Majumder and Maiti (1989)
	oxoflaccidin	Majumder and Maiti (1989)
Coelogyne nitida	flavidin	Veerraju *et al.* (1989b)
	coelogin	Veerraju *et al.* (1989b)
	imbricatin	Veerraju *et al.* (1989b)
	flaccidin	Veerraju *et al.* (1989b)
Coelogyne ochracea	ochrolide	Bhaskar *et al.* (1989)
	flavidin	Veerraju *et al.* (1989b)
	coelogin	Veerraju *et al.* (1989b)
	coeloginin	Veerraju *et al.* (1989b)
	ochrone A	Bhaskar *et al.* (1991)
	ochrone B	Bhaskar *et al.* (1991)

Table 4.11 (*Contd*)

Coelogyne ovalis	coelogin	Sachdev and Kulshreshtha (1986)
	flavidin	Sachdev and Kulshreshtha (1986)
	flavidinin	Sachdev and Kulshreshtha (1986)
	coeloginin	Sachdev and Kulshreshtha (1986)
	imbricatin	Sachdev and Kulshreshtha (1986)
Cymbidium aloifolium	cymbinodin-A	Barua *et al.* (1990)
	cymbinodin-B	Ghosh *et al.* (1992)
Cypripedium calceolus	cypripedin	Schmalle and Hausen (1979)
Dendrobium amoenum	amoenumin	Veerraju *et al.* (1989a)
	imbricatin	Veerraju *et al.* (1989b)
	flaccidin	Veerraju *et al.* (1989b)
Dendrobium herbaceum	imbricatin	Veerraju *et al.* (1989b)
Dendrobium pierardii	flavidin	Veerraju *et al.* (1989b)
Eria spp.	erianthridin	Majumder and Joarder (1985)
Otochilus fusca	flavidin	Majumder *et al.* (1982); Majumder, Laha and Datta (1982)
	isoflavidinin	Majumder, Laha and Datta (1982)
	iso-oxoflavidinin	Majumder, Laha and Datta (1982)
Otochilus porecta	flavidin	Majumder, Laha and Datta (1982)
	isoflavidinin	Majumder, Laha and Datta (1982)
	iso-oxoflavidinin	Majumder, Laha and Datta (1982)
Pholidota articulata	iso-oxoflavidinin	Majumder, Laha and Datta (1982)
	flavidin	Majumder *et al.* (1982); Majumder, Laha and Datta (1982)
	isoflavidinin	Majumder, Laha and Datta (1982)
Pholidota imbricata	imbricatin	Majumder and Sarkar (1982); Veerraju *et al.* (1989b)
	flavidin	Veerraju *et al.* (1989b)

Table 4.12 Orchid biphenanthrenes and biphenanthrene ethers

Species	*Compounds*	*Authors*
Arundina bambusifolia	4,4′-dimethoxy-2,2′,7,7′-tetrahydroxy-9,9′,10,10′-tetrahydro-[1,1′-biphenanthrene] (flavanthrin, blestriarene A)	Majumder and Ghosal (1993)
Bletilla striata	4,4′-dimethoxy-2,7,7′-trihydroxy-9,9′,10,10′-tetrahydro-[1,2′-biphenanthrene ether] (blestrin A)	Bai *et al.* (1990)
	4,7′-dimethoxy-2,5′,7-trihydroxy-9,9′,10,10′-tetrahydro-[1,2′-biphenanthrene ether] (blestrin B)	Bai *et al.* (1990)
	4,4′-dimethoxy-2,2′,7,7′-tetrahydroxy-9,9′,10,10′-tetrahydro-[1′,3-biphenanthrene] (blestrianol A)	Bai *et al.* (1991)
	8-(4-hydroxybenzyl)-4′,5-dimethoxy-2,2′,7,7′-tetrahydroxy-9,9′,10,10′-tetrahydro-[1′,3-biphenanthrene] (blestrianol B)	Bai *et al.* (1991)
	8-(4-hydroxybenzyl)-4′,5-dimethoxy-2,2′,7,7′-tetrahydroxy-9,10-dihydro-[1′,3-biphenanthrene] (blestrianol C)	Bai *et al.* (1991)

	4,4′-dimethoxy-2,2′,7,7′-tetrahydroxy-9,9′,10,10′-tetrahydro-[1,1′-biphenanthrene] (flavanthrin, blestriarene A)	Yamaki *et al.* (1989)
	4,4′-dimethoxy-2,2′,7,7′-tetrahydroxy-9,10-dihydro-[1,1′-biphenanthrene] (blestriarene B)	Yamaki *et al.* (1989)
	4,4′-dimethoxy-2,2′,7,7′-tetrahydroxy-[1,1′-biphenanthrene] (blestriarene C)	Yamaki *et al.* (1989)
Cirrhopetalum maculosum	4,4′-dimethoxy-2,2′,7,7′-tetrahydroxy-[1,1′-biphenanthrene] (cirrhopetalanthrin)	Majumder, Pal and Joarder (1990)
Eria flava	4,4′-dimethoxy-2,2′,7,7′-tetrahydroxy-9,9′,10,10′-tetrahydro-[1,1′-biphenanthrene] (flavanthrin, blestriarene A)	Majumder and Banerjee (1990)
Eulophia nuda	2,2′,7,7′-tetrahydroxy-4,4′,8,8′-tetramethoxy-[1,1′-biphenanthrene]	Tuchinda *et al.* (1988)
Spiranthes sinensis	spiranthesol	Tezuka *et al.* (1990)

Table 4.13 Stilbenes in the Polygonaceae

Species	*Compounds*	*Authors*
Polygonum cuspidatum	3,4′,5-trihydroxystilbene-3-*O*-β-D-glucopyronoside (piceid, polydatin)	Jayatilake *et al.* (1993)
	3,4′,5-trihydroxystilbene (resveratrol)	
Polygonum multiflorum	2,3,4′,5-tetrahydroxystilbene-2-*O*-β-D-(2″-*O*-galloyl)glucopyranoside	Nonaka, Miwa and Nishioka (1982)
	2,3,4′,5-tetrahydroxystilbene-2-*O*-β-D-(3″-*O*-galloyl)glucopyranoside	Nonaka, Miwa and Nishioka (1982)
	4′-methoxy-3,3′,5-trihydroxystilbene-3-*O*-β-D-glucopyranoside (rhapontin, rhaponticin)	Tsukida and Yokota (1954)
	2,3,4′,5-tetrahydroxystilbene-2-*O*-β-D-glucopyranoside	Hata, Kozawa and Baba (1975)
Rheum rhaponticum	3,4′,5-trihydroxystilbene (resveratrol) [*in vitro* enzymatic synthesis]	Rupprich and Kindl (1977)
Rheum spp.	3,5-dihydroxy-4′-methoxystilbene	Csupor (1970); Banks and Cameron (1971); Klimek (1973)
Rheum spp.	3,3′,4,5′-tetrahydroxystilbene (piceatannol)	Kashiwada, Nonaka and Nishioka (1984)
Rheum spp.	4′-methoxy-3,3′,5-trihydroxystilbene (rhapontigenin)	Kashiwada, Nonaka and Nishioka (1984)
Rheum rhaponticum	3,4′,5-trihydroxystilbene-4′-*O*-β-D-glucopyranoside (resveratroloside)	Nonaka, Minami and Nishioka (1977)
Rheum rhaponticum	3,5-dihydroxy-4′-methoxystilbene-3-*O*-β-D-glucopyranoside	Csupor (1970); Banks and Cameron (1971); Klimek (1973); Kashiwada, Nonaka and Nishioka (1984)

Rheum rhaponticum	4′-methoxy-3,3′,5-trihydroxystilbene-3-*O*-β-D-glucopyranoside (rhapontin, rhaponticin)	Rupprich, Hildebrandt and Kindl (1980) Hesse (1899) Kawamura (1938) Khafagy *et al.* (1974) Kimura and Hato (1952) Shibata and Takido (1952) Klimek (1973) Lang (1957) Neuhoff and Auterhoff (1954) Siegfried (1943) Siesto (1958) Tschirch and Chistofoletti (1905)
Rheum altaicum	4′-methoxy-3,3′,5-trihydroxystilbene-3-*O*-β-D-glucopyranoside (rhapontin, rhaponticin)	Klimek (1973)
Rheum cordatum	4′-methoxy-3,3′,5-trihydroxystilbene-3-*O*-β-D-glucopyranoside (rhapontin, rhaponticin)	Klimek (1973)
Rheum undulatum	4′-methoxy-3,3′,5-trihydroxystilbene-3-*O*-β-D-glucopyranoside (rhapontin, rhaponticin)	Shibata and Takido (1952) Gstirner and Holtzem (1949) Kimura and Hato (1952) Tsukida and Yoneshigo (1954)
Rheum emodi	4′-methoxy-3,3′,5-trihydroxystilbene-3-*O*-β-D-glucopyranoside (rhapontin, rhaponticin)	Youngken (1946)
Rheum collineanum	4′-methoxy-3,3′,5-trihydroxystilbene-3-*O*-β-D-glucopyranoside (rhapontin, rhaponticin)	Kimura and Hato (1952)

Table 4.13 *(Contd)*

Species	*Compounds*	*Authors*
Rheum compactum	4′-methoxy-3,3′,5-trihydroxystilbene-3-*O*-β-D-glucopyranoside (rhapontin, rhaponticin)	Kimura and Hato (1952)
Rheum spp.	4′-methoxy-3,3′,5-trihydroxystilbene-3′-*O*-β-D-glucopyranoside	Kashiwada, Nonaka and Nishioka (1984)
Rheum rhaponticum	3,3′,4,5′-tetrahydroxystilbene-3-*O*-β-D-glucopyranoside (astringin)	Gracza and Ruff (1984b) Kashiwada, Nonaka and Nishioka (1984)
Rheum spp.	3,3′,4,5′-tetrahydroxystilbene-3′-*O*-β-D-(6″-*O*-galloylglucopyranoside	Kashiwada, Nonaka and Nishioka (1984)
	3,3′,4,5′-tetrahydroxystilbene-3′-*O*-β-D-glucopyranoside	
	3,3′,4,5′-tetrahydroxystilbene-3′-*O*-β-D-xylopyranoside	
Rheum spp.	4′-methoxy-3,3′,5-trihydoxystilbene-3′-*O*-β-D-(6″-*O*-galloyl)glucopyranoside	Kashiwada, Nonaka and Nishioka (1984)
Rheum spp.	4′-methoxy-3,3′,5-trihydoxystilbene-3′-*O*-β-D-(2″-*O*-galloyl)glucopyranoside	Kashiwada, Nonaka and Nishioka (1984)
Rheum spp.	4′-methoxy-3,3′,5-trihydoxystilbene-3′-*O*-β-D-(2″-*O*-*p*-coumaryl)glucopyranoside	Kashiwada, Nonaka and Nishioka (1984)
Rheum rhaponticum	3,4′,5-trihydoxystilbene-4′-*O*-β-D-(6″-*O*-galloyl)glucopyranoside	Nonaka, Minami and Nishioka (1977); Kashiwada, Nonaka and Nishioka (1984)

Table 4.14 Stilbenoids in the Vitaceae

Species	*Compounds*	*Authors*
Ampelopsis brevipedunculata var. *hancei*	3,4′,5-trihydroxystilbene-4′-*O*-β-D-glucopyranoside (resveratroloside)	Oshima, Ueno and Hikino (1993)
	3,4′,5-trihydroxystilbene-3-*O*-β-D-glucopyranoside (piceid)[a]	Oshima, Ueno and Hikino (1993)
	ampelopsin A	Oshima *et al.* (1990)
	ampelopsin B	Oshima *et al.* (1990)
	ampelopsin C	Oshima *et al.* (1990)
	pallidol	Oshima, Ueno and Hikino (1993)
	miyabenol C	Oshima, Ueno and Hikino (1993)
	ampelopsin D	Oshima, Ueno and Hikino (1993)
	ampelopsin E[a]	Oshima, Ueno and Hikino (1993)
	ampelopsin H	Oshima, Ueno and Hikino (1993)
Vitis cognetiae	3,4′,5-trihydroxystilbene (resveratrol)	Langcake and Pryce (1976)
Vitis riparia	3,4′,5-trihydroxystilbene (resveratrol)	Langcake (1981)
Vitis riparia	3,5-dimethoxy-4′-hydroxystilbene (pterostilbene)	Langcake (1981)
Vitis riparia	α-viniferin	Langcake (1981)
Vitis riparia	ϵ-viniferin	Langcake (1981)

Table 4.14 *(Contd)*

Species	*Compounds*	*Authors*
Vitis vinifera	3,4′,5-trihydroxystilbene (resveratrol)	Langcake and Pryce (1976, 1977a,b,c); Langcake, Cornford and Pryce (1979); Fritzemeier and Kindl (1981); Hoos and Blaich (1988); Stein and Hoos (1984); Langcake (1981)
Vitis vinifera	3,5-dimethoxy-4′-hydroxystilbene (pterostilbene)	Langcake, Cornford and Pryce (1979); Langcake (1981)
Vitis vinifera	α-viniferin	Pryce and Langcake (1977); Langcake (1981)
Vitis vinifera	ε-viniferin	Langcake and Pryce (1977a,b,c); Langcake (1981)
Vitis vinifera	3,4′,5-trihydroxystilbene-3-*O*-β-D-glucopyronoside (piceid)	Waterhouse and Lamuela-Raventos (1994)
Cissus antarctica	3,4′,5-trihydroxystilbene (resveratrol)	Langcake and Pryce (1976)
Cissus pallida	pallidol	Khan *et al.* (1986)
Parthenocissus quiquefolia	3,4′,5-trihydroxystilbene (resveratrol)	Langcake and Pryce (1976)
Parthenocissus tricuspidata	tricuspidatol-A	Lins *et al.* (1991)
Rhoicissus rhomboidea	3,4′,5-trihydroxystilbene (resveratrol)	Langcake and Pryce (1976)

[a] *cis-* and *trans*-isomers

Table 4.15 Stilbenoids in *Hydrangea macrophylla*

Subspecies or variety	*Compounds*	*Authors*
var. *macrophylla*	3,4′-dihydroxybibenzyl (lunularin)	Gorham (1977a)
var. *macrophylla*	3,4′-dihydroxystilbene-2-carboxylic acid (hydrangeic acid)	Billek and Kindl (1962a,b,c); Gorham (1977a); Kindl (1971b); Kindl (1971b); Pryce (1971)
var. *macrophylla*	3,4′-dihydroxybibenzyl-2-carboxylic acid (lunularic acid)	Pryce (1971); Gorham (1977a)
var. *macrophylla*	3-(4-hydroxyphenyl)-8-hydroxydihydro-isocoumarin (hydrangenol)	Hashimoto *et al.* (1987); Billek and Kindl (1962a,b,c); Asahina and Asano (1930a,b); Asahina and Miyake (1916); Ibrahim and Towers (1960, 1962); Asen, Cathey and Stuart (1960); Kindl (1971b); Gorham (1977a); Pryce (1971)
var. *macrophylla*	hydrangenol-8-*O*-β-D-glucopyranoside	Hashimoto *et al.* (1987); Ueno (1937); Ibrahim and Towers (1960, 1962); Kindl (1971b); Asen, Cathey and Stuart (1960); Billek and Kindl (1962a,b,c); Pryce (1971)
subsp. *serrata*	hydrangenol-8-*O*-β-D-glucopyranoside	Hashimoto *et al.* (1987)
subsp. *serrata*	macrophylloside A	Hashimoto *et al.* (1987)
subsp. *serrata*	macrophylloside B	Hashimoto *et al.* (1987)
subsp. *serrata*	macrophylloside C	Hashimoto *et al.* (1987)
var. *thunbergii*	3-(4-hydroxyphenyl)-8-hydroxydihydro-isocoumarin (hydrangenol)	Hashimoto *et al.* (1987); Suzuki *et al.* (1977, 1979a, 1981); Akada *et al.* (1979); Yagi *et al.* (1972)

Table 4.15 (*Contd*)

Species	*Compounds*	*Authors*
var. *thunbergii*	3-(4-methoxyphenyl)-8-hydroxydihydroisocoumarin (hydrangenol monomethyl ether)	Suzuki *et al.* (1979a)
var. *thunbergii*	3-(3-hydroxy-4-methoxyphenyl)-8-hydroxydihydroisocoumarin (phyllodulcin)	Hashimoto *et al.* (1987); Suzuki *et al.* (1977, 1979a, 1981); Yagi *et al.* (1972, 1977); Akada *et al.* (1979); Asahina and Asano (1931); Arakawa and Nakazaki (1959); Arakawa (1960)
var. *thunbergii*	3-(3,4-dimethoxyphenyl)-8-hydroxydihydroisocoumarin (phyllodulcin monomethyl ether)	Suzuki *et al.* (1979a)
var. *thunbergii*	hydrangenol-8-*O*-β-D-glucopyranoside	Hashimoto *et al.* (1987); Suzuki *et al.* (1978, 1979a)
var. *thunbergii*	phyllodulcin-8-*O*-β-D-glucopyranoside	Hashimoto *et al.* (1987); Suzuki *et al.* (1977, 1979a); Yagi *et al.* (1972)
var. *thunbergii*	hydrangea glucoside A	Hashimoto *et al.* (1987); Yagi *et al.* (1972)
var. *thunbergii*	hydrangea glucoside B	Hashimoto *et al.* (1987); Yagi *et al.* (1972)
var. *thunbergii*	hydrangea glucoside C	Yagi *et al.* (1972)
var. *thunbergii*	glucoside I	Suzuki *et al.* (1979b)
var. *thunbergii*	glucoside II	Suzuki *et al.* (1979b)

Table 4.16 Stilbenoids in *Cannabis sativa* (Cannabinaceae)

Compounds	*Authors*
3,4′,5-trihydroxybibenzyl (dihydroresveratrol)	Elferaly (1984)
3,4′-dihydroxy-5-methoxybibenzyl	Kettenes-Van den Bosch and Salemink (1978); Crombie and Crombie (1978, 1982)
3,3′-dihydroxy-4,5′-dimethoxybibenzyl	Kettenes-Van den Bosch and Salemink (1978); Crombie and Crombie (1978)
3,3′-dihydroxy-4,5′-dimethoxy-2-isopentenylbibenzyl (canniprene)	Crombie and Crombie (1978, 1982); Kettenes-Van den Bosch and Salemink (1978)[a]
3,5-dihydroxy-7-methoxy-9,10-dihydrophenanthrene (cannithrene-1, cannabidihydrophenanthrene)	Crombie and Crombie (1982); Crombie, Crombie and Jamieson (1979)
4,5-dihydroxy-3,7-dimethoxy-9,10-dihydrophenanthrene (cannithrene-2)	Crombie and Crombie (1982)
cannabispirone (cannabispiran)	Crombie and Crombie (1982); El-Feraly *et al.* (1977); Bercht *et al.* (1976); Ottersen *et al.* (1976)
cannabispiradienone	Crombie and Crombie (1982); Crombie, Crombie and Jamieson (1979)
cannabispirenone (dehydrocannabispiran)	Crombie and Crombie (1982); El-Feraly *et al.* (1977); Bercht *et al.* (1976)
β-cannabispiranol	Crombie and Crombie (1982); Boeren *et al.* (1977)
isocannabispiran	ElSohly and Turner (1982)
cannabispirenone B	Kettenes-Van den Bosch and Salemink (1978); Crombie and Crombie (1982)
α-cannabispiranol	Crombie and Crombie (1982)

[a] Original structure revised to that of canniprene.

Table 4.17 Stilbenoids in other families

Species	*Compounds*	*Authors*
Empetraceae		
Empetrum nigrum	4-methoxy-3,3′,5-trihydroxybibenzyl	Arriaga-Giner, Wollenweber and Dörr (1993)
	3,3′-dihydroxy-4,5-dimethoxybibenzyl	
	5-hydroxy-2,3,4-trimethoxy-9,10-dihydrophenanthrene	Wollenweber *et al.* (1992)
	4,5-dihydroxy-2,3-dimethoxy-9,10-dihydrophenanthrene	Wollenweber *et al.* (1992)
Ericaceae		
Gaylussacia baccata	3,5-dihydroxystilbene-2-carboxylic acid-3-*O*-β-D-glucopyranoside (gaylussacin)	Askari, Worthen and Schimiza (1972)
Gaylussacia frondosa		
Asteraceae		
Helichrysum umbraculigerum	3,5-dihydroxy-4-geranylbibenzyl	Bohlmann and Hoffmann (1979)
	3,5-dihydroxy-4-prenylbibenzyl-2-carboxylic acid (and various 5-esters)	Bohlmann and Hoffmann (1979)
	3,5-dihydroxy-4-geranylbibenzyl-2-carboxylic acid	Bohlmann and Hoffmann (1979)
	3,5-dihydroxy-4-geranylstilbene-2-carboxylic acid	Bohlmann and Hoffmann (1979)
	3,5-dihydroxy-4-geranylstilbene	Bohlmann and Hoffmann (1979)
	structure X	Bohlmann and Hoffmann (1979)
Euphorbiaceae[a]		
Micrandropsis scleroxylon	2,6-dihydroxy-1,7-dimethyl-9,10-dihydrophenanthrene (micrandrol B)	De Alvarenga and Gottlieb (1974)

	2,6-dihydroxy-1,7-dimethylphenanthrene (micrandrol A)	De Alvarenga and Gottlieb (1974)
	2,6-dihydroxy-7-methyl-1-methylthio-phenanthrene (micrandrol C)	De Alvarenga *et al.* (1981)[b]; De Alvarenga and Gottlieb (1974)
	micrandrol D	De Alvarenga *et al.* (1981)
Sagotia raemosa	6-hydroxy-7-methoxy-1,2-dimethylphenanthrene (micrandrol E)	De Alvarenga, Gottlieb and Magalhues (1976)
Sagotia racemosa	1,2-dimethyl-6-hydroxy-7-methoxy-9,10-dihydrophenanthrene (micandrol-F)	De Alvarenga, Gottlieb and Magalhues (1976)
Papaveraceae		
Papaver somniferum	3,7-dimethoxy-5-hydroxyphenanthrene	Gombos *et al.* (1974)
	3,6-dimethoxy-4-hydroxyphenanthrene (α-thebaol)	Reisch *et al.* (1974)
Papaver bracteatum	3,6-dimethoxy-4-hydroxyphenanthrene (α-thebaol)	Küppers *et al.* (1976)
	3,4,6-trimethoxyphenanthrene (*O*-methyl-α-thebaol)	Theuns *et al.* (1985)
Betulaceae		
Alnus spp.	3-hydroxy-5-methoxystilbene (pinosylvin monomethyl ether)	Asakawa (1970, 1971); Suga, Iwata and Asakawa (1972)
Alnus crispa	3,5-dihydroxystilbene (pinosylvin)	Clausen, Reichardt and Bryant (1986)
	3-hydroxy-5-methoxystilbene (pinosylvin monomethyl ether)	Bryant *et al.* (1983); Clausen *et al.* (1986)
Alnus pendula	3,5-dihydroxystilbene (pinosylvin)	Asakawa (1971); Suga, Iwata and Asakawa (1972)
	3,5-dimethoxystilbene (pinosylvin dimethyl ether)	Asakawa (1970, 1971)
Alnus sieboldiana	stilbene	Asakawa (1971)

Table 4.17 (*Contd*)

Species	*Compounds*	*Authors*
Alnus viridis	4,5-dihydroxy-3-methoxystilbene	Favre-Bonvin, Jay and Wollenweber (1978)
Phytolaccaceae		
Petiveria alliacea	stilbene	Adesogan (1974)
Fagaceae		
Nothofagus spp.	3,5-dihydroxystilbene (pinosylvin)	Hillis and Orman (1962); Hillis and Inoue (1967)
	3,4′,5-trihydroxystilbene (resveratrol)	Hillis and Inoue (1967)
Palmaceae		
Phoenix dactylifera	3,4′,5-trihydroxystilbene (resveratrol)	Fernandez, Pedro and Seoane (1983)
	4-methoxy-3,3′,5,5′-tetrahydroxystilbene[c]	Fernandez, Pedro and Seoane (1983)
Myristicaceae		
Knema austrosiamensis	3,4′-dimethoxy-5-hydroxystilbene	Gonzalez *et al.* (1993)
	3,5-dihydroxy-4′-methoxystilbene	Gonzalez *et al.* (1993)
Virola cuspidata	3,4′,5-trimethoxystilbene	Blair *et al.* (1969)
Virola elongata	3,4′,5-trimethoxystilbene[c]	MacRae and Towers (1985)
Anacarrdiaceae		
Rhus pontifica	4′-methoxy-3,3′,5-trihydroxystilbene (rhapontigenin)	Kawamura (1938)
Didiereaceae		
Allaudia dumosa	(−)-androyol	Rasamoelisendra *et al.* (1991)
	(+)-androyol	Rasamoelisendra *et al.* (1991)
	(−)-6-*C*-methylandroyol	Rasamoelisendra *et al.* (1991)
	(+)-6-*C*-methylandroyol	Rasamoelisendra *et al.* (1991)

Guttiferae		
Clusia paralycola	paralycolin A[d]	Delle Monache *et al.* (1987)
Berberidaceae		
Epimedium grandiflorum var. *thunbergianum*	7-hydroxy-2,3,4,6-tetramethoxy-9,10-dihydrophenanthrene-7-*O*-glucoside (icariside A_1)	Miyase *et al.* (1987)
	2,5-dihydroxy-3,4,6-trimethoxy-9,10-dihydrophenanthrene-2-*O*-glucoside (icariside A_2)	Miyase *et al.* (1988)
	4,7-dihydroxy-2,3,6-trimethoxy-9,10-dihydrophenanthrene-4-*O*-glucoside (icariside A_3)	Miyase *et al.* (1988)
	4,5-dimethoxy-3,3′,4′-trihydroxybibenzyl-3′-*O*-glucoside (icariside A_4)	Miyase *et al.* (1988)
	icariside E_3	Miyase *et al.* (1988)
	2,3-dihydroxy-4,6,7-trimethoxy-9,10-dihydrophenanthrene-2-*O*-glucoside (icariside A_5)	Miyase and Ueno (1991)
	3′,5-dihydroxy-3,4,4′-trimethoxybibenzyl-5-*O*-glucoside (icariside A_6)	Miyase and Ueno (1991)
Algae		
Polysiphonia urceolata	3,3′-dibromo-4,4′,5,5′-tetrahydroxybibenzyl	Kurata, Amiya and Nakano (1976)

[a] phenanthrenes from the Euphorbiaceae are thought to be diterpenoids (De Alvarenga *et al.*, 1981)
[b] revised structure
[c] *cis-* and *trans* isomers
[d] thought to be derived from a 2-prenylbiphenyl

B-series combretastatins and other bibenzyls from *Combretum* species (Table 4.2). Bibenzyls and phenanthrenes were first isolated from this genus by Letcher and Nhamo (1972a) and Letcher, Nhamo and Gumiro (1972). Further examples, named combretastatin (**171**) and combretastatins B-1 to B-4 (**172–173**), were described by Pettit and colleagues (Pettit *et al.*, 1982, 1987; Pettit, Cragg and Singh, 1987; Pettit, Smith and Singh, 1987). A more recent study of *Combretum apiculatum* has revealed the presence of eight bibenzyls in this species (Malan and Swinny, 1993). Some of these *Combretum* bibenzyls show interesting biological properties (Chapter 7).

Another interesting group of bibenzyls with plant growth regulatory properties are batatasins III, IV and V (**167–170**) from *Dioscorea* species (Table 4.3). These were first discovered as dormancy factors in tubers of *D. opposita* (= *D. batatas*) by Hashimoto, Hasegawa and Kawarada (1972) and Hashimoto *et al.* (1974), and their effects on plant growth and metabolism were investigated in some detail (Chapter 7).

Batatasin III (**167**) and its 3′-*O*-methyl derivative (**168**) have also been found in several orchids (Table 4.8). Other orchid bibenzyls include the dihydro derivatives of pinosylvin dimethyl ether (**165b**) and resveratrol (**166a**), cirrhopetalidinin (**179c**), cirrhopetalidin (**179b**), cirrhopetalinin (**179a**), gigantol (**180**), aloifols I (**181**) and II (**182**), crepidatin (**183**), cumulatin (**184**) and erianin (**185**). Arundin (**186**) and arundinin (**187**) from *Arundina bambusifolia* are examples of 4-hydroxybenzyl derivatives of bibenzyls, other variations of which occur in *Bletilla striata*.

The liverwort bibenzyl, lunularin (**6**), has been found in *Hydrangea macrophylla* and *Morus laevigata*. Two further bibenzyls, 2,3′,4-trihydroxybibenzyl and dihydrooxyresveratrol (**166b**), occur in this and other *Morus* species. Dihydroresveratrol (**166a**) and three other bibenzyls are found in *Cannabis sativa*. One of these other bibenzyls, canniprene (**188**) contains an isopentenyl side group. Prenylated bibenzyls were also described from *Helichrysum umbraculigerum* and *Glycyrrhiza* species (3,5-dihydroxy-4-isopentenylbibenzyl, 3-hydroxy-4-isopentenyl-5-methoxybibenzyl, glepidotin C (**189**), gancaonins R (**190**) and S (**191**)).

Three bibenzyl glycosides have been found in higher plants. *Chlorophytum arundinaceum* contains 2,2′,4,4′-tetrahydroxybibenzyl-2-*O*-xylopyranoside, and *Epimedium grandiflorum* var. *thunbergianum* synthesizes icarisides A_4 (**192**) and A_6 (**193**).

4.3 BIBENZYL-2-CARBOXYLIC ACIDS

The first bibenzyl-2-carboxylic acid to be described was the widespread liverwort constituent lunularic acid (**18**). This compound has also been found in higher plants, in *Allium* species and in *Hydrangea macrophylla*. Two further bibenzyl-2-carboxylic acids (**194** and **195**) were reported in *Helichrysum umbraculigerum* by Bohlmann and Hoffmann (1979), together with the stilbene equivalent of the latter compound. Two prenylated bibenzyl-2-carboxylic acids, amorfrutins A (**196**) and B (**197**), were found in *Amorpha fruticosa* by Mitscher *et al.* (1981). Amorfrutin was also isolated from *Glycyrrhiza acanthocarpa* by Ghisalberti, Jefferies and McAdam (1981).

4.4 SPIRANS

Cannabispirone (= cannabispiran) (**198a**) and cannabispirenone-A (= dehydrocannabispiran) (**199a**) were isolated from *Cannabis sativa* of Indian and South African origin by two separate groups in 1976 (Ottersen *et al.*, 1976; El-Feraly *et al.*, 1977; Bercht *et al.*, 1976). The isomeric iso-cannabispiran (**198b**) was later identified in Panamanian *Cannabis sativa* by ElSohly and Turner (1982), and cannabispirenone-B (**199b**) was found in Mexican and Thailand *Cannabis* by Kettenes-Van den Bosch and Salemink (1978) and Crombie and Crombie (1982). Both of these groups of authors pointed to the probable biosynthetic relationship between the cannabispirans and 3,4′-dihydroxy-5-methoxybibenzyl, another constituent of *Cannabis* (Table 4.16). This suggestion was supported by the finding of cannabispiradienone (**200**), a probable intermediate in the pathway (Crombie and Crombie, 1982; Crombie, Crombie and Jamieson, 1979), together with related 9,10-dihydrophenanthrenes (**299**, **300**). Further spirans found in *Cannabis* include α- and β-cannabispiranols (**201a**, **201b**) (Bercht *et al.*, 1976; Boeren *et al.*, 1977; Crombie and Crombie, 1978, 1982) and an acetyl cannabispiranol (Shoyama and Nishioka, 1978). The synthesis and biosynthesis of these compounds are discussed in more detail in Chapter 5. Recently a phenanthrene derivative with a spirolactone ring, blespirol, has been isolated from the orchid *Bletilla striata* (Yamaki *et al.*, 1993b).

4.5 STILBENES

The unsubstituted *trans*-stilbene has been identified as a constituent of *Alnus sieboldiana* (Asakawa, 1970) and *Petiveria alliacea* (Adesogan, 1974). Most plant stilbenes have the *trans*- (*E*-) configuration, and this can be assumed unless the *cis*- (*Z*-) isomer is specifically mentioned. The majority of stilbenes have the 3,5-substitution pattern in one ring indicative of the polymalonate origin of that ring (Chapter 5).

The simplest angiosperm stilbene is 3,5-dihydroxystilbene (pinosylvin, **143a**), most commonly associated with the genus *Pinus* (Table 3.1). In angioperms it occurs in *Dalbergia sissoo* (Table 4.5) and *Nothofagus* species and together with its mono- and dimethyl ethers in *Alnus* species (Table 4.17). 3,4′,5-Trihydroxystilbene (resveratrol, **144a**) is more widely distributed, occurring in *Scirpus* species and *Veratrum grandiflorum* in the monocots (Table 4.1) and in various members of the Fabaceae (Table 4.5),

Moraceae (Table 4.6), Myrtaceae (Table 4.7), Polygonaceae (Table 4.13), Vitaceae (Table 4.14), Fagaceae and Palmae (Table 4.17). It is also the precursor of many of the oligostilbenes, exclusively so in the Dipterocarpaceae (Table 4.4). A number of methyl ethers of resveratrol are also found in angiosperms. 3,5-Dimethoxy-4′-hydroxystilbene (pterostilbene, **202a**) was found in sandalwood (*Pterocarpus* species) by King *et al.* (1953), in *Vitis* species by Langcake (1981) and in *Pterolobium hexapetalum* by Kumar *et al.* (1988). 3,5-Dihydroxy-4′-methoxystilbene (**202b**) occurs in *Clavisia racemosa* (Table 4.6) and *Rheum* species (Table 4.13). 3,4′,5-Trimethoxystilbene (**202c**) has been reported from *Pterolobium hexapetalum* (Fabaceae, Table 4.5), *Virola cuspidata* and *V. elongata* (Myristicaceae, Table 4.17). An unusual compound which occurs with resveratrol in *Bauhinia racemosa* is pacharin (**203**). This was suggested to be derived from resveratrol by ring closure (Anjaneyulu *et al.*, 1984), although the substitution pattern and the presence of a methyl side chain suggest that this needs further investigation.

The 3-*O*-glucoside of resveratrol, piceid (**144b**) is a common constituent of *Eucalyptus* species (Table 4.7) and also occurs in *Ampelopsis brevipedunculata* (Table 4.14). This species produces the 4′-*O*-glucoside, resveratroloside (**144c**) which has also been found in rhubarb (*Rheum rhaponticum*, Table 4.13). Two more unusual glycosides of resveratrol are the 3-*O*-rutinoside from *Terminalia sericea* (Bombardelli, Martinelli and Mustich, 1975), and the 4′-*O*-(6″-*O*-galloyl)glucoside, again from rhubarb (Nonaka, Minami and Nishioka, 1977; Kashiwada, Nonaka and Nishioka, 1984). The 3-rutinoside of 4′,5-dimethoxy-3-hydroxystilbene is a constituent of *Guibourtia coleosperma* (Steynberg *et al.*, 1988).

The tetrahydroxystilbene piceatannol (3,3′,4,5′-tetrahydroxystilbene, **147a**) is also widespread in angiosperms. Piceatannol was originally found in *Picea* species (Table 3.2) and in *Vouacapoua* species (King *et al.*, 1956). It is common in the Fabaceae (*Cassia, Centrolobium, Intsia, Laburnum, Maackia, Pericopsis, Schotia* and *Vouacapoua* species, Table 4.5), and is also found in *Morus* and *Rheum* species, and in several monocot genera including *Saccharum, Festuca* and *Scirpus*. In the last, it forms a dimer, scirpusin B (**252**), and a mixed resveratrol/piceatannol dimer, scirpusin A (**251**). A prenylated derivative of piceatannol occurs in groundnuts (*Arachis hypogaea*). The 3-*O*-glucoside of piceatannol, astringin (**147b**) is known from *Eucalyptus* and *Angophora* (Myrtaceae, Table 4.7), *Guibourtia coleosperma* (Table 4.5) and rhubarb (*Rheum rhaponticum*, Table 4.13). The 3′-*O*-glucoside occurs in rhubarb (*Rheum rhaponticum*) together with the 3′-*O*-xyloside and the 3′-*O*-(6″-*O*-galloyl)glucoside (Table 4.13). Oxyresveratrol (2,3′,4,5′-tetrahydroxystilbene, **204**) was found in members of the Moraceae (Table 4.6) and, with its 3-*O*-glucoside, in *Veratrum grandiflorum* (Table 4.1).

Rhapontigenin (4′-methoxy-3,3′,5-trihydroxystilbene, **205a**) is a constituent of *Rheum* species (Table 4.13) and *Rhus pontifica* (Table 4.17), but its 3-*O*-glucoside, rhapontin (**205b**), is the major constituent not only of *Rheum* species but also of *Polygonum multiflorum* (Tsukida and Yokota, 1954). It has also been reported to occur in *Eucalyptus sideroxylon* (Hillis and Hasegawa, 1962b) and *Guibourtia coleosperma*, in which the 3-*O*-rutinoside is also present (Steynberg *et al.*, 1988). The 3′-*O*-glucoside and its 2″-*O*- and 6″-*O*-galloyl and 2″-*O*-*p*-coumaryl derivatives have been found in *Rheum* species. 2,3,4′,5-Tetrahydroxystilbene-2-*O*-glucoside and its 2″-*O*- and 3″-*O*-galloyl derivatives were found in *Polygonum multiflorum* (Hata, Kozawa and Baba, 1975; Nonaka, Miwa and Nishioka, 1982).

Cassia species (Fabaceae, Table 4.5) contain 2,2′,3,3′-tetrahydroxystilbene (**206**, *C. siamea*), 3′-hydroxy-3,4,5′-trimethoxystilbene (**207a**) and 3′,4-dihydroxy-3,5′-dimethoxystilbene (**207b**, *C. didymobotrya*).

A few pentahydroxystilbenes are found in angiosperms. 3,3′,4,5,5′-Pentahydroxystilbene and its 3-*O*-glucoside are reported to occur in *Eucalyptus astringens* and *E. sideroxylon* (Table 4.7), and the *cis*- and *trans*-isomers of the aglycone occur in *Schotia brachypetala* (Table 4.5). 3,3′,4,4′,5-Pentahydroxystilbene was reported from *Vouacapoua americana* (King *et al.*, 1956). *Cis*- and *trans*-4-methoxy-3,3′,5,5′-tetrahydroxystilbene occur in the date palm (*Phoenix dactylifera*), together with resveratrol (Fernandez, Pedro and Seoane, 1983). Among the stilbenoids produced by *Combretum caffrum* (Table 4.2) are the A-series combretastatins. These *cis*-stilbenes include combretastatin A-2 (**175**, 3,4′-dimethoxy-3′-hydroxy-4,5-methylenedioxystilbene), combretastatin A-3 (**176**, 3′,5-dihydroxy-3,4,4′-trimethoxystilbene) and combretastatin A-4 (3′-hydroxy-3,4,4′,5-tetramethoxystilbene, **177**). Combretastatin A-1 is the hexa-substituted 2,3-dihydroxy-3′,4,4′5-tetramethoxystilbene (**174**). *C. caffrum* is so far the only stilbene-containing species in a genus which produces mainly bibenzyls, phenanthrenes and 9,10-dihydrophenanthrenes (Table 4.2).

Prenylated stilbenes occur in a number of angiosperm families, most notably the Fabaceae (Table 4.5). 3,5-Dihydroxy-4-geranylstilbene (amorphastilbol, **208**) was found in *Amorpha* species by Kemal *et al.* (1979). Groundnuts (*Arachis hypogaea*) produce several prenylated stilbenes as phytoalexins (Chapter 7). These include the *cis*- and *trans*- isomers of 4-isopentenyl-3,4′,5-trihydroxystilbene (arachidin II, **210**), 4-(3-methyl-but-1-enyl)-3,4′,5-trihydroxystilbene (arachidin III, **211**), 4-(3-methyl-but-1-enyl)-3,3′,4′,5-tetrahydroxystilbene (arachidin I, **209**) and 3-isopentadienyl-3′,4,5′-trihydroxystilbene. *Derris* species contain 3,5-dimethoxy-4-isopentenylstilbene (**212a**) and 4-isopentenyl-3,4′,5-trimethoxystilbene (**212b**). *Lonchocarpus longistylus* produces another series of prenylated stilbenes, the longistylines (**214–217**), two of which contain two isopentenyl side chains, while *L. nicou* produces the related pyran, lonchocarpene (**213**).

Two geranyl stilbenes are found in species of *Chlorophora* (Moraceae,

Table 4.6). These are 4-geranyl-3,4′,5-trihydroxystilbene (**218a**) and 4-geranyl-2′,3,4′,5-tetrahydroxystilbene (chlorophorin, **218b**). A dimer of chlorophorin has also been found in *C. excelsa*. *Helichrysum umbraculigerum* (Asteraceae, Table 4.17) contains amorphastilbol (**208**) and related stilbene-2-carboxylic acids (see below).

4.6 STILBENE-2-CARBOXYLIC ACIDS

The first stilbene-2-carboxylic acid to be identified was hydrangeic acid (**223**), which can be converted into the phenyldihydroisocoumarin hydrangenol (**222a**). Such stilbene-2-carboxylic acids are rare in plants, although they may be unstable or bound intermediates in stilbenoid synthesis. Pinosylvic acid (3,5-dihydroxystilbene-2-carboxylic acid, Figure 1.2) has been found as the 3-*O*-glucoside (gaylussacin, **219**) in *Gaylussacia* species (Ericaceae, Table 4.17). The other plant stilbene-2-carboxylic acids are prenylated. These are 3,5-dihydroxy-4-geranylstilbene-2-carboxylic acid (**220**) and a derivative with a hydroxylated side chain (**221**) from *Helichrysum umbraculigerum* (Bohlmann and Hoffmann, 1979), and the 4- and 6-isopentenyl derivatives of 3-hydroxy-5-methoxystilbene-2-carboxylic acid from *Cajanus cajan* (Cooksey *et al.* 1982).

4.7 PHENYLDIHYDROISOCOUMARINS

Compounds of this type have so far been found only in *Hydrangea macrophylla* and its various subspecies and varieties (Table 4.15). Hydrangenol (**222a**), phyllodulcin (**224a**), their monomethyl ethers (**222b**, **224b**), their 8-*O*-glucosides (**222c**, **224c**) and a range of glucosides of a more substituted phenyldihydrolisocoumarin (macrophyllosides A, B and C, **226a**, **226b** and **226c**; Hashimoto, Tori and Asakawa, 1987) have been described (Table 4.15). A number of glucosides of α-hydroxybibenzyl-2-carboxylic acids (**225a**, **225b**, **227a**, **227b**) have also been found in *Hydrangea* (Suzuki *et al.*, 1979a, b; Yagi *et al.*, 1972). The variety *thunbergii* ('Amacha' in Japan) is used to make a sweet ceremonial tea. Phyllodulcin and its derivatives contribute to the sweet taste of this tea, and phyllodulcin is reported to be sweeter than sucrose. *Hydrangea* has been used in several investigations of the biosynthesis of these compounds (Chapter 5).

4.8 STILBENE OLIGOMERS

Since the first description of an oligomeric stilbene derivative, hopeaphenol (**245**) from *Balanocarpus* (= *Hopea*) *heimii* by Coggan *et al.* (1965) and Coggan, King and Wallwork (1966), the number of reported structures has risen to 26. Most of these are derivatives of resveratrol and include dimers (ϵ-viniferin, **228**; balanocarpol, **229**; pallidol, **230**; tricuspidatol-A, **231**;

ampelopsins A, **232**, B, **233** and D, **235**), trimers (ampelopsins C, **234** and E (*cis* and *trans*), **236**; α-viniferin, **238**; distichol, **239**; canaliculatol, **240**; copalliferol A, **241**; copalliferol B, **242**; stremonoporol, **243**; and miyabenol C, **244**) and tetramers (hopeaphenol, **245**; vaticaffinol, **246**; miyabenol A, **249**; miyabenol B, **250**, kobophenol A, **247**; kobophenol B, **248**; and ampelopsin H, **237**). These are found in the Vitaceae (Table 4.14), Fabaceae (Table 4.5) and, especially, the Dipterocarpaceae (Table 4.4). Some resveratrol oligomers found in angiosperms and the Gnetales were recently reviewed by Sotheeswaran and Pasupathy (1993). There are also a number of piceatannol dimers (scirpusin B, **252**; cassigarols A–D, **253–256**) and a mixed piceatannol/reveratrol dimer, scirpusin A (**251**). A chlorophorin (**218b**) dimer, excelsaoctaphenol (**257**), was found in *Chlorophora excelsa* (Moraceae, Table 4.6). The possibility of higher oligomers occurring in plants is suggested by the detection of γ-viniferin, a partially characterized higher resveratrol oligomer, in *Vitis vinifera* (Langcake and Pryce, 1977a, b).

4.9 STILBENE CONDENSED TANNINS

Structures of this type (where a stilbene is linked directly to a flavonoid) have been isolated from three plant sources, *Alluaudia dumosa* (Didieriaceae, Table 4.17), Rhodesian copalwood, *Guibourtia coleosperma* and *Sophora* species (Fabaceae, Table 4.5). Androyol (**258a**) and methylandroyol (**258b**) from the former are resveratrol/3-methoxyflavone conjugates of which both the (+)- and (−)- enantiomers were found in extracts of the bark (Rasamoelisendra *et al.*, 1991). The sapwood of *G. coleosperma* contains a number of piceatannol/4′,7-dihydroxyflavan-3,4-diol derivatives, two of which are dimers (**259**, **260**) and two trimers (one molecule of piceatannol and two of the flavanoid; **261**, **262**). Three prenylated flavanones containing resveratrol residues, sophoraflavanones H, I and J, were found in the roots of *Sophora moorcroftiana* by Shirataki *et al.* (1991). Sophoraflavanone I and its cyclic isomer (with a lavandulyl group), leachianone C, were also isolated from the roots of another species, *S. leachiana*, by Iinuma *et al.* (1991). Sophoraflavanone I was given the trivial synonym leachianone B.

4.10 9,10-DIHYDROPHENANTHRENES AND PHENANTHRENES

Two biosynthetic pathways may lead to nitrogen-free phenanthrenes and 9,10-dihydrophenanthrenes in higher plants. Phenanthrenes such as those found in *Juncus* (juncunone, **263**; juncunol, **264**; juncusol, **265**; effusol and their 9,10-dihyro equivalents) and *Micrandropsis* (micrandrols A–F, e.g. **266–268**) with a methyl side chain may be terpenoid in origin. The origin of the unsubstituted phenanthrene found in wheat and *Pterolobium hexapetalum* is not known. The stilbenoid nature of other phenanthrenes is suggested by their substitution patterns and by the co-occurrence of other stilbenoids in the same plants.

The main families which produce phenanthrenes and 9,10-dihydrophenanthrenes are the Combretaceae (Table 4.6), the Dioscoreaceae (Table 4.6) and the Orchidaceae (Tables 4.9 and 4.10). *Combretum* species produce a large number of phenanthrenes, most with five substituents. With the exception of 7-hydroxy-2,4,6-trimethoxyphenanthrene (Ta IV, **270**), which has also been found in *Tamus communis*, none of these have trivial names. In the Dioscoreaceae both *Dioscorea* species and *Tamus communis* produce phenanthrenes (Table 4.6). The most famous of the *Dioscorea* phenanthrenes is batatasin I (**269**), one of the dormancy factors originally identified in *Dioscorea opposita* (= *D. batatas*). Among the other nine phenanthrenes and dihydrophenanthrenes in this genus only one (Ta VI, **271**) also occurs in *Tamus communis*. The orchid phytoalexin hircinol (**274b**) occurs in *D. rotundata*. Two 9,10-dihydrophenanthrenes and nine phenanthrenes have been found in *Tamus communis*, including batatasin I (**269**), Ta VI (**271**) and a phenanthraquinone, Ta II (**272**).

A large number of phenanthrenes and 9,10-dihydrophenanthrenes are produced by some members of the Orchidaceae (Tables 4.9 and 4.10). The first 9,10-dihydrophenanthrene, orchinol (**273**), was isolated as a phytoalexin from the European orchids *Orchis mascula* and *O. militaris* infected with *Rhizoctonia repens* by Gäumann, Hardegger and coworkers. The similar phytoalexins loroglossol (**274a**) and hircinol (**274b**) were later found in *Loroglossum hircinum*. The biological activity of these compounds is discussed in Chapter 7. One of the most common orchid dihydrophenanthrenes is coelonin (**275**), occurring in six species. These include *Bletilla striata* and *Eulophia nuda*, two species which contain 4-hydroxybenzyl derivatives of coelonin and other phenanthrenes (e.g. **276–278**). Other orchid phenanthrenes include cirrhopetalin (**279**), nudol (**280**), confusarin (**281**) and confusaridin (**282**).

Another group of stilbenoid derivatives which are particularly characteristic of some orchids are the phenanthropyrans and phenanthroquinones (Table 4.11). Examples of the former are flavidin (**283**), coelogin (**284**), imbricatin (**285**) and amoenumin (**286**), while orchid phenanthraquinones include ochrones A (**287**) and B (**288**), bulbophyllanthrone (**289**), cypripedin (**290**) and Ta II (**272**). An unusual phenanthraquinone of uncertain biosynthetic origin is latinone (**291**) from *Dalbergia latifolia* (Fabaceae, Table 4.5). Lonchocarpene (**213**) is a phenanthropyrone, also from the Fabaceae (*Lonchocarpus nicou*). An unusual dihydrophenanthropyrone, paralycolin A (**292**) from *Clusia paralycola*, was suggested by Delle Monache *et al.* (1987) to be derived from a 2-prenylbiphenyl.

Phenanthrene dimers, whether C-C linked or ether linked, are also found in the Orchidaceae (Table 4.12). Most of these structures have been found in *Bletilla striata* (blestrins A (**293a**) and B (**293b**), blestrianols A (**294**) and B (**295**), blestriarene A (= flavanthrin, **296**) and blestriarene B (**297**)). Blestriarine A (**296**) and other dimers, including cirrhopetalanthrin (**298**), have been found in *Arundina bambusifolia, Cirrhopetalum maculosum, Eria flava* and *Eulophia nuda*.

Other families contain species which produce 9,10-dihydrophenanthrenes. Cannithrene-1 (cannabidihydrophenanthrene, **299**) and cannithrene-2 (**300**) are constituents of *Cannabis sativa* (Cannabinaceae, Table 4.16). The prenylated dihydrophenanthrenes gancaonin V (**301**) and gancaonin U (**302**) were found in *Glycyrrhiza uralensis*, a member of a family (Fabaceae, Table 4.5) which is otherwise devoid of substituted phenanthrenes. Two 9,10-dihydrophenanthrenes were found on the surface of leaves of *Empetrum nigrum* (Empetraceae, Table 4.17) by Wollenweber *et al.* (1992). *Epimedium grandiflorum* var. *thunbergianum* (Berberidaceae, Table 4.17) produces a series of glucosylated 9,10-dihydrophenanthrenes, icarisides A_1 (**303**), A_2 (**304**), A_3 (**305**) and A_5 (**306**).

Papaver somniferum and *P. bracteatum* (Papaveraceae, Table 4.17) contain both alkaloid and nitrogen-free phenanthrenes. The latter include thebaol (**308a**) and *O*-methylthebaol (**308b**).

4.11 DIBENZOXEPINS

A dibenzoxepin, pacharin (**203**) has been reported from *Bauhinia racemosa* (Fabaceae) by Anjaneyulu *et al.* (1984). 2,8-Dihydroxy-1,7-dimethyl-6-vinyl-10,11-dihydrodibenz[b,f]oxepin is a cytotoxic constituent of *Juncus effusus* (Della Greca *et al.*, 1993). Although these authors suggested biosynthesis via oxidative coupling of the aromatic rings of a dihydrostilbene for this compound, the presence of methyl and/or vinyl substituents in both dibenzoxepins points to the possibility of a terpenoid origin.

5

Biosynthesis and chemical synthesis of stilbenoids

5.1 INTRODUCTION

The first steps in the biosynthesis of stilbenoids were outlined in Chapter 1, and the relationship between the biosynthesis of stilbenoids and that of flavonoids is illustrated in Figure 1.2. The initial cyclization steps have received considerable attention in recent years, and these studies have benefited greatly from the application of the techniques of molecular biology. Later stages in the biosynthesis of most of the newly discovered stilbenoids are still largely based on guesswork, since there have been few radio-tracer studies performed in the last 10 years. The emphasis on chemistry rather than biochemistry is also reflected in the advance of synthetic methods to various classes of stilbenoids.

5.2 BIOSYNTHETIC PATHWAYS

Three pathways to stilbenes were proposed in the earlier literature. One of these involved the condensation of a phenylacetic acid with a phthalic acid (Seshadri, 1957) and allowed for the existence of stilbene-2-carboxylic acids (Figure 5.1a). Robinson (1955) proposed an entirely polyketide origin (Figure 5.1b), based on the predominance of the 3,5- substitution pattern of natural stilbenes. An earlier suggestion by Birch and Donovan (1953) envisaged the mixed phenylpropanoid/polyketide pathway (Figure 5.1c), which was later shown by radio-tracer studies to be correct. At the time this route was proposed only one stilbene-2-carboxylic acid, hydrangeic acid (**223**), was known, and compounds of this type are still relatively rare in plants (Chapters 2 and 4).

Incorporation of label from both acetate and phenylpropanoids into hydrangenol (**222a**) in *Hydrangea macrophylla* was obtained by Ibrahim and Towers (1960, 1962) and Billek and Kindl (1961, 1962b). In the experi-

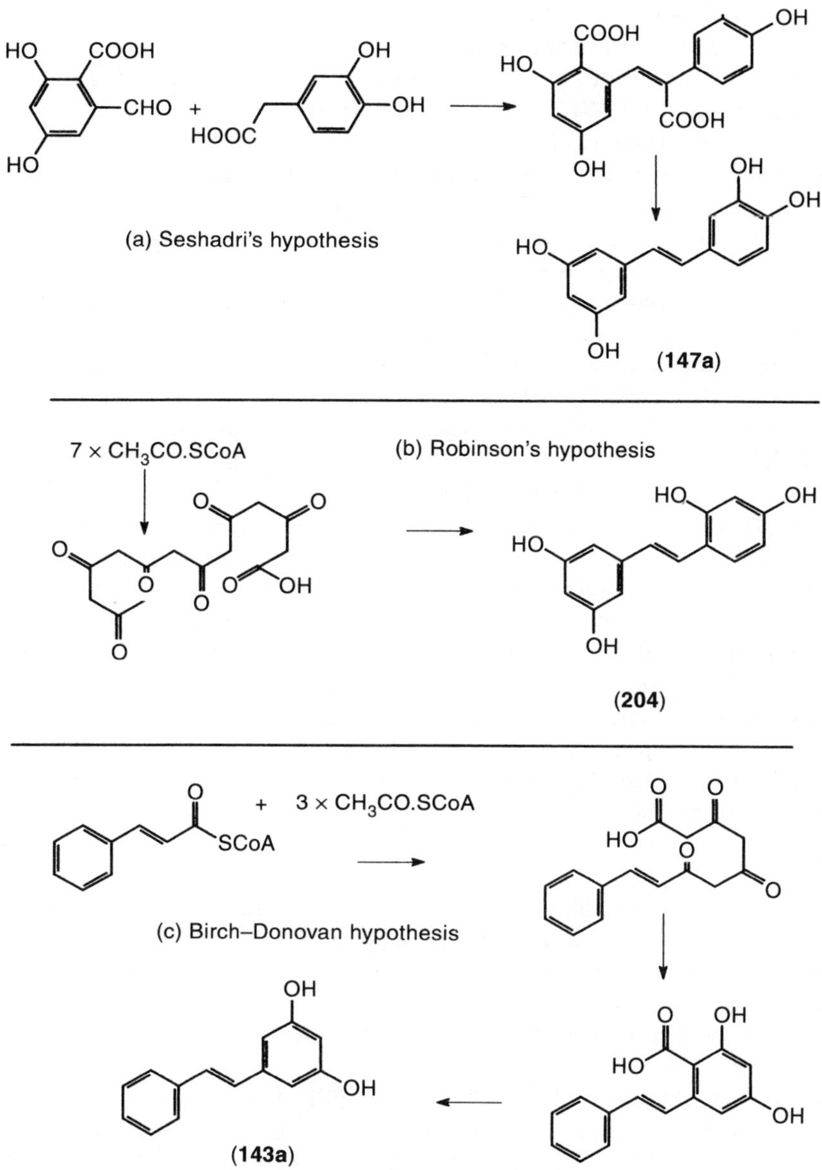

Figure 5.1 Possible biosynthetic routes to stilbenes.

ments of Ibrahim and Towers (1960), phenylalanine was a better precursor than tyrosine. Acetate and tyrosine were shown to be incorporated into both hydrangenol and hydrangeic acid by isolated *Hydrangea* chloroplasts by Kindl (1971a). Better incorporation of tyrosine or *p*-coumaric than of phenylalanine or cinnamic acid would suggest that hydroxylation at the 4′ position occurs before cyclization. Experiments with the variety *thunbergii* (Yagi *et al.*, 1977) further suggest that phyllodulcin is derived from hydrangenol by hydroxylation at the 3′ position and *O*-methylation at C-4′. These authors also obtained incorporation into phyllodulcin of label from tritiated hydrangea glucoside A (**225a**) and suggested that this was an intermediate in hydrangenol biosynthesis. This hypothesis requires confirmation using ^{14}C-labelled precursors to establish whether the α'-hydroxylated hydrangea glucosides are indeed intermediates in hydrangenol synthesis or later derivatives of hydrangenol.

Billek and Schimpl (1962, 1966) obtained labelling in the A ring of oxyresveratrol after feeding *Morus alba* with 1-^{14}C-acetate. Incubation with 3-^{14}C-*p*-coumaric acid resulted in label appearing in the B ring (Figure 5.2). Further evidence for the Birch and Donovan pathway was obtained when radiolabelled acetate and phenylpropanoids were incorporated into pinosylvin (**143a**), pinosylvin monomethyl ether (**143b**), rhapontin (**205b**) and piceid (**144b**) in *Pinus* species (Billek and Ziegler, 1962; Rudloff and Jorgensen, 1963) and *Eucalyptus sideroxylon* (Hillis and Hasegawa, 1962a). Langcake and Pryce (1977b) obtained incorporation of label from phenylalanine and malonate into the expected rings of resveratrol in grapevine. In this case phenylalanine was a better precursor than tyrosine, suggesting that hydroxylation at the 4′-position of resveratrol may occur post-cinnamate formation. Better incorporation of phenylalanine than of tyrosine into stilbenes was reported for *Eucalyptus* species by Hillis and Isoi (1965a), although tyrosine was more readily converted to sugars and acetate. Stilbene biosynthesis in *Eucalyptus* was also studied in some detail by Hillis and Ishikura (1970) and Hillis and Yazaki (1971). Synthesis of *p*-coumaric acid may involve a membrane-bound enzyme system at the endoplasmic reticulum in which phenylalanine ammonia lyase (PAL) is tightly coupled to the hydroxylating monooxygenase (Kindl, 1985). Further hydroxylation, *O*-methylation or prenylation of the A or B rings probably occurs after the aldol condensation of the polyketide precursor (Rupprich, Hildebrandt and Kindl, 1980; Kindl, 1985).

Incorporation of label from acetate and phenylalanine into the liverwort bibenzyls lunularic acid (**18**) and lunularin (**6**) was obtained by Pryce (1971a,b, 1972a,b) with *Lunularia cruciata* and by Gorham (1977b) with *Conocephalum conicum*. Hydrangenol (**222a**), although not found in liverworts, was also incorporated into lunularic acid in these liverworts, suggesting the capacity for the reduction of stilbenoids to bibenzyls in the

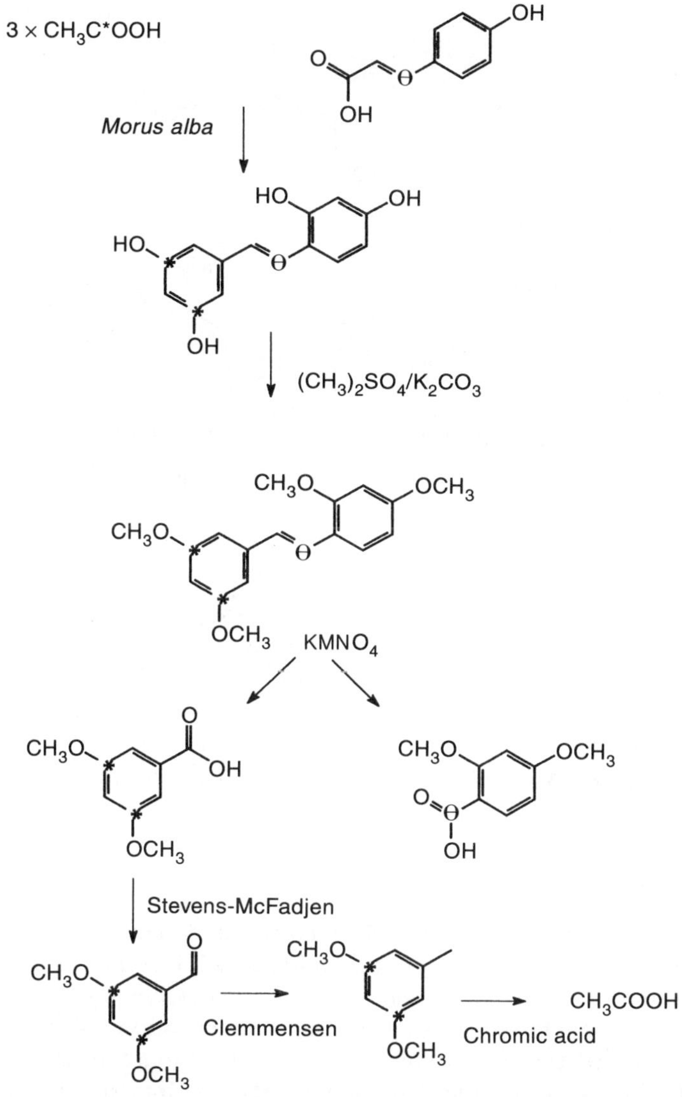

Figure 5.2 Incorporation of 1-^{14}C-acetate (*) and 3-^{14}C-*p*-coumaric acid (θ) into oxyresveratrol by *Morus alba* and the subsequent degradation.

Hepaticae. However, the discovery of prelunularic acid (**87**) in *Marchantia polymorpha* by Ohta *et al.* (1983, 1984a,b) and Abe and Ohta (1984), and its ready conversion to lunularic acid, suggested a more direct route to liverwort bibenzyls (Figure 5.3) involving *p*-hydroxyphenylpropionic acid (Kindl, 1985).

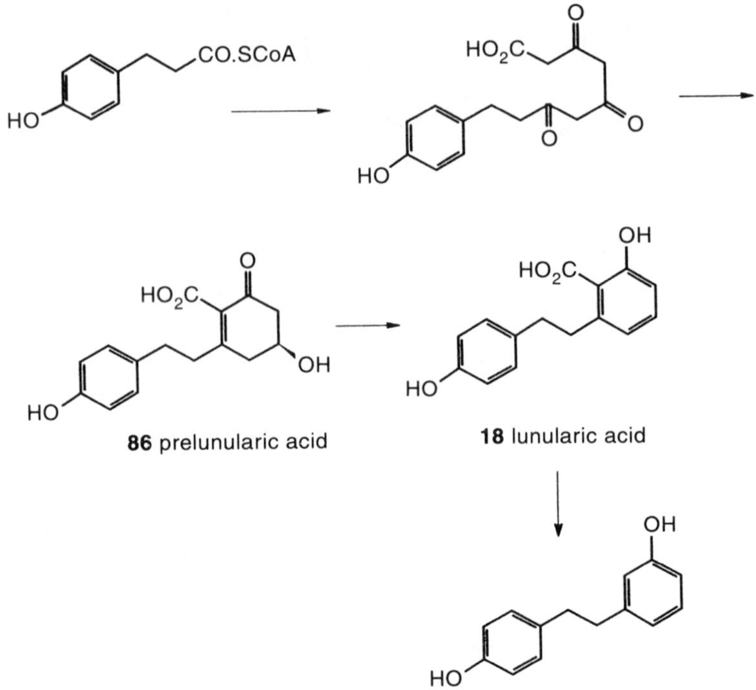

86 prelunularic acid

18 lunularic acid

6 lunularin

Figure 5.3 Biosynthesis of lunularic acid and lunularin.

There are nine types of bis(bibenzyl)s so far found in the Hepaticae. Their biosynthetic pathways are not yet fully elucidated, but the possible routes are illustrated in Figure 5.4. The mode of cyclization into macrocyclic bis(bibenzyl)s are inferred to occur through dimerization into perrottetins (Type I) followed by cyclization either with the oxygen of the hydroxyl groups (Types A–D) or direct C-C bond formation into riccardins (Types E–G) (Fig. 2.1). Compounds belonging to Type H derive from ether formation between rings B and D, followed by C-C bond formation at C-6 and C-2′ positions. A computational study (Keserú and Nógrádi, 1992) suggests that the possible conformations of bis(bibenzyl)s are constrained by the strain energies involved.

The formation of stilbene oligomers was reviewed by Sotheeswaran and Pasupathy (1993). They suggest that oligomers with an oxygen heterocycle are synthesized via ϵ-viniferin (Figure 5.5), whereas nonheterocyclic oligomers result from a different mechanism (Figure 5.6).

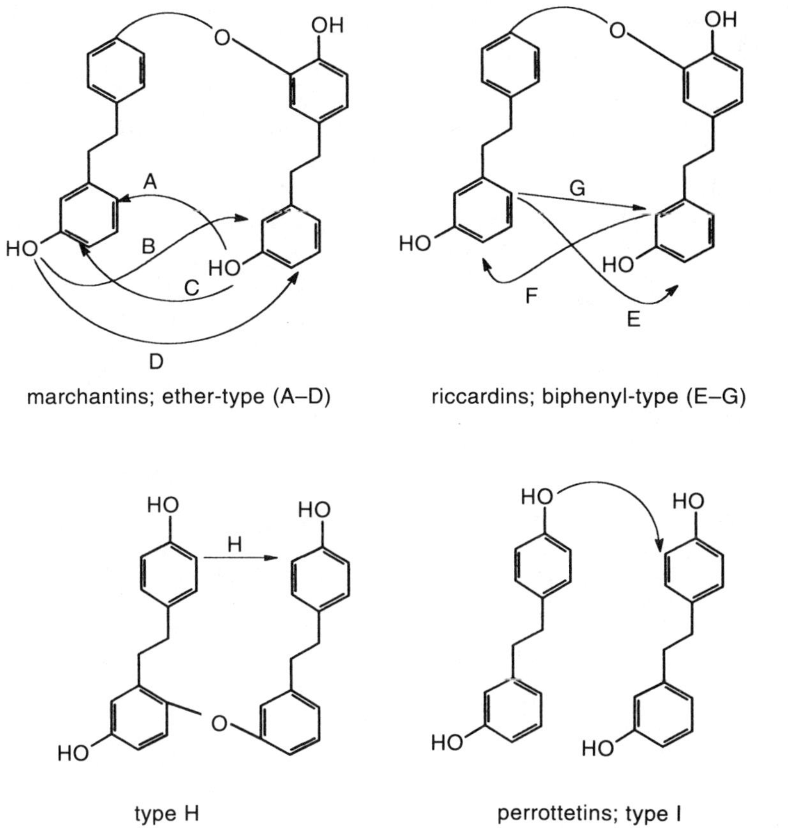

Figure 5.4 Possible cyclizations in bis(bibenzyl)s.

Bibenzyl and 9,10-dihydrophenanthrene biosynthesis in angiosperms has been studied by Kindl and coworkers. The role of coumaric acids in the pathway to orchinol (**273**) and related compounds was suggested by Stoessl (1982), but experiments by Fritzemeier and Kindl (1983) indicated a route involving the initial formation of bibenzyls from *m*-hydroxyphenylpropionic acid (Figure 5.7). In *Cannabis sativa* the co-occurrence of bibenzyls, 9,10-dihydrophenanthrenes and spirans led Crombie, Crombie and Jamieson (1979) and Crombie and Crombie (1982) to suggest the involvement of the latter in the cyclization of bibenzyls to dihydrophenanthrenes (Figure 5.8). In *Dioscorea* there appear to be two branches of the stilbenoid pathway (Fritzemeier, Kindl and Schlosser, 1984), one leading from *p*-coumaric acid to the stilbene resveratrol (**144a**) and then to phenanthrenes, and the other involving *m*-hydroxyphenylpropionic acid and bibenzyl

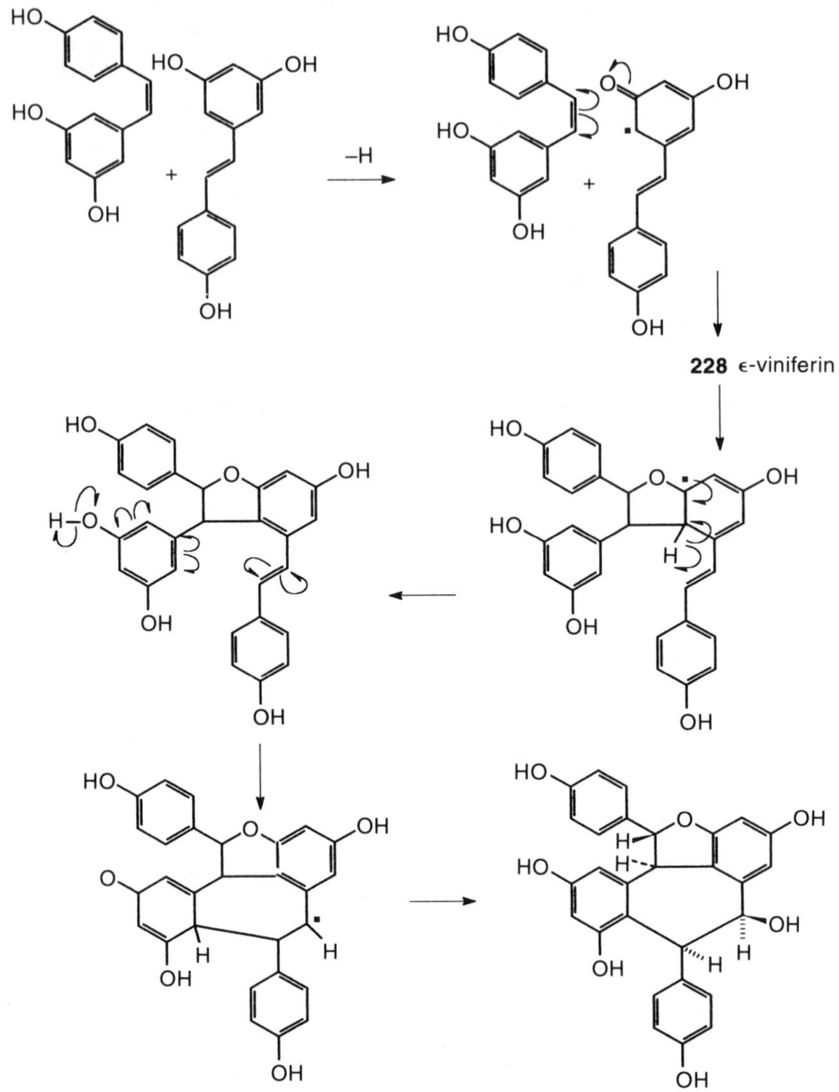

229 balanocarpol

Figure 5.5 Biosynthesis of heterocyclic resveratrol oligomers.

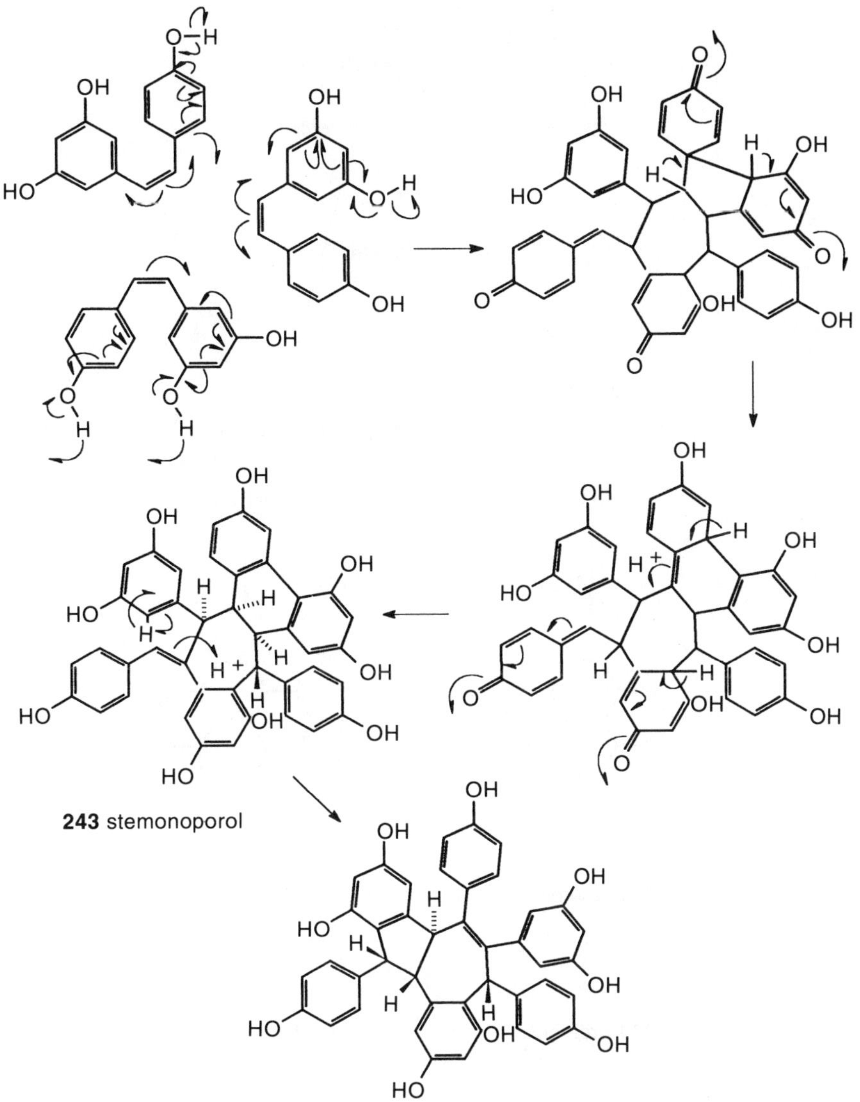

243 stemonoporol

241 copalliferol A

Figure 5.6 Biosynthesis of nonheterocyclic resveratrol oligomers.

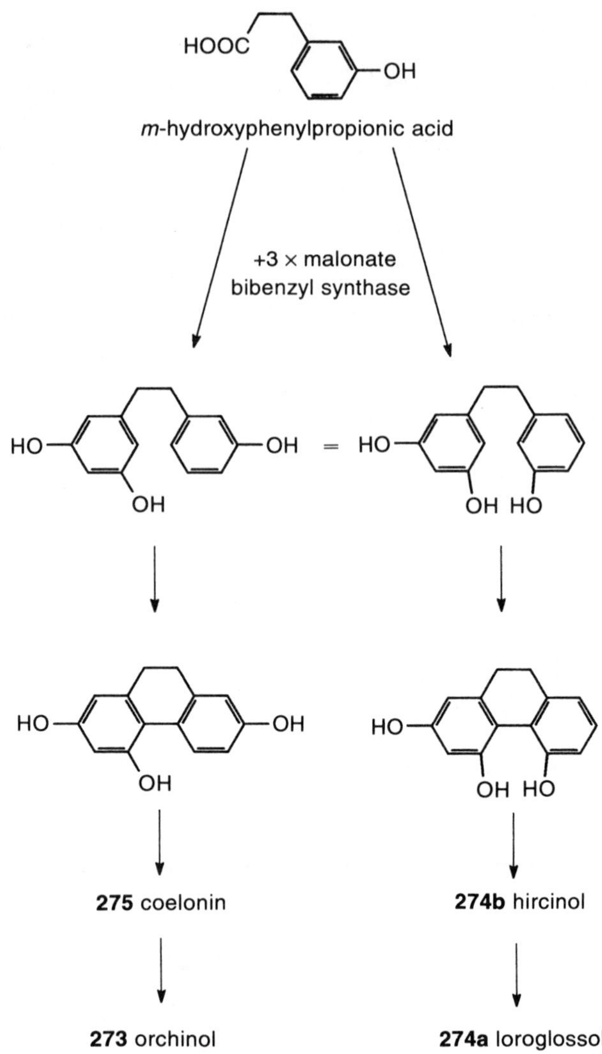

Figure 5.7 Biosynthesis of orchid bibenzyls and 9,10-dihydrophenanthrenes.

formation, with subsequent cyclization to 9,10-dihydrophenanthrenes. In the laboratory phenanthrenes may be obtained from stilbenes by the action of UV light, but the biosynthetic mechanism of phenanthrene formation is not known. The origin of the 4-hydroxybenzyl-substituted stilbenoids has not been established by radio-tracer experiments. Majumder and Ghosal (1994) suggested that arundin (**186**) might be derived from one malonate and three phenylpropanoid units (Figure 5.9).

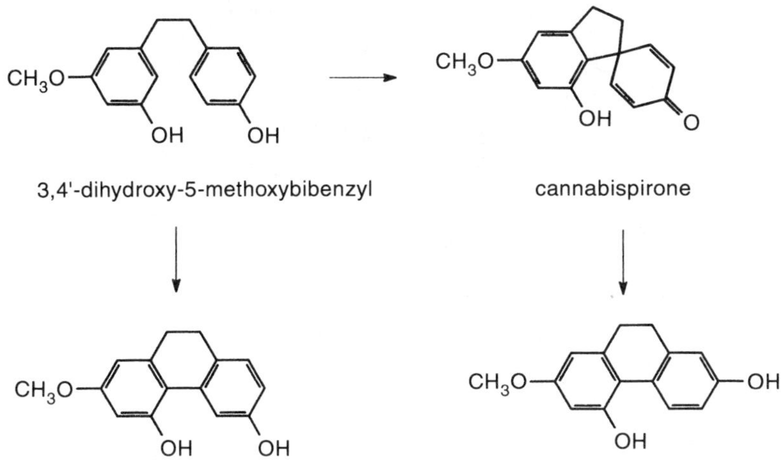

Figure 5.8 Possible involvement of cannabispirone in the biosynthesis of 9,10-dihydrophenanthrenes in *Cannabis*.

5.3 ENZYMES OF STILBENOID BIOSYNTHESIS

Hillis and Ishikura (1969a) extracted and partially purified an enzyme from *Eucalyptus sideroxylon* which converted cinnamoyl triacetate to *trans*-pinosylvin (**143a**). Enzyme activity was stimulated by Mg^{2+} ions (2–10 mol m^{-3}) and had a pH optimum of 8.2. A pinosylvin-forming stilbene synthase has also been detected in extracts of *Pinus sylvestris* seedlings (Schöppner and Kindl, 1979; Fliegmann *et al.*, 1992), and most of the activity appeared to be associated with membranes. Fliegmann *et al.* (1992) obtained expression of the cloned stilbene synthase in *Escherichia coli*, but suggested that the enzyme was a bibenzyl synthase (dihydropinosylvin synthase) because of an apparent preference, *in vitro*, for phenylpropionyl-coenzyme A rather than cinnamoyl-coenzyme A. Improved techniques for the expression of the enzyme in *E. coli* removed the influence of bacterial factors that altered the substrate specificity of the stilbene synthase (Schanz, Schroder and Schroder, 1992) and confirmed that cinnamoyl-coenzyme A was the preferred substrate.

A different stilbene synthase catalyses the formation of resveratrol in *Rheum rhaponticum* (Rupprich and Kindl, 1977; Rupprich, Hildebrandt and Kindl, 1980), *Picea excelsa* (Rolfs and Kindl, 1984), *Pinus griffithii* (Kindl, 1985), *Arachis hypogaea* (Rolfs, Fritzemeier and Kindl, 1981; Schöppner and Kindl, 1984) and *Vitis* species (Fritzemeier and Kindl, 1981). The most detailed studies have been performed on the resveratrol-forming

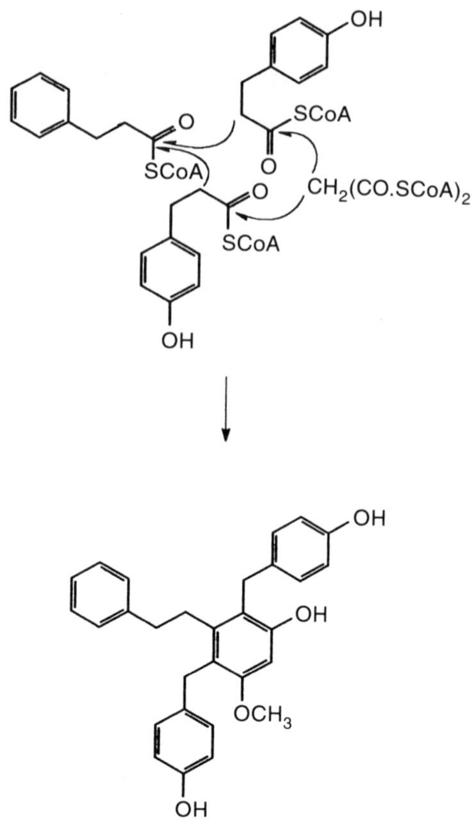

186 arundin

Figure 5.9 Possible biosynthesis of arundin from one malonate and three phenylpropanoid units, as suggested by Majumder and Ghosal (1993).

stilbene synthase in cell suspension cultures of *Arachis hypogaea* (Rolfs and Kindl, 1984; Schöppner and Kindl, 1984; Vornam, Schön and Kindl, 1988). It is a dimer composed of subunits with a molecular weight of about 47 kDa (sodium dodecyl sulphate (SDS) gel electrophoresis) and has an isoelectric point of 5.8. With a K_m of 2 mmol m^{-3}, *p*-coumaroyl-coenzyme A (*p*-coumaryl-CoA) is about six times better as a substrate than cinnamoyl-CoA (i.e. the reverse of the preference of the pinosylvin-forming stilbene synthase), but neither caffeoyl-CoA nor feruloyl-CoA were good substrates (Rupprich, Hildebrandt and Kindl, 1980). The K_m for malonyl-CoA is higher (380 mmol m^{-3}) for the synthase from rhubarb than from groundnut (10 mmol m^{-3}). The pH optimum for the resveratrol-forming stilbene synthase lies between 7.6 and 8.4 (Kindl, 1985). Stilbene synthase from grapevine is also a dimer, but with smaller subunits (43 kDa), and a pI of 5.4 (Liswidowati *et al.*, 1991). Both groundnut and grapevine stilbene synthases had a low K_m for *p*-coumaroyl-CoA (2 mmol m^{-3}), indicating tight binding of this substrate. Stilbene synthase incubated with malonyl-CoA alone exhibits reversible decarboxylase activity (Kindl, 1985). Incubation of stilbene synthase with labelled CO_2 and acetyl-CoA results in the formation of labelled malonyl-CoA. No free stilbene-2-carboxylic acid was produced by the purified stilbene synthase. Activity of the stilbene synthase from *Cissus antarctica* was reduced by thiol group inhibitors and required the presence of dithiothreitol (Fritzemeier and Kindl, 1981).

Bibenzyl synthases from both orchids (Fritzemeier and Kindl, 1983) and yams (Fritzemeier, Kindl and Schlosser, 1984) catalyse the formation of 3,3′,5-trihydroxybibenzyl from *m*-hydroxyphenylpropionyl-CoA and malonyl-CoA. A partially purified bibenzyl synthase preparation from *Epipactis palustris* also catalysed the conversion of phenylpropionyl-CoA to dihydropinosylvin (**165a**), but was less active with cinnamoyl- or *p*-coumaroyl-CoA (Gehlert and Kindl, 1991). Bibenzyl synthase from *Bletilla striata* is a dimer of 46 kDa subunits and exhibited the highest activity with dihydro-*m*-coumaroyl-CoA as substrate (Reinecke and Kindl, 1994). Dihydrocinnamoyl-CoA and dihydro-*p*-coumaroyl-CoA produced half the activity of dihydro-*m*-coumaroyl-CoA, while cinnamoyl-CoA and *p*- and *m*-coumaroyl-CoAs were much less active.

The enzymes involved in post-cyclization modification of stilbenoids are not well characterized. *O*-Methyltransferases showing specificity for the 4′-hydroxyl of piceatannol were detected in extracts of *Rheum rhaponticum* (Rupprich, Hildebrandt and Kindl, 1980) and in cell cultures of *Picea excelsa* (Rolfs and Kindl, 1984).

5.4 INDUCTION OF STILBENE SYNTHESIS

Although many stilbenoids are constitutive plant products, quite a large number are only synthesized following induction by biotic or abiotic factors. Indeed, many inducible stilbenoids were discovered during investigations into the responses of plants to infection or chemical or physical damage. The role of stilbenoids as phytoalexins will be considered in more detail in Chapter 7, but there are a number of reports which deal specifically with environmental factors affecting stilbenoid synthesis.

Cultured cells of *Vitis vinifera* responded to the presence of cell wall fragments of *Botrytis cinerea*, or to a soluble glucan elicitor, by switching from general protein synthesis to the synthesis of proteins specific to induced fungal resistance (Liswidowati *et al.*, 1991). Principal among these proteins were resveratrol-forming stilbene synthase and phenylalanine ammonia lyase. Increased amounts of resveratrol were synthesized by induced cells and secreted into the culture medium. Treatment of grapevine leaves with fungal elicitors resulted in increased activity of cinnamate *p*-hydroxylase and cinnamoyl-CoA ligase in addition to stilbene synthase and PAL (Fritzemeier and Kindl, 1981). In cultured cells of *Arachis hypogaea* stilbene synthase activity showed two peaks, one immediately after transfer to new medium and one in mid-log phase. During the stationary phase of growth (after about 10 days) stilbene synthase could be induced by fungal elicitors or other factors, and the increased activity was shown to be a result of *de novo* protein synthesis (Kindl, 1985). Induction occurs at the level of transcription, resulting in rapid (within 30 min) but transient increases in mRNA specific for the stilbene synthase. The resveratrol-forming stilbene synthases from rhubarb and *Picea excelsa* were found to be constitutive, i.e. not affected by various elicitors, but pinosylvin-forming stilbene synthase could be induced in young pine seedlings by inoculation with pathogenic fungi or with purified fungal elicitors (Gehlert, Schwekendiek and Kindl,1991; Schwekendiek, Pfeffer and Kindl, 1992), or by treatment with ozone (Rosemann, Heller and Sandermann, 1991).

In the orchid *Epipactis palustris* increased concentrations of 1,7-dihydroxy-5-methoxy-9,10-dihydrophenanthrene and 3,3′-dihydroxy-4′,5-dimethoxybibenzyl were found after wounding the rhizomes (Gehlert and Kindl, 1991). These increases were accompanied by increased activity of bibenzyl synthase, and were related to the extent of mycorrhizal infection. Similar results were obtained with *Bletilla striata* (Reinecke and Kindl, 1994).

In addition to fungal infection (or inoculation with fungal elicitors), stilbenoid synthesis may be induced by UV light. In *Vitis vinifera* the action spectrum for the UV induction of resveratrol and viniferin synthesis showed a peak at 260–270 nm, with little effect below 220 nm or above 310 nm (Langcake and Pryce, 1977b). Sunlight, with little radiation below 300–320 nm, was not effective in inducing resveratrol synthesis. Irradiation of the lower (abaxial) rather than the upper (adaxial) leaf surfaces was necessary for induction. Induction by UV light or infection with *Plasmopara viticola* led to the sequential formation of resveratrol, the dimer ε-viniferin and then the trimer α-viniferin (with *P. viticola* only).

Five species from the family Vitaceae (*Vitis sylvestris, V. riparia, Cissus antarctica, C. rhombifolia* and *Parthenocissus quinquefolia*) produced resveratrol-forming stilbene synthase after irradiation with UV light at 254 nm (Fritzemeier and Kindl, 1981). Cinnamate-4-hydroxylase and PAL activities were also induced concomitantly with stilbene synthase activity, all three enzymes showing maximum activity about 15 h after induction. Only a slight increase in cinnamate-2-hydroxylase was detected, and catalase and malate dehydrogenase activities were not affected.

Both infection with *Helminthosporium carbonum* (Ingham, 1976; Keen and Ingham, 1976; Rolfs, Fritzemeier and Kindl, 1981) and irradiation with UV light at 260–270 nm but not red or far-red light (Fritzemeier *et al.*, 1983) induced resveratrol and isopentenylresveratrol formation in groundnut (*Arachis hypogaea*). The peak of resveratrol production preceded that of its isopentenyl derivative. The time of maximum stilbene synthase and PAL activities were influenced by both the intensity and duration of the irradiation, and with the PAL peak preceding that of stilbene synthase (which in turn preceded maximum stilbene accumulation). Four different resveratrol synthase sequences were identified in *Arachis hypogaea* cell cultures by Lanz, Schröder and Schröder (1990). One (RS2) was never expressed, but expression of the others (RScDNA, RS1, RS3) was induced by elicitors from *Phytophthora megasperma*, yeast extract, dilution of the cultures and UV light (254 nm). These elicitors had different effects on the expression of the three sequences.

5.5 MOLECULAR BIOLOGY OF RESVERATROL-FORMING STILBENE SYNTHASE

DNA sequences for several members of the resveratrol synthase gene family from *Arachis hypogaea* have been published (Lanz, Schröder and Schröder, 1990; Schröder, Brown and Schröder 1988; Schröder and Schröder, 1990). In addition to the clones mentioned above, Lanz, Schröder and Schröder (1990) also found another (RS4) which was not functional. Two genomic clones (pGSG10 and pGSGll) had large areas of nearly identical sequences with the former containing the 5′ end and the latter the 3′ end of the resveratrol synthase genes (Schröder, Brown and Schröder, 1988). There was a single intron. The size of the predicted protein (42.7 kDa) was in good agreement with that of the extracted enzyme (43 kDa on polyacrylamide gels).

Comparisons with chalcone synthase sequences (Schröder, Brown and Schröder 1988; Schröder and Schröder, 1990) revealed substantial similarities, as might be expected of enzymes that use the same substrates but yield different products (resveratrol from resveratrol synthase and naringenin from chalcone synthase). There was little homology with 2-ketoacyl–acyl carrier protein (ACP) synthases, except for a short sequence which may be the active centre of the condensing reaction in the 2-ketoacyl-ACP synthase. Lanz *et al.* (1991) changed the six conserved cysteines in both resveratrol synthase and chalcone synthase to alanines by site-directed mutagenesis. While most of the changes led to reduced activities, only one (cysteine 169) was essential in both enzymes, and this may be the site of action of cerulenin, an antibiotic that reacts with cysteine at the active sites of condensing enzymes. The sequences around cysteine 169 were different from those at the active sites of other condensing enzymes such as polyketide synthases. Functional differences between the two enzymes were revealed at positions 135 and 195. Since chalcone synthase in cooperation with a reduced nicotine adenine dinucleotide phosphate (NADPH)-dependent reductase can produce a deoxychalcone, Schröder and Schröder (1990) suggested that a similar cooperation between a stilbene synthase and a reductase might result in the synthesis of stilbene-2-carboxylic acids.

Pinosylvin synthase cDNA was identified in *Pinus sylvestris* seedlings treated with *Botrytis cinerea* (Schwekendiek, Pfeffer and Kindl, 1992). The calculated mass of the predicted peptide (42.6 kDa) was very similar to that of the groundnut enzyme (42.7 kDa; Schröder, Brown and Schröder, 1988). While there were similarities with resveratrol synthase, there were differences (insertions of three amino acid residues near the *N* terminal, a glutamyl residue at position 90 and differences in the charged residues in positions 81–89) which were thought to be related to the different substrates of the two enzymes (*p*-coumaroyl and cinnamoyl-CoA).

Since stilbene synthase could use precursors commonly found in most plants, and is the only enzyme required to initiate the pathway to various stilbene phytoalexins, it is an attractive target for plant transformation, especially if it can be introduced into crop plants. This has been achieved with resveratrol synthase from grapevine being expressed in transformed tobacco plants (Hain *et al.*, 1993). These tobacco plants acquired increased resistance to infection by *Botrytis cinerea*, and the characteristics of resveratrol synthase induction were similar to those found in grapevine.

5.6 CHEMICAL SYNTHESIS OF STILBENES

A variety of methods are available for the synthesis of most types of stilbenoids. Some are generally applicable to the synthesis of a range of structures, whereas others are quite specific, and a few follow routes related to the biosynthetic pathways (Harris and Carney, 1966, 1967; Takeuchi *et al.*, 1976, 1980.) An example of the latter is the condensation of 8-phenyloct-7-ene-2,4,6-trione and its dihydro derivative to pinosylvin and dihydropinosylvin (Birch, Cameron and Rickards, 1960).

Stilbenes may be prepared by the condensation of the sodium salt of a phenylacetic acid with a benzaldehyde in acetic anhydride (Figure 5.10), during which any free hydroxyls are acetylated. The reaction yields both *cis*- and *trans*-isomers of stilbene-α-carboxylic acids. These can be decarboxylated (e.g. with copper and quinoline) to the corresponding stilbenes.

An alternative is the Wittig reaction, in which a benzaldehyde is reacted with benzyl triphenylphosphonium halide, examples of which are shown in Figure 5.11. Hydroxyls can be protected during preceding and subsequent reaction as benzyl, *p*-bromophenacyl, methoxyethoxymethyl or trimethylsilyl ethers or other derivatives (e.g. Crombie, Crombie and Jamieson, 1980; Reimann, 1969, 1970, 1971). There are many variations

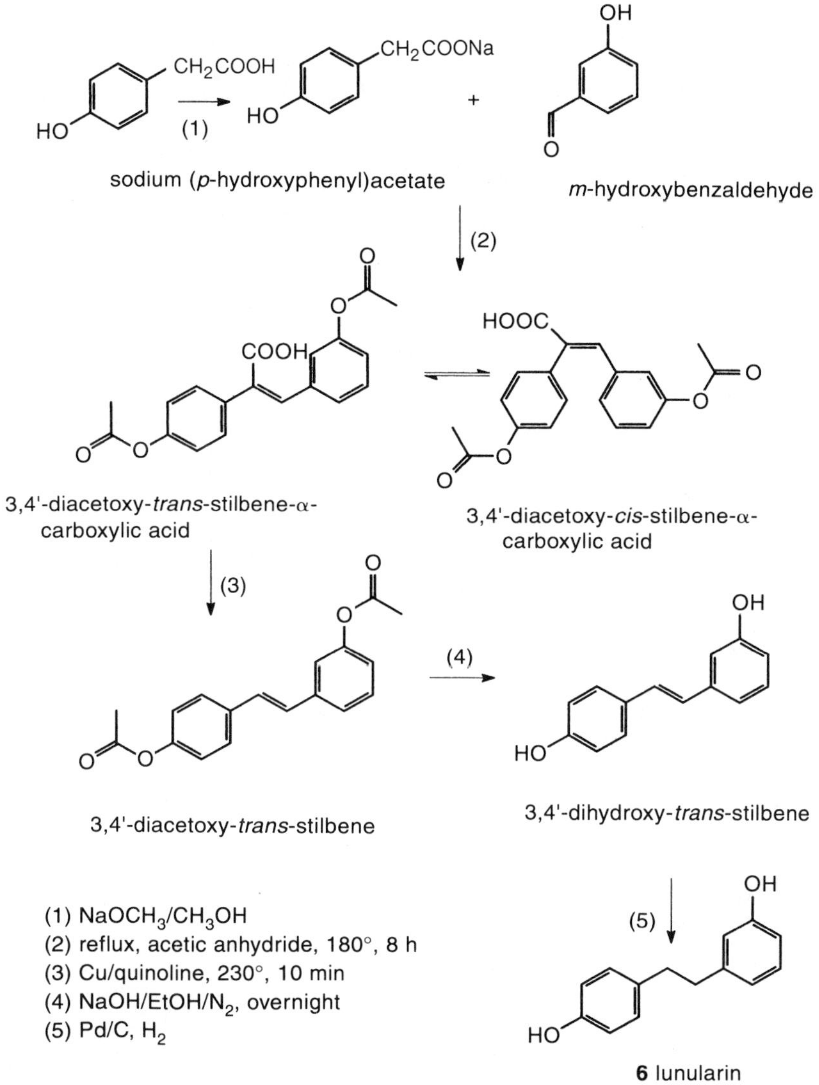

Figure 5.10 Synthesis of lunularin from sodium (*p*-hydroxyphenyl)acetate and *m*-hydroxybenzaldehyde.

on this method, including the use of lithium catalysts (*n*-butyllithium in benzene: Cardona *et al.*, 1986; phenyllithium in ether: Reimann, 1970) to generate the Wittig reagents, or sodium hydride/tetrahydrofuran (NaH/THF) to facilitate condensation. The use of diethylbenzylphosphonates with NaH and ethylene glycol dimethyl ether during the conden-

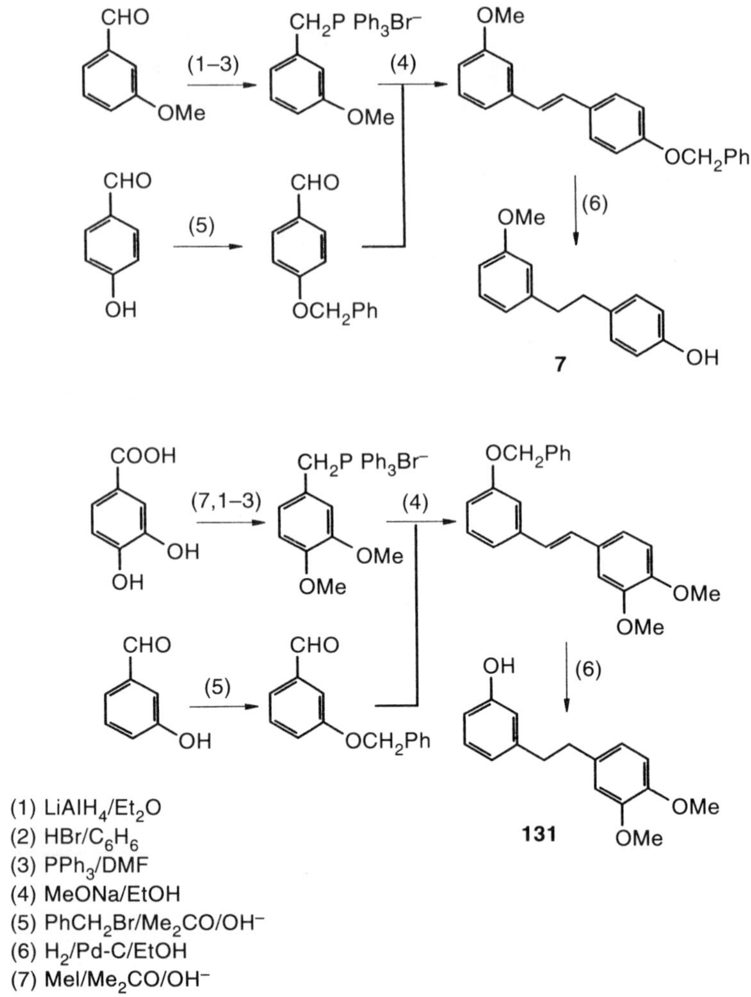

Figure 5.11 Wittig synthesis of 3-hydroxy-2-methoxybibenzyl (**7**) and 3,4-dimethoxy-3′-hydroxybibenzyl (**131**).

sation reaction with a benzaldehyde gave exclusively the *trans*-isomer (Wheeler and Battle de Pabon, 1965; Bachelor, Loman and Snowdon, 1970), whereas triphenylphosphonium salts produce mixtures of *cis*- and *trans*-isomers.

Other methods of more limited use for the production of symmetrically substituted stilbenes are the condensation of two molecules of a benzaldehyde in the presence of ethylene glycol and sodium hydroxide (Tadros and Ekladuis, 1950; Tadros, Ekladuis and Sakla, 1954), of two molecules of substituted thiobenzaldehydes with a copper catalyst (Laarhoven, Nivard and Havinga, 1961), or of two molecules of veratrol in the presence of chloral hydrate and acetic acid. Fletcher *et al.* (1971) condensed substituted toluenes and benzaldehydes in the presence of sodium hydroxide. Benzyl cyanides have been used with resorcinol or ethyl benzoates to

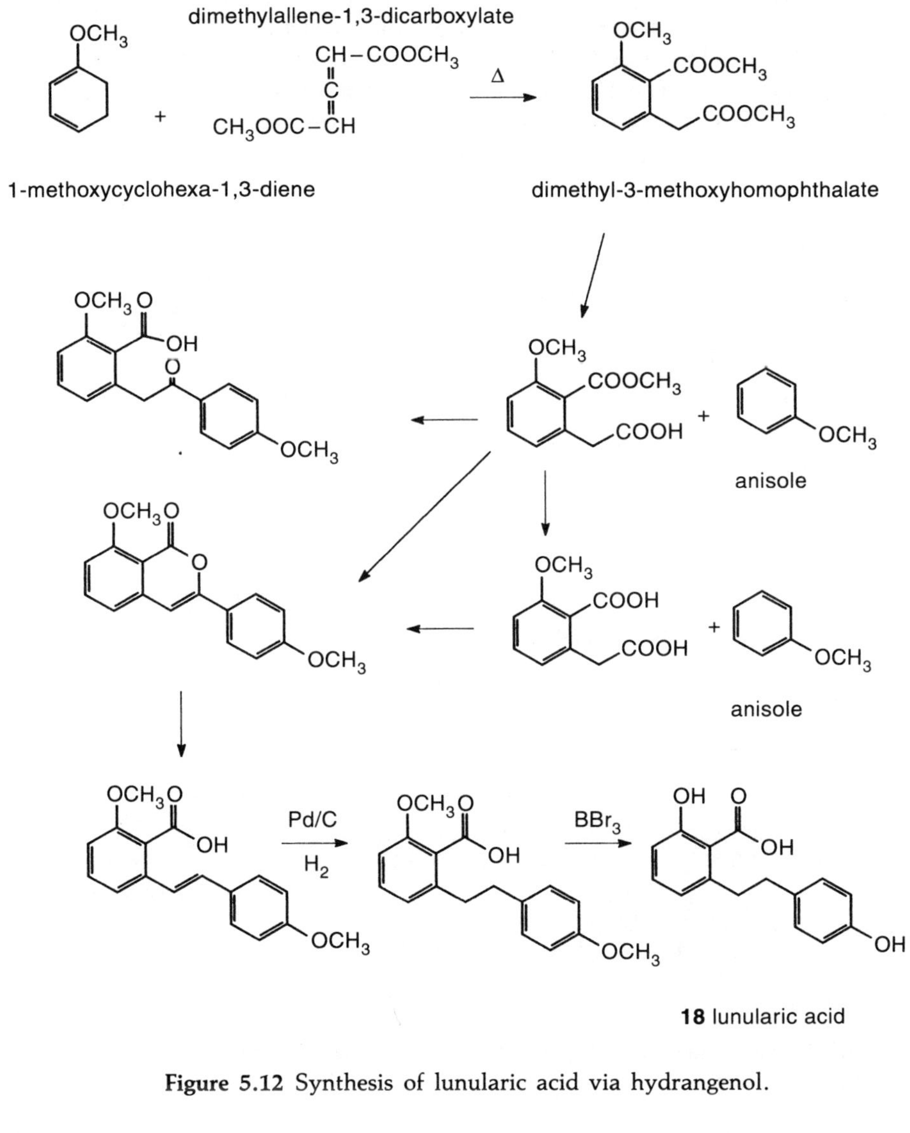

Figure 5.12 Synthesis of lunularic acid via hydrangenol.

prepare oxyresveratrol (**204**; Mongulsuk, Robertson and Towers, 1957) and lunularin (**6**; Hopkins and Perold, 1974).

Cyclization of *cis*-stilbenes to phenanthrenes is facilitated by UV light (254 nm) and iodine (e.g. Letcher, 1973; Bhandari *et al.*, 1985). In the synthesis of the phenanthropyran coeloginin, Sargent and Stanojevic (1984) used an iodinated diisopropoxy benzaldehyde and a methoxycarbonylbenzyltriphenylphosphonium salt as precursors in the Wittig reaction. Cyclization to a phenanthrene and palladium-catalyzed reduction to a 9,10-dihydrophenanthrene with H_2 was followed by treatment with boron trichloride to yield coeloginin directly. Synthesis of juncunone and the micrandrols was described by Carvalho and Sargent (1984).

Crombie and Jamieson (1982) used two routes to the prenylated bibenzyl, canniprene (**188**). The first involved a conventional Wittig synthesis of 5-benzoyloxy-3′-hydroxy-3,4′-dimethoxybibenzyl, followed by produc-

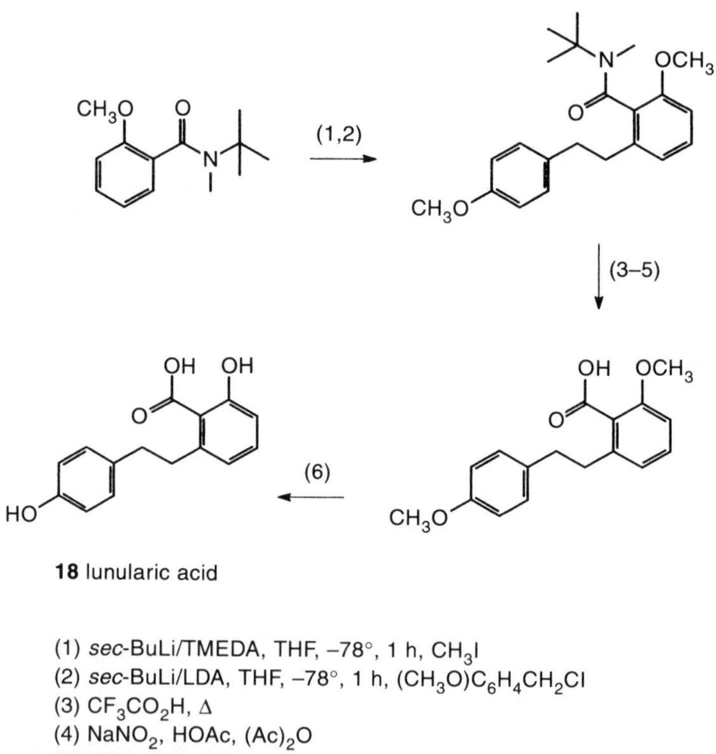

18 lunularic acid

(1) *sec*-BuLi/TMEDA, THF, –78°, 1 h, CH_3I
(2) *sec*-BuLi/LDA, THF, –78°, 1 h, $(CH_3O)C_6H_4CH_2Cl$
(3) CF_3CO_2H, Δ
(4) $NaNO_2$, HOAc, $(Ac)_2O$
(5) KOH, C_2H_5OH, Δ
(6) BBr_3, CH_2Cl_2, –78°

Figure 5.13 Synthesis of lunularic acid from *N*-*tert*-butyl-*N*-methyl-2-methoxybenzamide.

tion of the dimethylallyl ether of the free hydroxyl and Claisen rearrangement and hydrolysis to the canniprene. In the second route isovanillin was converted to the dimethylprop-2-ynyl ether and subsequently to 2-prenylated isovanillin. This could be used in a Wittig synthesis to produce canniprene directly.

An alternative route to 3,5-dihydroxybibenzyl (dihydropinosylvin) was used by Crombie, Crombie and Firth (1988). 3,5-Dimethoxybenzoic acid was treated with sodium and liquid ammonia (−70°C) to yield a bis-anion. This was then alkylated with phenethyl bromide and the product decarboxylated with lead tetraacetate/copper acetate to give the dimethoxybibenzyl. This was demethylated with BBr_3, in dichloromethane.

Several methods have been developed for the synthesis of the bibenzyl-2-carboxylic acid, lunularic acid (**18**). One of these is the reduction of hydrangenol/hydrangeic acid (Valio, Burdon and Schwabe, 1969), but the synthesis of hydrangenol (Asahina and Asano, 1930a,b) was originally inefficient and extraction from *Hydrangea* was preferred (Hashimoto, Tori and Asakawa, 1988). A ten-step synthesis of hydrangenol from phenyl β-chloropropionate was described by Huneck and Schreiber (1977). Condensation of 1-methoxycyclohexa-1,3-dione with dimethylallene-1, 3-dicarboxylate (Figure 5.12) gave dimethyl-3-methoxyhomophthalate in 42% yield, and hydrolysis of this and reaction of the products with anisole yielded hydrangeic acid dimethyl ether, and hence lunularic acid with a yield of 21% (Arai, Kamikawa and Kubota, 1972; Arai *et al.*, 1973). A higher yield was obtained by Reitz and Massey (1990), starting with *N-tert*-butyl-*N*-methyl-2-methoxybenzamide (Figure 5.13). A number of synthetic routes to bibenzyls and related compounds were described by Eicher (1990).

Synthesis of oligostilbenes has been attempted by oxidative dimerization of 4-hydroxystilbene and resveratrol with horseradish peroxidase and hydrogen peroxide (Langcake and Pryce, 1977c). The products differed from natural stilbene oligomers in the positions at which the monomers coupled, but the synthetic viniferin still exhibited antifungal activity.

6

Analysis and identification

6.1 INTRODUCTION

The procedures for isolation and characterization of unknown compounds in plant material are usually carried out on a larger scale than those used in quantitative analysis. Plant stilbenoids vary in polarity from highly polar glycosides to less polar, highly methylated derivatives. A variety of polar and nonpolar solvents can be used for extraction, depending on the nature of compounds being investigated. A first fractionation may involve phase partition (for large-scale extracts) or solid-phase purification. The latter allows a variety of preliminary fractionations to be made by changing the pH and/or the polarity of the eluant. The use of acidic extractants and solvents is best avoided where there is a danger of hydrolysis of acid-labile glycosides or compounds such as prelunularic acid (Abe and Ohta, 1984). Highly hydroxylated stilbenoids can be easily oxidized or polymerized, and are best isolated as derivatives, normally the acetyl esters. The first stage in the isolation of an unknown compound is usually column chromatography on silica or Sephadex LH-20®. This will sometimes give fractions from which crystalline compounds can be obtained, but further chromatographic procedures are usually required to isolate minor components in pure form, or prior to further chromatographic separation. Of the vast array of techniques that have been used in the isolation and identification of plant stilbenoids, a few selected examples are presented in the following sections.

6.2 PAPER AND THIN-LAYER CHROMATOGRAPHY

A wide variety of chromatographic techniques, including paper chromatography and electrophoresis, and thin-layer chromatography have all been used for the separation of naturally occurring stilbenes (Franc, Kovar and Mikes, 1966; Hemingway, Hillis and Bruerton, 1970; Hillis and Ishikura, 1968; Lindstedt and Missiorny, 1951, 1952; Sinsheimer and Smith, 1967).

Useful solvent systems for paper (PC) and thin-layer (TLC) chromatography, together with colour reactions of a number of stilbenes to diazotized aniline spray reagents, potassium permanganate and phosphomolybdic acid, were described by Hillis and Ishikura (1968). Spraying TLC plates with Fast Blue B or 10% ceric sulphate in 10% sulphuric acid, followed by heating, are useful methods for some compounds. Most stilbenes and phenanthrenes can be seen as fluorescent spots under UV light at 254 or 366 nm, and the colour or intensity of the spot may change after fuming with ammonia. Thin-layer plates incorporating fluorescent dyes can also be useful, the stilbenoids being viewed as dark or fluorescent spots against a bright background under UV illumination. A technique used for the detection of antifungal substances on thin-layer plates involves spraying the dried, developed plates with a suspension of fungal spores suspended in a suitable medium. Species of *Cladosporium* are most often used for this purpose. TLC data for a number of stilbenoids are shown in Table 6.1.

Table 6.1 Data from thin-layer chromatography (TLC) for selected stilbenoids

	TLC on silica gel G	
Compound	*Mobile phase*	*Rf*
Lunularic acid	EtOAc : $CHCl_3$: HOAc (15 : 5 : 1)	0.50
	benzene : MeOH : HOAc (20 : 4 : 1)	0.33
Lunularin	EtOAc : $CHCl_3$: HOAc (15 : 5 : 1)	0.64
	benzene : MeOH : HOAc (20 : 4 : 1)	0.33
4-Hydroxystilbene	MeOH : $CHCl_3$: petroleum ether (2 : 4 : 7)	0.87
Pinosylvin	EtOAc : $CHCl_3$: HOAc (15 : 5 : 1)	0.73
	MeOH : $CHCl_3$: petroleum ether (2 : 4 : 7)	0.71
Pinosylvin dimethyl ether	EtOAc : $CHCl_3$: HOAc (15 : 5 : 1)	0.79
Resveratrol	MeOH : $CHCl_3$: petroleum ether (2 : 4 : 7)	0.36
Batatasin I	$CHCl_3$: HOAc (19 : 1)	0.49
Batatasin III	$CHCl_3$: HOAc (19 : 1)	0.13
Oxyresveratrol	MeOH : $CHCl_3$: petroleum ether (2 : 4 : 7)	0.21
Chlorophorin	MeOH : $CHCl_3$: petroleum ether (2 : 4 : 7)	0.39
Pterostilbene	MeOH : $CHCl_3$: petroleum ether (2 : 4 : 7)	0.52
Astringin	MeOH : $CHCl_3$: petroleum ether (2 : 4 : 7)	0.04
	benzene : MeOH : HOAc (45 : 8 : 4)	0.08
Rhapontin	MeOH : $CHCl_3$: petroleum ether (2 : 4 : 7)	0.08
	benzene : MeOH : HOAc (45 : 8 : 4)	0.15

6.3 GAS–LIQUID CHROMATOGRAPHY

Most plant stilbenoids are insufficiently volatile for direct gas–liquid chromatography. The more volatile (less polar) methyl, acetyl or trimethylsilyl ether and/or ester derivatives are often used for the analysis of otherwise nonvolatile hydroxystilbenes (Gorham, 1977a; Hemingway, Hillis and Bruerton, 1970; Pryce, 1971a,b; Sinsheimer and Smith, 1967). Trimethylsilyl derivatives are easy to prepare, but methyl or acetyl derivatives are more useful for combined gas chromatography/mass spectroscopy. Detection is routinely by flame-ionization, but Sinsheimer and Smith (1967) have shown that *trans*-stilbene and the methyl, acetyl and trimethylsilyl derivatives of 4-hydroxy- and 4,4′-dihydroxystilbene are electron-capturing. These compounds had greater affinity for electrons than chlorobenzene, and the chloromethyldimethylsilyl derivative of 4-hydroxystilbene and the di(chloromethyl-dimethylsilyl) derivative of 4,4′-dihydroxystilbene could be detected using an electron-capture detector with a sensitivity 40 and 70 times (respectively) that for *trans*-stilbene. Electron-capture detection of derivatives of other, naturally occurring hydroxystilbenes has not been widely employed. In combination with capillary column gas chromatography, this technique should be capable of decreasing the limit of detection by several orders of magnitude. Retention times for a number of stilbenoids are given in Table 6.2.

6.4 HIGH-PERFORMANCE LIQUID CHROMATOGRAPHY

Efficient separations of underivatized stilbenoids can be performed by high-performance liquid chromatography (HPLC). Several different separation mechanisms have been employed for this purpose, including gel permeation, normal-phase partition on silica and reverse-phase partition on nitrile or C18 columns (Aguamah *et al.*, 1981; Akada *et al.*, 1979; Charpentier and Cowles, 1981; Cooksey *et al.*, 1988; Gracza and Ruff, 1984a; Hashimoto *et al.*, 1979; Tyukavkina *et al.*, 1979).

Langcake, Cornford and Pryce (1979) determined pterostilbene concentrations in grapevine leaves by reverse-phase HPLC on a 200 × 4.5 mm i.d. column of 5 μm Hypersil ODS® (octadecyl silyl) eluted with 50% acetonitrile in water. As an internal standard, *trans*-3,4-methylenedioxy-4′-hydroxystilbene was used, and detection was by UV absorbance at 300 nm. For resveratrol and the viniferins the column was eluted with 35% acetonitrile and detection was at 288 nm, and ε-viniferin was used as an external standard. Toscano Underwood and Pearce (1991) employed a 100 × 8 mm i.d. Radial Pak® cartridge column of Novapak C18® to separate piceatannol (astringenin), isorhapontigenin and their glucosides in extracts of *Picea sitchensis* bark. The eluant was a gradient of methanol in an aqueous solution of ethylenediamine tetraacetic acid (EDTA);

Table 6.2 Gas–liquid chromatographic data for selected stilbenoids

Compound	*Derivative*	*Stationary phase*	*Mobile phase* (N_2, *ml min*$^{-2}$)	*Temperature* (°C)	*Rt* (*min*)
Lunularic acid	TMSi	3% SE-30	60	243	12.1
	permethylated	1% OV-17	60	187	22.2
Lunularin	TMSi	3% SE-30	60	243	4.6
	dimethyl ether	1% OV-17	60	187	4.9
	diacetate	1% OV-17	60	187	16.1
4-Hydroxystilbene	TMSi	2% OV-17	50	195	0.27[a]
	methyl ether	3% SE-52	60	183	1.32
	acetate	3% SE-52	60	183	2.42
Pinosylvin	TMSi	3% SE-30	60	220	10.8
	TMSi	2% OV-17	50	195	0.58[a]
Pinosylvin dimethyl ether	TMSi	3% SE-30	60	220	7.6
Resveratrol	TMSi	2% OV-17	50	195	2.13[a]
Batatasin I	TMSi	1.5% OV-17	35	250	23.3
Batatasin III	TMSi	1.5% OV-17	35	250	3.5
Pterostilbene	TMSi	2% OV-17	50	195	2.36[a]
	TMSi	2% SE-54	50	215	0.94[a]
Piceatannol	TMSi	2% OV-17	50	195	3.58[a]
	TMSi	2% SE-54	50	215	2.38[a]

[a] Relative to 4,4′-dihydroxystilbene bis(trimethylsilylether)

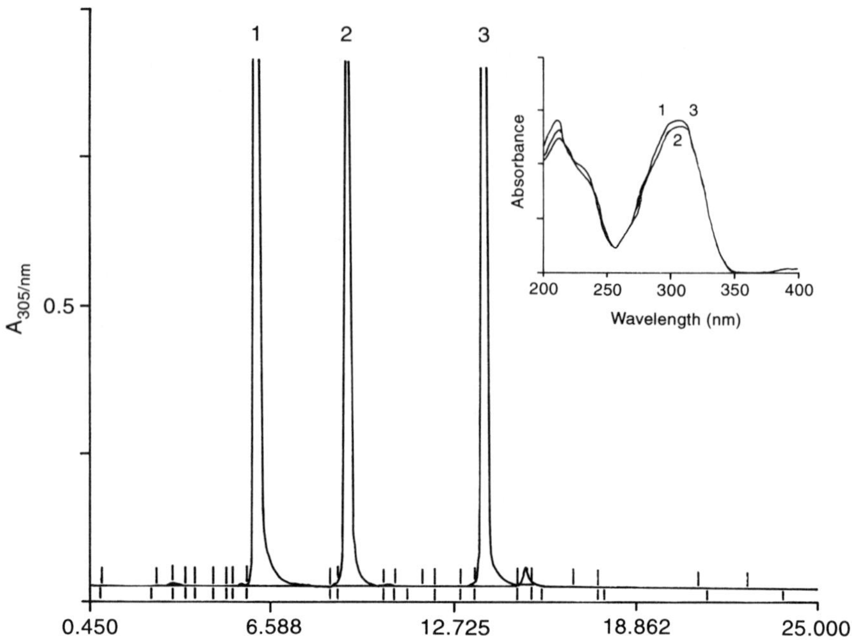
1
2
3
A305/nm
0.5
0.450
6.588
12.725
18.862
25.000
Time (min)
Absorbance
Wavelength (nm)
200
250
300
350
400

(a)

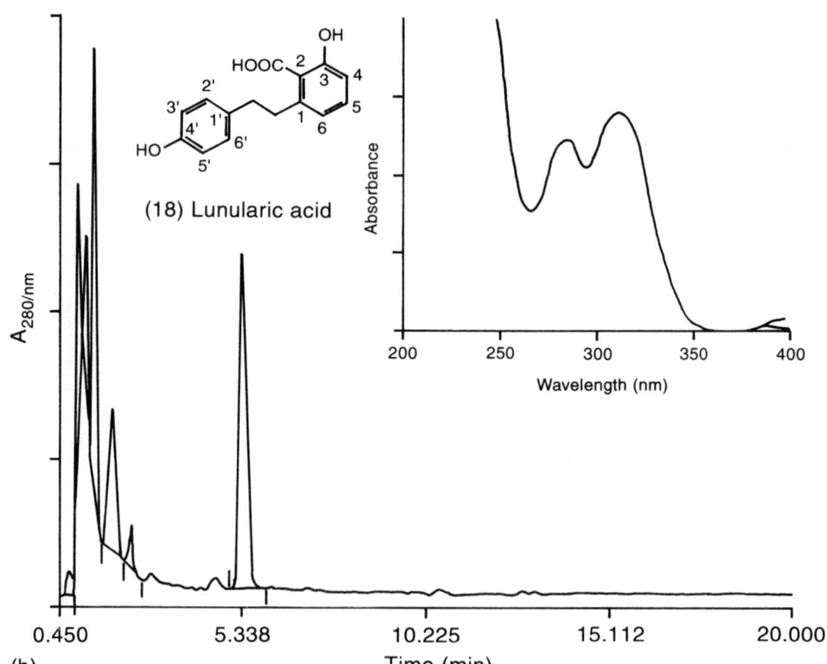
OH
HOOC
(18) Lunularic acid
A280/nm
0.450
5.338
10.225
15.112
20.000
Time (min)
Absorbance
Wavelength (nm)
200
250
300
350
400

(b)

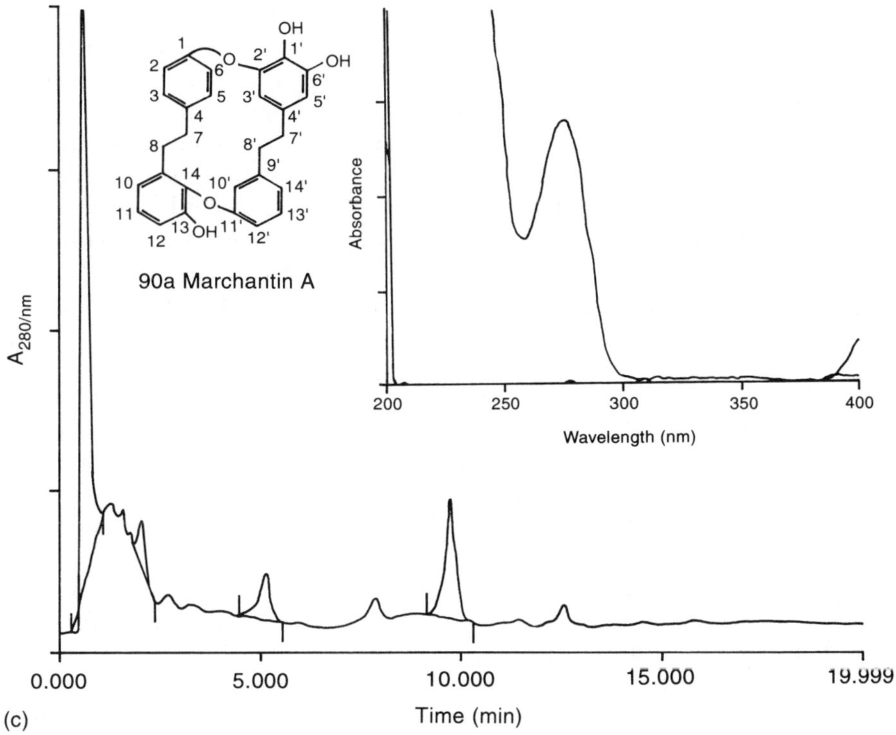

Figure 6.1 High performance liquid chromatography of stilbenoids with 'in-flight' wavelength scanning. All separations were performed on a 150 × 4.6 mm column of Spherisorb 5C8 eluted with methanol in 0.1% H_3PO_4 (linear 40–85% methanol gradient over 20 min). UV absorbance detection at 280 nm or 305 nm with a Kontron 430 UV absorbance detector + Kontron 450 data system. Inset: In-flight wavelength scans from 200 to 400 nm. (a) Synthetic pinosylvins; 1, pinosylvin; 2, pinosylvin monomethyl ether; 3, pinosylvin dimethyl ether. (b) Lunularic acid in an extract from the liverwort *Conocephalum conicum*. (c) Marchantin A (peak 4) from the liverwort *Marchantia polymorpha*.

$10\,mg\,l^{-1}$) and UV detection was at 325 nm. An external standard of isorhapontin was used for quantitative measurements. The use of a fully end-capped stationary phase and the minimum amount of prechromatographic purification limited the degradation of the labile stilbenes, and detection at a high wavelength reduced interference from other compounds in the extracts. EDTA was added to the eluant to reduce complex formation between astringin and traces of metal ions in the eluant.

Analysis of the resveratrol content of wine has recently received attention, because of its possible benefits to human health (see Chapter 7).

Mattivi (1993) used a Hypersil ODS reverse-phase column to measure mean concentrations of, respectively, 2.24 and 0.52 mg l^{-1} resveratrol in red and *rosé* wines from the Trentino region of Italy. In an analysis of Californian wines, Lamuela-Raventós and Waterhouse (1993) found concentrations of resveratrol ranging from <0.05 to 0.68 mg l^{-1}. A 150 × 3.9 mm column of 4-μm Novapak C18 was eluted with a gradient of acetonitrile in 200 mol m^{-3} orthophosphoric acid adjusted to pH 1.5 with ammonium hydroxide. UV absorbance in the range of 200 to 400 nm was monitored with a diode array detector, and the purity of the resveratrol peak was monitored by comparing its spectrum with that of a sample of pure resveratrol.

Pettit, Cragg and Singh (1987) used a chiral *N*-(3,5-dinitrobenzoyl)-D-phenylglycine bonded silica column eluted with hexane-propan-2-ol (9 : 1) to separate synthetic (+) and (−) epimers of 3-[(*tert*-butyl-dimethylsilyl) oxy]-combretastatin prior to Horeau esterification with 2-phenylbutyric anhydride to determine the configuration of natural combretastatin.

All stilbenoids can be detected by UV absorption because of the absorption of their benzene rings, although with varying response factors. Combined with simultaneous wavelength spectrum analysis in diode-array or 'in-flight' scanning detectors, HPLC can be a powerful analytical technique (Gorham, 1989; Lamuela-Raventós and Waterhouse, 1993; Figure 6.1). Fluorescence detection offers much greater sensitivity and selectivity for some compounds. HPLC interfaced, online, with multidimensional gas chromatography has been used for the analysis of stilbene hormones in corned beef (Chappell, Creaser and Shepherd, 1993), and this would be a powerful technique for the analysis of complex stilbenoid mixtures in plant extracts. The technical problems of coupling high-performance liquid chromatography and mass spectrometry are considerable, but such hyphenated techniques offer the ultimate in sensitivity and selectivity. Preparative-scale HPLC may be a useful preliminary for purification before more efficient chromatographic separations.

6.5 ULTRAVIOLET SPECTROSCOPY

Ultraviolet absorption maxima for a number of stilbenes in ethanolic solution are given in Hillis and Ishikura (1968). Two main bands can usually be distinguished at 308–336 nm and 281–313 nm, corresponding to the A and B benzene rings. Stilbene- and bibenzyl-2-carboxylic acids have two distinct bands at 260–280 nm and 300–312 nm in acidic alcoholic solvents. Increasing the pH shifts the absorbance of the lower wavelength band of stilbenes, bibenzyls and their -2-carboxylic acids to lower wavelengths. The UV absorbance may be affected by the solvent used, the pH and the presence of shift reagents such as aluminium chloride or boric acid/sodium acetate. UV absorbance data for a number of stilbenoids are shown in Table 6.3.

Table 6.3 Ultraviolet absorbance data for selected stilbenoids

Compound	*Solvent*	*Wavelength maxima (nm) [log ϵ]*
Lunularic acid	EtOH	281 [3.53], 287 [3.55], 310 [3.61]
	EtOH/NaOEt	300 [3.82]
Lunularin	MeOH	275, 279
4-Hydroxystilbene	EtOH	309, 322
	EtOH/NaOEt	323, 355
Pinosylvin	EtOH	303, 312
	EtOH/NaOEt	316
Pinosylvin dimethyl ether	MeOH	230, 237, 300, 309
	MeOH/NaOEt	230, 237, 300, 309
Resveratrol	EtOH	310, 323
	EtOH/NaOEt	324, 338(s)
Batatasin I	EtOH	261 [7.94], 283 [4.22], 295(s) [4.02] 307 [3.96], 328 [3.64], 344 [3.85], 362 [4.03]
	EtOH/KOH	258 [4.77], 334 [3.95], 353 [3.97], 371 [4.11]
Batatasin III	EtOH	208 [4.77], 226(s) [4.32], 275 [3.81], 282 [3.79]
	EtOH/KOH	215 [4.74], 242 [4.29], 284(s) [3.87], 292 [3.90]
Agrostophyllin	95% EtOH	226 [4.60], 268 [4.69], 300 [4.13]
	EtOH/NaOH	223 [4.47], 275 [4.72], 306 [4.21]
Hydrangenol	EtOH	285 [3.63], 315 [3.91]
ϵ-Viniferin	EtOH	218 [4.28], 235(s) [4.07], 297(s) [4.39] 306 [4.43], 320 [4.41], 336 [4.22]
Oxyresveratrol	EtOH	296(s), 305, 332
	EtOH/NaOEt	308(s), 317, 353
Chlorophorin	EtOH	296(s), 307, 335
	EtOH/NaOEt	307(s), 347, 358
Pterostilbene	EtOH	313, 324
	EtOH/NaOEt	324(s), 358
Astringin	EtOH	297, 331
	EtOH/NaOEt	307, 346
Rhapontin	EtOH	308(s), 329
	EtOH/NaOEt	240, 308, 327

6.6 NUCLEAR MAGNETIC RESONANCE SPECTROSCOPY

Simple ^{1}H and ^{13}C NMR techniques provide a large amount of structural information about stilbenoids, but require purification of the compound concerned. Detailed spectra for particular stilbenoids can be found in many of the references listed in Tables 2.1 to 4.17. More sophisticated techniques such as two-dimensional NMR, D_2-exchange and NOE difference spectroscopy, ^{1}H-^{1}H COSY, ^{13}C-^{1}H COSY, HMBC, distortionless enhancement of polarization transfer (DEPT) and insensitive nuclei enhanced by polarization transfer (INEPT) spectra may be required to determine the positions of the various substituents on the stilbene or phenanthrene skeleton (e.g. Pettit and Singh, 1987; Steynberg *et al.*, 1988; Tori *et al.*, 1985). ^{1}H and ^{13}C NMR data for lunularic acid, marchantin A and 6,7-dihydroxy-2,3,4-trimethoxy-9,10-dihydrophenanthrene are shown in Tables 6.4 to 6.6.

6.7 MASS SPECTROMETRY

Various forms of mass spectrometry (high- and low-resolution, chemical or electron-impact ionization or fast atom bombardment) have been used to determine the molecular weight and fragmentation patterns of stilbenoids.

Table 6.4 Nuclear magnetic resonance data for lunularic acid (**18**), from Hashimoto, Tori and Asakawa (1988)

Position	*^{1}H (400 MHz)*	*^{13}C (100 MHz)*
Ring A		
1		146.5 s
2		115.4 s
3		163.2 s
4	6.68 d	116.6 d
5	7.25 dd	134.8 d
6	6.78 d	123.8 d
Ring B		
1′		135.0 s
2′	7.00 d	130.8 d
3′	6.70 d	116.5 d
4′		156.8 s
5′	6.70 d	116.5 d
6′	7.00 d	130.8 d
Bridge		
α	3.14 t	39.2 t
α′	2.76 t	40.9 t
COOH		174.8 s

Table 6.5 Nuclear magnetic resonance data for marchantin A (**90a**), from Tori *et al.* (1985)

Position	^{1}H	^{13}C
Ring A		
1		152.9
2	6.58 d	121.2
3	6.93 d	129.5
4		139.0
5	6.93 d	129.5
6	6.58 d	121.2
Ring B		
9		136.1
10	7.02 dd	121.9
11	7.15 t	126.0
12	6.87 dd	114.3
13		148.6
14		139.6
Bridges		
7	2.96–3.01 m	35.2
8	2.96–3.01 m	30.2
7′	2.72–2.78 m	34.0
8′	2.72–2.78 m	35.4
Ring C		
1′		130.6
2′		146.4
3′	5.13 d	107.9
4′		132.4
5′	6.47 d	109.3
6′		144.1
Ring D		
9′		143.0
10′	6.85 t	115.4
11′		156.6
12′	6.55 ddd	112.0
13′	6.98 t	128.8

Details for particular compounds can be found in the references cited in Tables 2.1–4.17.

6.8 X-RAY CRYSTALLOGRAPHY

X-ray crystal structure analysis also provides valuable structural information. Pettit *et al.* (1982) gave details of the X-ray crystal structure of

Table 6.6 Nuclear magnetic resonance data for 6,7-dihydroxy-2,3,4-trimethoxy-9,10-dihydrophenanthrene (**178**), from Pettit, Cragg and Singh (1987)

Position	^{1}H *(400 MHz)*	^{13}C *(100.6 MHz)*
Ring A		
1	6.639	108.92
2	3.885	152.63
3	3.882	142.69
4	3.771	152.34
4a		121.86
4b		130.73
5	7.904	115.68
6		144.51
7		143.89
8	6.739	115.36
8a		134.85
8b		125.27
Ring C		
9	2.682 m	31.35
10	2.692 m	31.52
C2-methoxyl		56.31
C3-methoxyl		60.89
C4-methoxyl		60.53

combretastatin. The crystal structure of combretastatin A-1 was determined by X-ray crystallography combined with various computer techniques (Pettit *et al.*, 1987). The NMR and crystallographic data were then used to design a total chemical synthesis of combretastatin A-1. The structures of various phenanthrenes and 9,10-dihydrophenanthrenes from *Combretum caffrum* were also established by a combination of high-resolution NMR and X-ray crystallography (Pettit *et al.*, 1988a,b).

6.9 CHEMICAL SYNTHESIS AND DEGRADATION

Chemical degradation of stilbenes and identification of the reaction products as a method of structural determination has largely been superseded by modern spectrometry, but can still provide useful confirmatory evidence. Much of the earlier work on stilbenes employed oxidation of derivatives with potassium permanganate or chromium trioxide, or ozonolysis, to cleave the molecule at the ethene bridge. The resulting benzoic acids or benzaldehydes could then be compared with known compounds (e.g. Erdtman, 1939a,b; Kawamura, 1938; Späth and Schläger, 1940; Takaoka, 1940c).

Ultimate proof of chemical structure is usually provided by chemical synthesis from well-characterized precursors. Details of several synthetic routes are given in Chapter 5. A detailed account of the synthesis of combretastatin A-1, for example, is given by Pettit *et al.* (1987).

6.10 ELUCIDATION OF THE STRUCTURE OF MARCHANTIN A

As an illustrative example, a description of the elucidation of the structure of marchantin A (**90a**) follows. The ^1H NMR spectrum of **90a** (Table 6.5) contains signals of four benzylic methylenes, signals of three OH protons that disappear on addition of D_2O and thirteen protons on benzene rings. Methylation of marchantin A (**90a**) with methyl iodide gave a trimethyl ether (**90b**) and acetylation gave a triacetate (**90c**), indicating the presence of three phenolic hydroxyl groups. Treatment of **90a** with methylene iodide in dimethylsulphoxide in the presence of cupric oxide gave a methylene dioxide (**90d**), suggesting the presence of two vicinal phenolic hydroxyls.

The MS spectrum displayed an intense molecular ion peak at m/z 440 and the molecular formula was determined to be $C_{28}H_{24}O_5$ by high-resolution MS. Thus the degree of unsaturation was 17. The four aromatic rings account for 16 out of these 17, and this was confirmed in the ^{13}C NMR spectrum, showing 24 sp^2 carbon peaks. Therefore, **90a** should have at least one carbocycle. The remaining two oxygen atoms were attributable to ether linkages since the IR spectrum of **90a** contained neither carbonyl nor hydroxyl absorption bands. Hence **90a** was a cyclic bis(bibenzyl) in which the two bibenzyl units were linked by two ether oxygen atoms.

The degradation of **90b** using sodium in liquid ammonia gave two compounds, **7** and **131**, each synthesized as shown in Figure 6.2. These results clearly show that **90a** was a macrocyclic compound.

The 400-MHz NMR spectrum had enough resolution to analyse the fine details, due to couplings of the four aromatic rings. The protons due to four

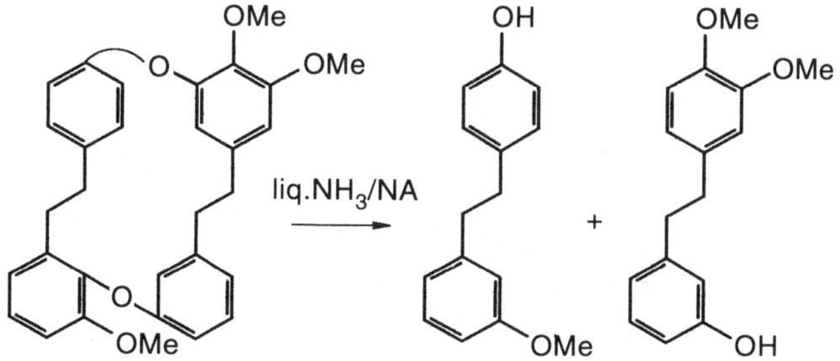

90b **7** **131**

Figure 6.2 Degradation of marchantin A trimethyl ether (**90b**).

independent aromatic rings were assigned by decoupling experiments and the substitution pattern of the four benzene rings were deduced. The NOE experiments displayed the relations between the ortho functional groups. When the benzyl methylenes H-7′, -8′ were irradiated, the NOEs into H-3′, -5′, -10′ and -14′ were observed. Similarly, the NOEs between H-7 and -8 and H-3, -5 and -10 were detected. Moreover, the position of the hydroxyl groups were determined by the NOEs of the methyl ether **90b**.

Although marchantin A and its derivatives are viscous gums, the trimethyl ether **90b** furnished crystals suitable for X-ray analysis. The characteristic proton of H-3, resonating in the extraordinarily higher field (δ 5.13), is due to the paramagnetic effects from rings A and D. This proton is sandwiched between two aromatic rings A and D. The complete structure of marchantin A was thus established.

^{1}H- and ^{13}C-NMR spectra of marchantins and their derivatives have been completely assigned using NOE difference, ^{13}C-^{1}H COSY and long-range proton selective decoupling spectra (Tori *et al.*, 1985).

7

Biological activity

7.1 ANTIBACTERIAL ACTIVITY

A list of stilbenoids reported to have mostly weak antibacterial activity is presented in Table 7.1. Unfortunately, the antibacterial assays are not always reported in as much detail, or as rigorously, as the evidence for the structural elucidations of the compounds in question.

Bryophytes are not usually damaged by bacteria, fungi, insects, snails or slugs. The common thalloid liverwort, *Marchantia polymorpha*, a member of the Marchantiaceae (Marchantiales), grows on wet soil. Extracts of this species have inhibitory activity against Gram-positive bacteria, produce allergenic contact dermatitis and are diuretic. The bisbibenzyl marchantin A (**90a**) shows antibacterial activity against *Acinetobacter calcoaceticus* (minimum inhibitory concentration (MIC) 6.25 μg/ml), *Alcaligenes faecalis* (100), *Bacillus cereus* (12.5), *B. megaterium* (25), *B. subtilis* (25), *Cryptococcus neoformans* (12.5), *Enterobacter cloacae, Escherichia coli, Proteus mirabilis, Pseudomonas aeruginosa, Salmonella typhimurium* (100) and *Staphylococcus aureus* (3.13–25). 4-Hydroxy-3′-methoxybibenzyl has recently been identified as an antibacterial constituent of the New Zealand liverwort *Plagiochila stephensoniana* (Lorimer, Perry and Tangney, 1993).

Resveratrol (**144a**) and oxyresveratrol (**204**) were reported to be inactive against *E. coli, S. aureus, P. aeruginosa* and *Candida albicans* at 120 and 100 μg ml^{-1} respectively (Gerber, 1986). Gollapudi *et al.* (1989), writing of the antibacterial activity of glepidotin C, concluded that 'this level of activity makes it unworthy of further study as a potential agent for treatment of human bacterial diseases'. A similar conclusion was reached by Adesanya, Ogundana and Roberts (1989) for the role of demethylbatatasin IV and dihydroresveratrol as antibacterial agents in yams. Takasugi *et al.* (1987) reported activity against some bacteria for various yam bibenzyls and phenanthrenes. The results obtained by these workers with dihydropinosylvin (activity against *Agrobacterium tumefaciens, Xanthomonas campestris* and *Bacillus subtilis*, but not against *Pseudomonas* species)

Table 7.1 Stilbenoids reported to have antibacterial activity

Compounds	*Authors*
BIBENZYLS	
Batatasin III	Yamaki *et al.* (1989)
3-*O*-methylbatatasin IV	Yamaki *et al.* (1989)
Dihydropinosylvin	Fagboun *et al.* (1987)
Demethylbatatasin IV	Fagboun *et al.* (1987)
Dihydroresveratrol	Adesanya, Ogundana and Roberts (1989)
3,5-dihydroxy-4-(3-methyl-2-butenyl) bibenzyl	Mitscher *et al.* (1983)
Glepidotin C	Gollapudi *et al.* (1989)
3,3′-dihydroxy-2,6-bis (4-hydroxybenzyl)-5-methoxy-bibenzyl	Takagi, Yamaki and Inoue (1983)
3,3′-dihydroxy-5-methoxy-2,5′,6-tris-(4-hydroxybenzyl)bibenzyl	Takagi, Yamaki and Inoue (1983)
4′,7-dihydroxy-2-methoxy-9,10-dihydrophenanthrene	Takagi, Yamaki and Inoue (1983)
Blestriarene A	Yamaki *et al.* (1989)
Blestriarene B	Yamaki *et al.* (1989)
Blestriarene C	Yamaki *et al.* (1989)
BIS(BIBENZYL)S	
Marchantin A	Asakawa (1990a,b)
STILBENE	
Amorphastilbol	Mitscher *et al.* (1985)
OLIGOSTILBENES	
Distichol	Sultanbawa, Surendrakumar and Bladon (1987)
ϵ-viniferin	Sultanbawa, Surendrakumar and Bladon (1987)
	Sotheeswaran *et al.* (1985)
Miyabenol A	Kawabata *et al.* (1989)
Kobophenol A	Kawabata *et al.* (1991)
Kobophenol B	Kurihara *et al.* (1991)
Copalliferol A	Sultanbawa, Surendrakumar and Bladon (1980)
	Samaraweera, Sotheeswaran and Sultanbawa (1982)
	Sotheeswaran *et al.* (1983)
Copalliferol B	Geewananda *et al.* (1986)
Vaticaffinol	Sotheeswaran *et al.* (1985)

suggest that activity may be (bacterial) species-specific, although Fagboun *et al.* (1987) obtained good inhibition of *B. cereus*, *S. aureus*, *P. aurigenosa* and *E. coli* with this bibenzyl. Pinosylvin (**143a**) is reported to be 7–30 times more inhibitory to the growth of *Salmonella typhimurium* and *Staphylococcus aureus* than phenol (Frykholm, 1945).

7.2 ANTIFUNGAL ACTIVITY

A variety of stilbenoids have been shown to inhibit fungal growth or spore germination. In many cases the fungal species are not those which might be expected to be natural pathogens of the plant from which the tested compound was isolated. A particular example is the widespread use of the TLC bioassay with *Cladosporium* species (e.g. Bailey and Burden, 1973). Although this is a rapid and convenient screening method, the spectrum of activity against other fungi may be very different. An interesting variation to this technique is to incorporate a fungicide into the thin-layer plates and screen for antidotes to the fungicide. Tahara *et al.* (1993) identified 3,5-dihydroxy-4-methoxystilbene from *Polygonum lapathifolium* as an antidote to benzimidazole fungicides in this way. Results of fungal inhibition studies are often expressed as the dose required to give 50% inhibition (ED_{50} values), or the minimum inhibitory concentration (MIC) in p.p.m. ($\mu g\ ml^{-1}$), rather than in molar concentrations.

The liverwort bisbibenzyl, marchantin A, exhibited antifungal activity against *Alternaria kikuchiana*, *Aspergillus fumigatus* (MIC 100 μg/ml), *A. niger* (25–100), *Candida albicans*, *Microsprorum gypseum*, *Penicillium chrysogenum* (100), *Pyricularia oryzae* (12.5), *Rhizoctonia solani* (50), *Saccharomyces cerevisiae*, *Sporothrix schenckii* (100), *Trichophyton mentagrophytes* (3.13) and *T. rubrum* (100) (Asakawa, 1990a,b). High concentrations of lunularic acid (**18**) inhibited fungal spore germination (Pryce, 1972a,b), but had little effect on growth of mycelia in liquid culture (Thomas, 1977). Other compounds, probably bis(bibenzyl)s, gave major antifungal bands in extracts of *Conocephalum conicum* assayed on thin-layer plates with *Cladosporium cucumerinum* (Gorham, 1975). 4-Hydroxy-3′-methoxybibenzyl has recently been identified as an antifungal constituent of the liverwort *Plagiochila stephensoniana* (Lorimer, Perry and Tangney, 1993).

Dietrich and Valio (1973) reported that hydrangenol at $1\ mol\ m^{-3}$ inhibited the growth of *Pythium* by 63%. Malan and Swinny (1993) found that representatives of the three classes of stilbenoids present in *Combretum apiculatum* (a bibenzyl, a phenanthrene and a dihydrophenanthrene) all inhibited the growth of *Penicillium expansum* when applied as 20-μg spots on TLC plates. The 4-hydroxybenzyl derivatives of bibenzyls and 4,7-dihydroxy-2-methoxy-9,10-dihydrophenanthrene isolated from the orchid *Bletilla striata* by Takagi, Yamaki and Inoue (1983) showed very little activity against the two fungi tested (*Candida albicans* and *Trichophyton mentagrophytes*), although other orchid stilbenoids are phytoalexins (see below).

Some stilbenoids are normal constituents of heartwood and may help to protect the wood from attack by fungi, insects and other organisms. This is true for pinosylvin (**143a**) (Basset *et al.*, 1967; Coutts, 1970; Erdtman, 1949b; Erdtman and Rennerfeldt, 1944; Frykholm, 1945; Gibbs, 1972; Hemingway, McGraw and Barras, 1977; Rennerfeldt, 1949; Shrimpton and Watson, 1971), astringin (**147b**) and isorhapontin (**148b**) (Alcubilla-Martin, 1970; Shain and Hillis, 1971), oxyresveratrol (**204**) and pterostilbene (**202a**) (King *et al.*, 1953; Bevan, 1958; Takasugi *et al.*, 1978). Growth of the wood-rotting fungi *Polyporus vaporarius, Coniophora cerebella* and *Lentinus squamosus* was inhibited by 0.005–0.02% pinosylvin or its monomethyl ether in agar culture (Erdtman and Rennerfeldt, 1944), but the ability of these fungi to attack central or peripheral heartwood or softwood could not be correlated with the distribution of pinosylvins in different parts of the wood. Pinosylvin from *P. radiata* inhibited the growth of *Amylostereum arcolatum* (Coutts, 1970). In roots of Corsican pine (*Pinus nigra* var. *maritima*) concentrations of pinosylvin and pinosylvin monomethyl ether (**143b**) were related to infection by *Heterobasidion annosum* (Prior, 1976). Dubrovinskaya, Yakavleva and Tokin (1974) reported ultrastructural changes in hyphae of *Phytophthora infestans* caused by pinosylvin monomethyl ether.

In addition to their normal presence in pine heartwood, pinosylvin and pinosylvin monomethyl ether may also accumulate in sapwood in response to fungal or insect attack. Diseased sapwood of red pine (*P. resinosa*) was found to contain more pinosylvins than healthy sapwood (Jorgensen, 1961). Lodgepole pine (*P. contorta*) seedlings infected with the blue-stain fungus *Europhium clavigerum* or with *Dendroctonus ponderosae* also contained more pinosylvins (Shrimpton and Watson, 1971; Shrimpton, 1973). Shain (1967) observed that infection of sapwood of loblolly pine (*P. taeda*) with the fungus *Fomes annosus* resulted in the formation of a 'reaction zone' (i.e. a hypersensitive reaction) which inhibited the further spread of the fungus. This was accompanied by the accumulation of pinosylvin, which may inhibit enzymes produced by the fungus (Lyr, 1961). Just as attack by bark beetles can inoculate elm trees with dutch elm disease (*Ceratocystis ulmi*), the southern pine beetle (*Dendroctonus frontalis*) can infect pines with the blue-stain fungus *Ceratocystis minor*. Loblolly pine sapwood inoculated with *C. minor* produced pinosylvins, and possibly also resveratrol (Hemingway, Mcgraw and Barras, 1977). The concentrations of the stilbenes declined in later stages of incubation. Some fungi (e.g. *Peniophora pseudo-pini* and *Tympanis hypopodia*) are able to metabolize pinosylvin and pinosylvin monomethyl ether (Loman, 1970a; Lyr, 1962a,b,c), and *Fomes annosus* isolates were able to tolerate the presence of pinosylvin (Gibbs, 1972). Production of pinosylvins is also induced in the sapwood of Monterey pine (*P. radiata*) by inoculation with *Amylostereum chailletii* following attack by the European wood wasp, *Sirex noctilio*. Inoculation occurs following insertion of the wasp's ovipositor into the sapwood (Hillis and Inoue, 1968). Fungal infection of *P. sylvestris* induced the synthesis of pinosylvin-forming stilbene synthase (Gehlert, Schoeppner and Kindl, 1990). Pinosylvin-forming enzymes may also be induced by exposure to UV light or ozone (Rosemann, 1990; Rosemann, Heller and Sandermann, 1991; Schoeppner and Kindl, 1979).

Infection of the wood of Norway spruce (*Picea abies*) with *Fomes annosus* resulted in extensive rotting which spread in a longitudinal direction, because of the absence of fungitoxic stilbenes in the heartwood. Radial spread of the infection was inhibited by a reaction zone between the healthy and diseased wood, although no stilbene accumulation was detected (Shain and Hillis, 1971). Stilbenes in spruce are mainly found in the bark, where they may also have fungitoxic properties and help to prevent infection (Alcubilla-Martin, 1970). Woodward and Pearce (1988) investigated the role of astringin (**147b**) and isorhapontin (**148b**) and their aglycones, in resistance of excised spruce bark discs to six wood-decaying fungi. They found that the stilbenes were the major antifungal compounds present in bark, and that there was some relationship between sensitivity of the fungi to isorhapontin and their pathogenicity. The aglycones were more toxic than the glycosides and were released by hydrolysis of the latter with fungal β-glucosidase. The distribution, quantitative variation and association with disease resistance *in vitro* was examined in more detail by Toscano Underwood and Pearce (1991, 1992). Variation in susceptibility of the host was not directly correlated with stilbene content, although this did not exclude the possibility that the stilbenes were effective against some fungi.

Schultz *et al.* (1990, 1991) compared the effects of a number of synthetic stilbenes and bibenzyls on the growth of the white-rot fungus *Coriolus versicolor* and the brown-rot fungi *Gloeophyllum trabeum* and *Poria placenta* using both agar plate and soil block techniques. These studies revealed some relationships between structure and inhibitory activity, but mainly an effect of hydrophobicity. Bibenzyls were generally more active than the corresponding stilbenes. They suggested that *in vivo* methylation might increase the fungitoxicity of hydroxylated stilbenoids. Surprisingly, synthetic stilbene dimers and oligomers had little activity in contrast to reports of the activity of natural and synthetic oligomers of resveratrol (Langcake and Pryce, 1977b).

The problems of demonstrating the antifungal effects of stilbenes *in vivo* are illustrated by the experiments of Hart and Hillis (1974). Resveratrol (**144a**) and piceid (**144b**) were extracted from the heartwood of *Eucalyptus sideroxylon*, which was resistant to attack by *Polyporus versicolor* and

Poria monticola. Both heartwood extracts and pure resveratrol were inhibitory to the fungi in culture, but impregnation with resveratrol did not increase the decay resistance of cottonwood (*Populus deltoides*) blocks. Hathaway (1962) reported that these two stilbenes were cellulase inhibitors. The conflicting results of *in vitro* and wood impregnation experiments with regard to the effectiveness of stilbenes as antifungal compounds were considered by Hart and Shrimpton (1979). They demonstrated that pinosylvin (**143a**) was bound to lodgepole pine sawdust, and that addition of sawdust to malt extract broth containing pinosylvin reduced the inhibitory effect of the stilbene on the growth of *Coriolus versicolor*. Synergistic effects in complex mixtures of wood extractives were also considered. Schultz *et al.* (1991) found no synergism between synthetic stilbenes and bibenzyls in soil block tests, but did find an interaction with the preservative didecyldimethylammonium chloride. Hart and Shrimpton (1979) suggested that critical experiments to resolve the conflict between *in vitro* results and the apparently low activity of stilbenes applied to wood had not been conducted. In a detailed review of the role of stilbenes in disease resistance, mainly in *Pinus*, Hart and Shrimpton concluded that 'the dynamic resistance of sapwood to disease differs from the passive decay resistance of heartwood', and that 'stilbene accumulation may provide some unspecific resistance to fungal proliferation in tissues already resisting invasion'. Kemp and Burden (1986) reviewed the evidence that stilbenes may be considered as phytoalexins in sapwood. The stilbenes accumulate in narrow 'reaction zones' and are produced by dying parenchyma cells.

The role of stilbenoids as antifungal plant constituents has also been reviewed by Scheffer and Cowling (1966), Ingham (1972) and Stoessl (1970). It is not clear to what extent the constitutive stilbenoids of herbaceous species also have a role in protecting the plants against fungal and insect attack.

7.3 PHYTOALEXINS IN ANGIOSPERMS

Synthesis of a number of antifungal stilbenoids can be induced by infection with the appropriate organism, by isolated elicitors or by a number of other stimuli (such as UV light).

The orchid dihydrophenanthrenes orchinol (**273**), hircinol (**274b**) and loroglossol (**274a**) were identified as phytoalexins in the 1960s (Fisch and Arditti, 1972; Fisch, Flick and Arditti, 1973; Gäumann and Kern, 1959a,b; Gäumann and Hohl, 1960; Gäumann, Nuesch and Rumpau, 1960; Gäumann, 1963; Hardegger, Biland and Carrodi, 1963; Hardegger, Schellenbaum and Carrodi, 1963; Hardegger *et al.*, 1963; Ward, Unwin and Stöessl, 1975a,b; Urech *et al.*, 1963). Orchinol was more effective than hircinol in inhibiting the growth of *Candida lipolytica*, although both phenanthrenes were apparently degraded by the fungus (Fisch, Flick and

Arditti, 1973). These authors suggested that continuous production of slowly degraded phytoalexins would be most appropriate for maintaining a balance of infection in mycorrhizal associations, but did not test the effectiveness of orchinol and hircinol against mycorrhizal fungi. Although loroglossol was described as inactive against fungi (Fisch, Flick and Arditti, 1973; Hardegger, Schellenbaum and Carrodi, 1963), it did inhibit germination of spores of *Monilinia fruticola* and *Phytophthora infestans* (Ward, Unwin and Stössl, 1975a,b). Orchinol also inhibited growth of hyphae of a number of fungi, although with higher ED_{50} values than for spore germination (Ward, Unwin and Stössl, 1975b). Some related bibenzyls, stilbenes and phenanthrenes had similar or lower ED_{50} values than did orchinol.

Bibenzyls and phenanthrenes are reported to be phytoalexins in various species of yam (*Dioscorea*). Coxon, Ogundana and Dennis (1982) reported that hircinol (**274b**) and 7-hydroxy-2,4,6-trimethoxyphenanthrene, constitutive components of *Dioscorea rotundata*, inhibited germ tube growth, but not germination, of a range of fungi. Later dihydropinosylvin, batatasin IV and demethyl batatasin IV were identified as phytoalexins in this species after infection with *Bothyodiplodia theobromae* (Fagboun *et al.*, 1987). Takasugi *et al.* (1987) concluded that dihydropinosylvin was a phytoalexin in *D. opposita* (= *D. batatas*) infected with the bacterium *Pseudomonas cichorii*, while six preformed bibenzyls and phenanthrenes also had antifungal activity. The bibenzyls were more active than the phenanthrenes, and were more inhibitory to fungi than to bacteria. None of these compounds inhibited the growth of cultures of *Pseudomonas cichorii* or *P. solanacearum*. Adesanya, Ogundana and Roberts (1989) found that demethylbatatasin IV was the major phytoalexin in *D. bulbifera*, while dihydroresveratrol was found in *D. dumentorum*. Dihydroresveratrol was active against *Cladosporium cladosporoides*, and to a lesser extent against *Trichophyton mentagrophytes*, but not against a number of other fungi. *Dioscorea alata* accumulated dihydropinosylvin, batatasin IV, demethylbatatasin IV and batatasin III after infection with *Bothyodiplodia theobromae* or treatment with mercuric chloride (Cline *et al.*, 1989).

The stilbene phytoalexins include resveratrol (**144a**) and its derivatives in *Trifolium*, *Arachis* and members of the Vitaceae (Aguamah *et al.* 1981; Cooksey *et al.*, 1988; Ingham, 1976, 1978; Keen and Ingham, 1976; Langcake, 1981; Langcake and Pryce, 1976, 1977a,b,c; Pryce and Langcake, 1977). Resveratrol production was induced in several members of the Vitaceae by infection with *Botrytis cinerea* or exposure to 254-nm UV radiation (Langcake and Pryce, 1976). The antifungal activity of this stilbene was, however, rather weak, and 4-hydroxystilbene was considerably more effective in a number of assays. The resveratrol oligomers α- and ε-viniferin, also found in infected or irradiated vine leaves, were much more effective than resveratrol itself (Langcake and Pryce, 1977a,b).

γ-Viniferin showed slightly higher activity than resveratrol (Langcake and Pryce, 1977a). The resveratrol trimer canaliculatol and the tetramer vaticaffinol also have antifungal properties as revealed on TLC plates incubated with *Cladosporium cladosporoides* (Bokel *et al.*, 1988; Sotheeswaran *et al.*, 1987).

An even more active stilbene, pterostilbene (**202a**), was identified as a minor component of the phytoalexin mixture synthesized by vine leaves (Langcake, Cornford and Pryce, 1979). Both resveratrol (**144a**) and pterostilbene (**202a**) have also been detected in both infected and uninfected grape berries (Jeandet, Bessis and Gautheron, 1992; Jeandet, Sbaghi and Besses, 1992; Langcake, Cornford and Pryce 1979; Pezet and Pont, 1988a,b). Pezet and Pont (1990) and Pont and Pezet (1990) have confirmed that resveratrol is not an effective inhibitor of fungi, whereas pterostilbene is highly toxic at 0.5 mol m^{-3}. A more detailed study by Langcake (1981) of the responses of *Vitis vinifera* and *V. riparia* to infection with *Botrytis cinerea* and *Plasmopara viticola* suggested that pterostilbene, despite its high fungitoxicity, played no role in disease resistance. ϵ-Viniferin appeared to be the most effective phytoalexin in *V. riparia*, while α-viniferin was more important in the age-related resistance of *V. vinifera* to *Botrytis cinerea*. Whichever compound is the most effective, in practice the speed and intensity of resveratrol production is a reasonable indicator of *Botrytis cinerea* resistance in grapevine selection (Barlass, Miller and Douglas, 1987; Bavaresco and Eibach, 1987; Blaich and Bachmann, 1980; Dercks and Creasy, 1989; Langcake and McCarthy, 1979; Pool, Creasy and Frackleton, 1981; Stein and Blaich, 1985; Stein and Hoos, 1984). Calderon *et al.* (1993) found that a cellulase preparation from *Trichoderma viride* contained an elicitor that produced a hypersensitive response in cell suspension cultures of immature grape pericarp (cv. Monastrell), accompanied by the production of resveratrol. This response was not obtained with pectinase or pectolyase from other fungi, mycelial extracts or culture filtrates of *Botrytis cinerea*, or with grape cell wall fragments released by the cellulase. It was suggested that biocontrol of *Botrytis cinerea* infection in grapes by treatment with *Trichoderma viride* might be mediated by this hypersensitive response. Liswidowati *et al.* (1991) were able to induce stilbene synthase and hence resveratrol production by incubating grapevine cell suspension cultures with cell wall fragments from *Botrytis cinerea*. Induction of stilbene synthase production by UV light at 254 nm was reported for several members of the Vitaceae by Fritzemeier and Kindl (1981). Stilbene synthase genes isolated from grapevines have been transferred to and expressed in tobacco plants (Hain *et al.*, 1993). The transgenic tobacco plants and suspension cultures responded to an elicitor from *Phytophthora megasperma* by accumulating stilbene synthase mRNA. Transgenic tobacco plants that responded to infection with *B. cinerea* with a large and rapid accumulation of resveratrol were more resistant. The ability of fungi to overcome inhibition by stilbene phytoalexins may be related to the ability of the fungi to produce stilbene-oxidizing enzymes (Hoos and Blaich, 1988, 1990; Pezet, Pont and Hoang-Van, 1991).

Ingham (1976) used a TLC bioassay (Homans and Fuchs, 1970) with *Cladosporium herbarum* to identify *cis*- and *trans*-resveratrol as phytoalexins in African material of *Arachis hypogaea* infected with *Helminthosporium carbonum*. He obtained inhibition of mycelial growth of this fungus by *trans*-resveratrol with an ED_{50} of 50 $\mu g\,ml^{-1}$, indicating that it was slightly less active than the isoflavonoid phytoalexins medicarpin and sativan. Infected *Arachis hypogaea* of American origin produced the *cis*- and *trans*-isomers of 4-isopentenyl-3,4′,5-trihydroxystilbene (Keen and Ingham, 1976). Other prenylated stilbenes (Table 4.5) were later isolated by Aguamah *et al.* (1981) and Wotton and Strange (1985). The arachidins were shown to inhibit spore germination and hyphal growth of *Aspergillus flavus* (the organism responsible for aflatoxin contamination) and to be inversely correlated with the extent of infection of wounded peanut kernels. Later work identified 3-isopentadienyl-3′,4,5′-trihydroxystilbene as the major phytoalexin in kernels of 11 groundnut genotypes (Cooksey *et al.*, 1988). Callus cultures of *A. hypogaea* produced resveratrol and isopentenyl resveratrol in response to irradiation with UV light at 260–270 nm (Fritzemeier *et al.*, 1983). Two prenylated stilbene-2-carboxylic acids were reported as phytoalexins in *Cajanus cajan* by Cooksey *et al.* (1982).

Piceatannol (**147a**) was isolated from sugarcane (*Saccharum* spp.) infected with *Colletotrichum falcatum* (Brinker and Seigler, 1991). A later study (Brinker and Seigler, 1993) raised doubts about the effectiveness of piceatannol as a phytoalexin, and illustrated some of the problems with interpretation of the results of such experiments. Piceatannol concentrations only started to increase 45 h after infection, whereas earlier accumulation might have been expected for a phytoalexin. Furthermore, concentrations were higher in the susceptible than the resistant cultivar. The authors point out the problem of analysing small areas at the site of infection rather than bulk tissue.

Oxyresveratrol (**204**) was isolated, together with resveratrol (**144a**), piceid (**144b**) and oxyresveratrol-3-*O*-glucoside, from leaves of *Veratrum grandiflorum* treated with cupric chloride (Hanawa, Tahara and Mizutani, 1992). It was inhibitory to the growth of *Cladosporium herbarum* on thin-layer plates, whereas the glucosides were not. Oxyresveratrol had earlier been identified as a phytoalexin in *Morus alba* (Takasugi *et al.*, 1978).

The main stilbenoids implicated as phytoalexins in angiosperms are listed in Table 7.2. From the account given above it is clear that in some cases a mixture of stilbenoids may be associated with microbial infection. The composition of the mixture may depend on the genotype of the plant or

Table 7.2 Stilbene phytoalexins

Plant	*Stilbenes*	*Authors*
Orchids	orchinol, hircinol, loroglossol	Gäumann (1963); Gäumann and Kern, 1959a,b; Gäumann, Nuesch and Rumpau (1960); Urech *et al.* (1963); Hardegger, Biland and Carrodi (1963); Hardegger *et al.*, 1963; Ward, Unwin and Stössl (1975a,b); Fisch, Flick and Arditti (1973)
Vitis spp.	resveratrol, pterostilbene, α-viniferin, ϵ-viniferin	Langcake and Pryce 1976, 1977a,b,c Langcake 1981, Langcake, Cornford and Pryce (1979); Calderón *et al.* (1993)
Arachis hypogaea	3-isopentadienyl-3′,4,5′-trihydroxystilbene, arachidin I, arachidin II, arachidin III, resveratrol	Cooksey *et al.* (1988); Aguamah *et al.* (1981); Wotton and Strange (1985); Ingham (1976); Keen and Ingham (1976)
Cajanus cajan	3-hydroxy-6-isopentenyl-5-methoxystilbene-2-carboxylic acid, 3-hydroxy-4-isopentenyl-5-methoxystilbene-2-carboxylic acid	Cooksey *et al.* (1982)
Dioscorea spp.	batatasin IV, demethylbatatasin IV, dihydropinosylvin, dihydroresveratrol, batatasin III, hircinol, 7-hydroxy-2,4,6-trimethoxyphenanthrene	Fagboun *et al.* (1987) Adesanya, Ogundana and Roberts (1989) Cline *et al.* (1989); Coxon, Ogundana and Dennis (1982) Takasugi *et al.* (1987)
Saccharum spp.	piceatannol	Brinker and Seigler (1991, 1993)
Veratrum grandiflorum	resveratrol, oxyresveratrol, piceid, oxyresveratrol-3-*O*-glucoside	Hanawa, Tahara and Mizutani (1992)
Morus alba	oxyresveratrol	Takasugi *et al.* (1978)

the fungus, and in most cases interactions between components of these mixtures have not been tested in antifungal assays.

7.4 DORMANCY AND GROWTH INHIBITION IN PLANTS

Growth-regulating roles have been postulated for a number of bibenzyls (lunularic acid (**18**) and batatasins II–IV) and the phenanthrene, batatasin I. Lunularic acid, batatasins and other stilbenoids have been shown to affect photosynthesis and respiration (Iino, Hashimoto and Heber, 1978; Gorham and Coughlan, 1980; Nimmanpisut, Chudapongse and Ratanabanangkoon, 1976), adenosine triphosphatase activity (Janthasoot, Ratanabanangkoon and Chudaponyoe, 1977) and indoleacetic acid (IAA) oxidation (Gorham, 1978; Lee, Rock and Stoessl, 1978; Mato and Calvo, 1977).

7.4.1 Lunularic acid

Lunularic acid (**18**) was first identified as a dormancy factor in an accession of the thallose liverwort *Lunularia cruciata* (Valio, Burdon and Schwabe, 1969). It was also isolated, but not characterized, as a growth inhibitor in *Marchantia polymorpha* (Fries, 1964).

Lunularic acid inhibits the growth of liverwort gemmalings *in vitro*, and possibly also on the plant (Gorham, 1977a, 1978; Valio and Schwabe, 1970). Circumstantial evidence for inhibition of gemmalings in the gemmae cups of thallose liverworts came from experiments in which diffusion of extracellular products away from isolated, germinating gemmalings was prevented, resulting in inhibited growth. Separation of the two halves of the bilaterally symmetrical gemmalings also enhanced growth (Schwabe and Valio, 1970). A similar role has been ascribed to IAA (La Rue and Narayanaswami, 1957). Other stilbenes and bibenzyls, including lunularin and 4-hydroxybibenzyl, are more potent inhibitors of gemmaling growth (Gorham, 1978). Exogenously applied prelunularic acid was not as inhibitory to the growth of *Marchantia gemmalings* as was lunularic acid (Imoto and Ohta, 1985).

Lunularic acid was also reported to inhibit seed germination (Fries, 1964), although this was not confirmed in a later study (Thomas, 1977). Hashimoto, Tori and Asakawa (1988) reported that lunularic acid and a number of related compounds inhibited the germination and coleoptile elongation of rice. The most active compound was methyl hydrangeate. Lunularic acid was not very inhibitory to cress root growth (Huneck and Schreiber, 1972; Gorham, 1978), but some analogues were more effective. Growth of *Avena* seedlings appeared to be stimulated, although no statistical analysis of the results was presented (Huneck and Schreiber, 1972). In sufficient quantity, lunularic acid inhibited the elongation of *Avena* coleoptiles, but at low concentrations it had no effect, or promoted growth (Arai, Kamikawa and Kubota, 1972; Arai *et al.*, 1973; Fries, 1964; Huneck and Schreiber, 1972; Thomas, 1977). Added IAA may counteract the inhibitory effect of lunularic acid (Gorham, 1978). Lunularic acid enhances the oxidation of IAA by both horseradish peroxidase and lentil root homogenate in the absence of other phenols (Gorham, 1978; Mato and Calvo, 1977). When phenols such as 2,4-dichlorophenol are included in the reaction mixture, addition of lunularic acid and other polyhydroxystilbenoids inhibits IAA oxidation (Gorham, 1978).

Dormancy in the Israeli strain of *Lunularia cruciata* was associated with drought resistance (Schwabe and Nachmony-Bascomb, 1963). Suggestions of a structural similarity between the carboxylic acid and 4′-hydroxygroup of lunularic acid and equivalent groups of abscisic acid (Pryce, 1972a,b) led to an assessment of the effect of lunularic acid on stomatal opening in higher plants. Thomas (1977) was, however, unable to demonstrate significant effects of lunularic acid on transpiration rates of *Lycopersicon*, *Phaseolus* or *Apium*, or on stomatal aperture in epidermal strips of *Commelina communis*. Lunularic acid occurs together with abscisic acid (ABA) in *Hydrangea macrophylla* and *Allium* species (Pryce, 1971b; Goda and Sankawa, 1985), and the carboxylic acid group does not seem to be necessary for its inhibitory activity.

Both lunularic acid and some synthetic stilbenoids are inhibitory to the growth of unicellular and filamentous algae (Sheers, 1971; Thomas, 1977). The presence of lunularic acid in algae (Pryce, 1972a,b) has not been confirmed (Gorham, 1977a; Thomas, 1977). ABA was not detected in liverworts by Weiler (1979), but a more recent report suggests that it does occur in this group (B. V. Milborrow, quoted in Katekar, Venis and Geissler, 1993). An allelopathic role has also been suggested for lunularic acid, but Thomas (1977) was unable to detect the compound in the soil or liquid media in which liverworts had been growing.

Experiments in which lunularic acid is applied externally may be misleading, since there is evidence (Gorham, 1980) of binding of radiolabelled lunularic acid to cell walls, a phenomenon also observed with other stilbenoids (Hart and Shrimpton, 1979). Lunularic acid concentrations do not appear to be related to dormancy in other liverworts and are highest in actively growing thalli of *Conocephalum conicum* (Gorham, 1975). The highest concentrations in this and other thalloid liverworts were found near the growing apices. In most species lunularic acid accumulates in plants grown under high light intensities and long daylengths, and does not respond to a short far-red light break in thalli grown in short-day conditions. Most of the lunularic acid extracted from these liverworts may be in the form of prelunularic acid in the plants. In fractionation experiments the absolute amounts of lunularic acid + prelunularic acid were similar in the vacuoles and the cytoplasm (Imoto and Ohta, 1985). Cell cultures of *Marchantia polymorpha* accumulated lunularic acid in the stationary phase of growth, and possibly also in response to phosphate deficiency (Abe and Ohta, 1983).

Lunularic acid (**18**) and hydrangeic acids were reported by Katekar, Venis and Geissler (1993) to bind to the receptor for 1-*N*-naphthylphthalamic acid in maize coleoptiles and *Cucurbita pepo* hypocotyls. Analogues of lunularic acid with the carboxylic acid in the 3- position showed weaker binding than analogues with the carboxylic acid at the 2- position. Both compounds promoted IAA accumulation in membrane vesicles of *Cucurbita*, but did not inhibit the gravitropic response of cress roots. The nonphenolic analogues, bibenzyl- and stilbene-2-carboxylic acids, showed more consistent effects on IAA accumulation and inhibited gravitropism.

Lunularic acid at $<10\,\mathrm{mol\,m^{-3}}$ inhibited the de-differentiation of isolated, mature cells of the liverwort *Riella helicophylla* (Grotha and Schwabe, 1978). At higher concentrations a red-brown diffusible substance, possibly an oxidation product of lunularic acid, was observed in plastids and cell walls of nondividing cells. Lunularic acid inhibited photoreduction of methyl viologen in uncoupled spinach chloroplasts at $500\,\mathrm{mol\,m^{-3}}$, but phenol and several stilbenes and bibenzyls were more inhibitory (Gorham and Coughlan, 1980).

7.4.2 Effects of stilbenoids on higher plants

The batatasins (**169, 170, 269**) are reputed to be dormancy factors in the tubers of *Dioscorea* species (El-Olemy and Reisch, 1979a,b; Hashimoto, Hasegawa and Kawarada, 1972; Hashimoto, Tamura and Kawarada 1973; Hashimoto and Hasegawa, 1974; Hashimoto *et al.*, 1974; Hasegawa and Hashimoto, 1974; Hashimoto and Tajima, 1978; Saito *et al.*, 1976). Abscisic acid was also found in aerial bulbils of *Dioscorea opposita* (= *D. batatas*), but was not correlated with dormancy and did not prevent sprouting (Hasegawa and Hashimoto, 1973; Hashimoto and Tajima, 1978). Batatasin concentrations were related to dormancy, which could be induced by exogenous application of these compounds. Gibberellin (GA_3) suppressed the sprouting of bulbils and increased the concentrations of batatasins, but not of abscisic acid (Hasegawa and Hashimoto, 1974). Ireland and Passam (1985) found that exogenous GA_3 was more effective than the batatasins in prolonging dormancy of yam tubers, and that the concentrations of batatasins which were effective were impractical for commercial use.

Exogenous application of batatasins also inhibits the growth of liverworts and algae (Gorham, 1978; Thomas, 1977). Batatasin I inhibited the growth of oat coleoptiles (Hashimoto and Hasegawa, 1973), but the other batatasins were more inhibitory than batatasin I to lettuce seed germination and hypocotyl elongation, and the elongation of wheat coleoptile sections (Hashimoto and Tajima, 1978). Dihydropinosylvin, demethylbatatasin IV and batatasin IV, found in *Dioscorea alata*, inhibited the germination of *Sorghum bicolor* seeds, and root elongation (Cline *et al.*, 1989). Batatasin IV was the most inhibitory of these bibenzyls.

In spinach chloroplasts, CO_2-dependent O_2 evolution was inhibited by batatasins, as was the flow of electrons from water to methylviologen (Iino, Hashimoto and Heber, 1978). Photosystem-1-dependent cyclic photophosphorylation was uncoupled from electron transport, and electron transport from ascorbate to oxygen was stimulated. The proton conductivity of the thylakoid membranes was also increased. Batatasin I was more effective than batatasins II and V. Succinate-dependent O_2 uptake in potato tuber mitochondria was inhibited by the batatasins. In a separate study, Gorham and Coughlan (1980) found that a number of stilbenes and bibenzyls inhibited CO_2-dependent O_2 evolution and electron flow from water and dichlorophenolindophenol ($DCPIPH_2$) to methylviologen. Pinosylvin (**143a**) and its methyl ethers were the most inhibitory of the compounds tested. Martin and Bassham (1980) found that pine chloroplasts, and extracts of pine chloroplasts, inhibited CO_2 fixation by spinach chloroplasts, and this may be attributable to pinosylvins.

Phenanthrenes and 9,10-dihydrophenanthrenes such as orchinol (**273**) with a free hydroxyl at the 7- position inhibit IAA oxidation catalysed by horseradish peroxidase with 2,4-dichlorophenol as cofactor (Lee, Rock and Stoessl, 1978). Phenanthrenes without a free hydroxyl group were ineffective, but trihydroxyphenanthrenes were highly inhibitory. In the absence of 2,4-dichlorophenol batatasin III and 3,5-dimethoxy-3′-hydroxystilbene acted as cofactors, as was also reported for lunularic acid (**18**) and other bibenzyls and stilbenes by Gorham (1978), but phenanthrenes with similar substitution patterns (including orchinol) did not. Even in the presence of 2,4-dichlorophenol, batatasin II and 3,5-dimethoxy-3′-hydroxystilbene slightly (9–20%) increased IAA oxidation. Low concentrations of orchinol disappeared from the reaction mixture, after which the oxidation rate returned to the control value. Orchinol competed preferentially with IAA for the ferriporphyrin group of the horseradish peroxidase.

Two bibenzyl analogues inhibited the elongation of rice coleoptiles at 500 $\mu g\,ml^{-1}$ (Hashimoto *et al.*, 1987). Isorhapontin (**148b**) and astringin (**147b**) in extracts of Engelmann spruce inhibited seed germination and growth of various conifers (Taylor and Shaw, 1983). Orchinol, loroglossol, 3,4′-dihydroxy-5-methoxybibenzyl and batatasin III inhibited the growth of seedlings of the orchid *Cattleya aurantiaca* (Hills *et al.*, 1984). Ohwaki, Ogino and Shibano (1993) have reported that 3-hydroxy-5-methoxystilbene-2-carboxylic acid, from *Cajanus cajan*, inhibits the elongation of lettuce radicles.

7.5 EFFECTS OF STILBENOIDS ON ANIMALS

Pinosylvins are reported to be toxic to bacteria, fungi (Rennerfeldt, 1949; Scheffer and Cowling, 1966), insects, fishes (Erdtman, 1939a,b) and rats (Erdtman, 1949b). In rats pinosylvin (**143a**) is slightly more toxic than phenol, while the fish *Lebistes reticulatus* was killed by a 0.002% solution. Wood impregnated with pinosylvin is more resistant to attack by ants and termites (Wolcott, 1951, 1953; Sandermann and Dietrichs, 1957a,b). Piscicidal activity (against *Oryzia latipes*) was demonstrated for some methyl or ethyl ether analogues of hydrangeic acid, but not for other, nonmethylated bibenzyls and stilbenes, including lunularic acid (**18**) (Hashimoto, Tori and Asakawa, 1988). Pinosylvin monomethyl ether (**143b**) was found to be the most active of a number of nematicidal substances from *Pinus* species (Suga *et al.*, 1993). Pinosylvin monomethyl ether showed no antiacetylcholinesterase activity and thus acts differently from organophosphorus agents and carbamoyl insecticides.

Pinosylvin (**143a**) and pinosylvin monomethyl ether (**143b**) are potent feeding deterrents in *Alnus crispa* (Bryant *et al.*, 1983; Clausen, Reichardt and Bryant, 1986), limiting damage from snowshoe hares (*Lepus americanus*). Piceatannol (**147a**) and scirpusin B exhibited antifeeding activity in tests with the fall army worm, *Spodoptera frugiperda* (Powell, Bajaj and McLauchlin, 1987). Phyllodulcin (**224a**), in contrast, has an intense sweet taste and was isolated from leaves of *Hydrangea macrophylla* var. *thunbergii* used to make *Amacha* sweet tea (Asahina and Ueno, 1916; Maniwa, 1915, 1924). The macrophyllosides (**226**), also from a *Hydrangea macrophylla* variety (Hashimoto, Tori and Asakawa, 1987) were bitter-tasting and induced vomiting. Other phenyldihydroisocoumarins were tasteless. Chlorophorin (**218b**), from the West African tree *Chlorophora excelsa*, produces allergenic contact dermatitis.

Marchantin A (**90a**) had 5-lipoxygenase inhibitory activity – 89% at 10^{-5} mol, 94% at 10^{-6} mol, 45% at 10^{-7} mol, 16% at 10^{-8} mol) against leukotriene B_3 (LTB_3 = 5*S*,12*R*-dihydroxy-6,8,10,14-eicosatetraenoic acid); 99% at 10^{-5} mol, 97% at 10^{-6} mol, 70% at 10^{-7} mol, 40% at 10^{-8} mol against 5-HETE (= 5-hydroxy-6,8,11,14-eicosatetraenoic acid); and calmodulin inhibitory activity at ID_{50} 1.85 μg/ml (Asakawa, 1990a,b, Asakawa *et al.*, 1984). Marchantins D (**93**) and E (**94**), riccardin A (**110a**), perrottetins A (**46**) and D (**56**) and prenyl bibenzyls (**46, 51, 71, 79**) had calmodulin inhibitory activity (50% inhibitory dose, ID_{50}, 2.0–95.0 μg/ml). The simple bibenzyls (**1, 8, 16, 25**) also showed weak calmodulin inhibitory activity (ID_{50} 100 μg/ml) (Asakawa, 1990a,b). Marchantins D (**93**) and E (**94**), riccardin A (**110a**), perrottetin A (**46**), prenylbibenzyls (**11, 50**) also inhibited 5-lipoxygenase (4–76% at 10^{-6} mol) (Asakawa, 1990a,b; Asakawa *et al.*, 1991,a,b). Lunularic acid (**18**) found in almost all liverworts, has antihyaluronidase activity (IC_{50} 0.13 μmol m^{-3}). This activity is stronger than that of tranilast (*N*-3′,4′-dimethoxycinnamoylanthranilic acid), which is an antiallergenic agent developed in Japan for oral administration (Hashimoto, Tori and Asakawa 1988).

Low concentrations of oxyresveratrol (**204**) inhibit the respiration of rat liver mitochondria (Nimmanpisut, Chudapongse and Ratabanangkoon, 1976) and reduce mitochondrial ATPase activity (Janthasoot, Ratanabanangkoon and Chudaponyoe, 1977). Effects of resveratrol on lipid (Arichi *et al.*, 1982) and arachidonate metabolism (Kimura, Okuda and Arichi, 1985) have been described. Frankel, Waterhouse and Kinsella (1993) reported that resveratrol (**144a**) inhibited human low-density lipoprotein (LDL) oxidation. Resveratrol from *Polgygonum cuspidatum* was found to inhibit bovine protein–tyrosine kinase (Jayatilake *et al.*, 1993). The 3- and 4-*O*-glucosides were less inhibitory. Piceatannol (Deanin *et al.*, 1993) and some analogues (Thakkar, Geahlen and Cushman, 1993) are also selective inhibitors of $Fc_{\epsilon}RI$-induced protein–tyrosine phosphorylation and inhibit receptor-mediated signal transduction in RBL-2H3 rat tumour mast cells.

The biological activity ascribed to the Hepaticae is due mainly to terpenoids and aromatic compounds which are constituents of oil bodies. Many bis(bibenzyl)s possess cytotoxicity (Asakawa, 1990a,b). Riccardins A (**110a**) and B (**109a**) inhibited KB cells at a concentration of 10 and 12 μg/ml, respectively (Asakawa *et al.*, 1983a). The ED_{50} values for marchantin A (**90a**), B (**91**) and C (**92**) and perrottetin E (**122**) are recorded as 8.39, 10.0, 10.0 and 12.5 μg/ml, respectively (Asakawa, 1990a,b). The T/C (100%) values of marchantin A (**90a**) against P388 cells were estimated as 117% (Asakawa, 1990a,b).

The prenylated bibenzyl (**36**) has vasopressin antagonist activity (27 μg/ml). The monomethyl ethers of **51** and **52** showed similar activity (57 and 17 μg/ml) (Asakawa, 1990a,b; Asakawa *et al.*, 1991a). Marchantin A (**90a**) increases coronary blood flow (2.5 ml/min at 0.1 μg) (Asakawa, 1990b). This compound might be used as a coronary vasodilator.

Marchantin A (**90a**) and the related cyclic bis(bibenzyl)s are structurally similar to bis (bibenzyl)isoquinoline alkaloids such as tubocurarine, which are pharmacologically important muscle relaxing active drugs. Surprisingly, marchantin A (**90a**) and its trimethyl ether (**90b**) also had muscle relaxing activity (Hashimoto *et al.*, 1991b). Nicotine in frog Ringer solution effects maximum contraction of the rectus abdominis in frogs (RAF) at a concentration of 10^{-6} mol l^{-1}. After preincubation of marchantin A trimethyl ether (**90b**) (at a concentration of 2×10^{-7}–2×10^{-4} mol l^{-1}) in Ringer solution, nicotine (10^{-8}–10^{-4} mol l^{-1}) was added. At a concentration of 10^{-6} mol l^{-1}, the contraction of RAF decreased by about 30%. d-Tubocurarine exhibits similar effects to those described above with the use of acetyl choline. Although the mechanism of action of marchantin A (**90a**) and its methyl ether (**90b**) in effecting muscle relaxation is still unknown, it is interesting that these cyclic bis(bibenzyl)s from liverworts without nitrogen atoms cause a concentration-dependent decrease of contraction of RAF. Marchantin A and its trimethyl ether also had muscle relaxing activity *in vivo* in mice. Calculations by the computer program MM2 indicate that the conformation and the presence of an ortho hydroxyl group in **90a** and an ortho methoxyl group in **90b** contribute to the muscle relaxation activity (Taira *et al.*, 1990).

Stilbenes from higher plants also have effects on animal systems. The coronary vasodilatory and hypertensive effects of 3,3′-dihydroxy-α, β-diethylstilbene, piceatannol and related compounds were studied by Inamori *et al.* (1984a, 1985, 1987b, 1988a, 1989, 1990). Piceatannol and its dihydro derivative inhibited the antigen-induced release of histamine in human basophils *in vitro* (Inamori *et al.*, 1991b,c).

Little is known about the biological activities of the large number of bibenzyls and phenanthrenes found in orchids, but Sachdev and Kulshreshtha (1986) identified coelogin and flavidin as spasmolytic agents. Bibenzyls and dihydrophenanthrenes from *Bletilla striata* showed no activity against leukaemic P388 cells (Takagi, Yamaki and Inoue, 1983). Shan (1988) reported that resveratrol affected rabbit platelet aggregation. Ampelopsin E, from roots of the wild vine *Ampelopsis brevipedunculata* var. *hancei*, protected rat hepatocytes against carbon tetrachloride-induced damage. Kimura *et al.* (1983) found that piceid (**144b**) protected rats fed peroxidized oils from liver injury. Skinnider and Stoessl (1986) reported that pinosylvin and dehydroloroglossol inhibited the growth of human lymphoblastoid cell lines Molt and Raji. Isorhapontin (**148b**) and piceatannol (147a) from *Picea abies* displayed antileukaemic activity (Mannila and Talvitie, 1992). Several stilbene derivatives showed inhibitory activity against mouse leukaemia L1210 cultures, the most active being dihydroastringin and 3,3′,4,5′-tetrahydroxybibenzyl (Mannila, Talvitie and Kolehmainen, 1993; Talvitie, Mannila and Kolehmainen, 1992). Piceatannol (**147a**) had earlier been identified as the antileukaemic principle in the seeds of *Euphorbia lagascae* (Ferrigni *et al.*, 1984), and, together with scirpusin B, in seeds of *Scirpus maritimus* (Powell, Bajaj and McLauchlin, 1987). The 9,10-dihydrophenanthrenes from *Juncus roemerinus* are cytotoxic (Bhattacharyya and Miles, 1977; Cossey, Gunter and Mander, 1980; Miles *et al.*, 1977, 1978, 1981).

The genus *Combretum* includes about 250 species of tropical trees and climbers. Of these about 25 species are used for medicinal purposes in Africa and India. Their uses include treatments for leprosy, cancer, heart problems, scorpion stings and mental illness, and as wound dressings (Pettit *et al.*, 1987; Pettit, Cragg and Singh, 1987; Pettit, Smith and Singh, 1987). Screening of plant extracts in the murine P388 lymphocytic leukaemia system by the United States National Cancer Institute resulted in the identification of *Combretum molle* and *Combretum caffrum* as potential sources of antileukaemic substances. Using an assay based on the 9ASK astrocytoma system, Pettit, Cragg and Singh (1987) first identified combretastatin (**171**), an α-hydroxylated bibenzyl, as the active compound in extracts of *C. caffrum*. Combretastatin (**171**) also inhibited the growth of murine P388 cells. Later work identified combretastatin A-1 (**174**), a *cis*-stilbene, as one of the main antineoplastic constituents of this species, effective against P388 cells with an ED_{50} of 0.99 $\mu g\,ml^{-1}$ (Pettit *et al.*, 1987). Combretastatin A-1 (**174**) and the bibenzyl combretastatin B-1 (**172**) were potent inhibitors of microtubule assembly *in vitro*, and of the binding of colchicine to tubulin. Further *cis*-stilbenes (combretastatins A-2 and A-3) and bibenzyls (combretastatin B-2, B-3 and B-4) from the same source also inhibited the growth of P388 cells (Pettit and Singh, 1987; Pettit *et al.*, 1988a,b), and inhibited tubulin polymerization. Combretastatin analogues were synthesized by Mannila and Talvitie (1993) and also showed antileukaemic activity, but Malan and Swinny (1993) reported that 4,7-dihydroxy-2,3,6-trimethoxyphenanthrene only slightly inhibited murine 3T6 fibroblast cells at 1 $mmol\,m^{-3}$. No toxicity was shown for 2,7-dihydroxy-2,4,6-trimethoxy-9,10-dihydrophenanthrene and 4,4′-dihydroxy-3,5-dimethoxystilbene, also isolated from *Combretum apiculatum*. The activities of several *Combretum* stilbenoids in the P388 assay are listed in Table 7.3.

The astrocytoma assay which had been used to fractionate extracts of *C. caffrum* shows good selectivity for antimitotic agents and this, together

Table 7.3 Antineoplastic activity of *Combretum* stilbenoids

Compound	*P388 inhibition (ED_{50}, $g\,ml^{-1}$)*	*Authors*
Combretastatin	0.011	Pettit, Cragg and Singh (1987)
Combretastatin A-1	0.99	Pettit *et al.* (1987)
Combretastatin A-2	0.027	Pettit and Singh (1987)
Combretastatin A-3	0.026	Pettit and Singh (1987)
Combretastatin B-1	1.7	Pettit *et al.* (1987)
Combretastatin B-2	0.32	Pettit and Singh (1987)
9,10-DiHYDRO-PHENANTHRENES		
6,7-dihydroxy-2,3,4-trimethoxy-	2.2	Pettit *et al.* (1988a)
7-hydroxy-2,3,4,6-tetrahydroxy-	2.8	Pettit *et al.* (1988a)
2,7-dihydroxy-3,4,6-trimethoxy-	2.6	Pettit *et al.* (1988a)
PHENANTHRENE		
7-hydroxy-2,3,4,6-trimethoxy-	2.0	Pettit *et al.* (1988a)

with the structural similarities (Figure 7.1) of combretastatin, colchicine and its analogue 2-methoxy-5-(2′,3′,4′-trimethoxyphenyl)tropone (MTPT) led to tests of the antimitotic and antitubulin activity of combretastatin (Hamel and Lin, 1983). It was found to have high activity, causing a marked rise in the mitotic index of L1210 cells and competitively inhibited the binding of colchicine to tubulin. Like colchicine, it inhibited tubulin polymerization and stimulated tubulin-dependent GTP hydrolysis. Sackett (1993) compared combretastatin (**171**) with podophyllotoxin and steganacin, two other natural products that bind at the colchicine site of tubulin. Microtubule assembly, *in vitro* and *in vivo*, was found to be affected by both 3,3′-dihydroxy-α,β-diethylstilbene and 3,3′,4,5′-tetrahydroxystilbene in a study by Sakakibara *et al.* (1992).

Lin *et al.* (1988) examined the antimitotic activities of 39 combretastatin analogues and found that 19 compounds inhibited cell growth by 50% at 1 mmol m^{-3} or less. Fourteen compounds inhibited tubulin polymerization by at least 50% at stoichiometric drug concentrations. The most toxic compounds strongly inhibited both tubulin polymerization and colchicine binding. Combretastatin A-4 (**177**) was the most active compound, and has been the subject of further studies (Lin *et al.*, 1989; McGown and Fox, 1989, 1990; Pettit *et al.*, 1989). El-Zayat *et al.* (1993) found that combretastatin A-4 was selectively toxic to six out of eight human tumour cell lines

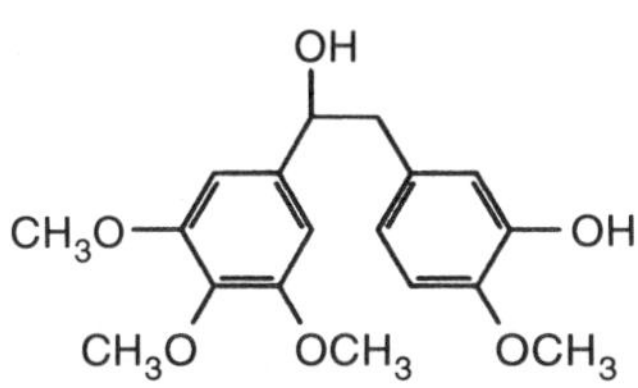

171 combretastatin

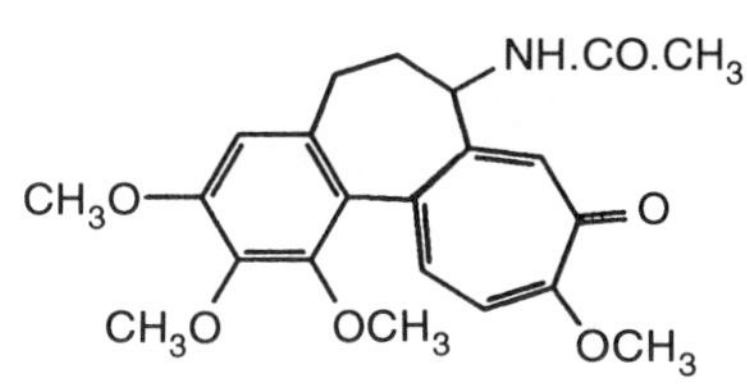

colchicine

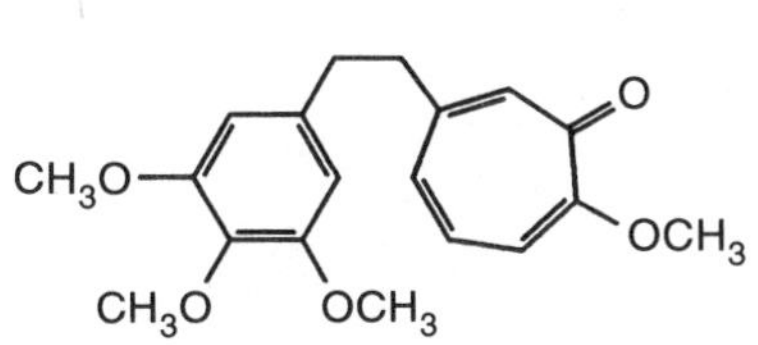

'combretatropone'

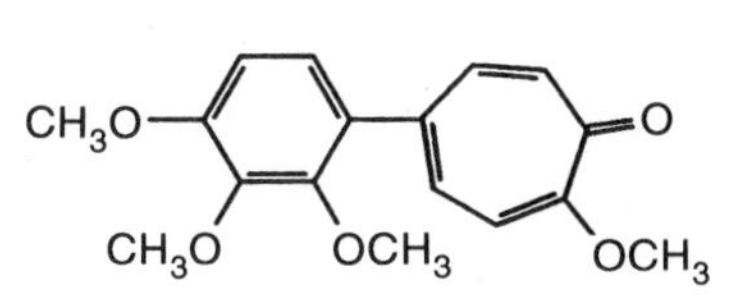

2-methoxy-5-(2',3',4'-trimethoxyphenyl) tropone (MTPT)

Figure 7.1 Relationship of combretastatin to other tubulin-binding compounds.

(i.e. it affected tumour cells more than it did normal cells) and to a variety of human tumours. They suggested further tests on *in vivo* preclinical model systems. McGown and Fox (1990) investigated the toxicity of combretastatins A-1 and A-4 to daunorubicin-resistant P388 cell lines. Combretastatin A-4 (**177**) was inhibitory but A-1 was not, indicating cross resistance for daunorubicin and combretastatin A-1 (**174**). Since multiple drug resistance is becoming more of a problem, McGown and Fox (1990) suggested that combretastatin A-4 might be used in the treatment of tumours resistant to anthracycline antibiotics. Andres *et al.* (1993) have synthesized a number of 'combretatropones' with a hybrid combretastatin/colchicine structure (Figure 7.1) and tested their interactions with tubulin.

Synthetic stilbenoids have a number of effects on plants and animals, including oestrogenic activity (e.g. diethylstilboestrol: Knorr, Lehr and Probst, 1956) and inhibition of membrane transport (4,4′-dinitrostilbene-2, 2′-disulphonate: Balke and Hodges, 1979; Knauf *et al.*, 1993; Marten *et al.*, 1993). The biological properties of synthetic stilbene derivatives have been described in detail by Inamori *et al.* (1984b,c, 1987a, 1988a,b, 1991a).

Interest in the resveratrol content of wine has been stimulated by the suggestion that it may reduce heart disease (Waterhouse and Lamnela-Raventos, 1994).

References

Abe, S. and Ohta, Y. (1983) Lunularic acid in cell suspension cultures of *Marchantia polymorpha*. *Phytochemistry*, **22**, 1917–20.

Abe, S. and Ohta, Y. (1984) The concentrations of lunularic acid and prelunularic acid in liverworts. *Phytochemistry*, **23**, 1379–81.

Adam, K. P. and Becker, H. (1994) Phenanthrenes and other phenolics from *in vitro* cultures of *Marchantia polymorpha*. *Phytochemistry*, **35**, 139–43.

Adesanya, S. A., Ogundana, S. K. and Roberts, M. F. (1989) Dihydrostilbene phytoalexins from *Dioscorea bulbifera* L. and *D. dumetorum* (Kunth.) Pax. *Phytochemistry*, **28**, 773–4.

Adesogan, E. K. (1974) Trithiolaniacin, novel trithiolane from *Petiveria alliacea*. *J. Chem. Soc. Chem. Commun.*, 906–7.

Adler, E. (1951) Sulfite pulping properties of sprucewood from unpeeled, floated logs. II. Sulfite cooking experiments. *Svensk Papperstidn.*, **54**, 477–82.

Aguamah, G. E., Langcake, P., Leworthy, D. P. *et al.* (1981) Two novel stilbene phytoalexins from *Arachis hypogaea*. *Phytochemistry*, **20**, 1381–3.

Ahluwalia, V. K., Dass, I. and Mehta, A. C. (1956) Crystalline components of certain Indian pines. *Curr. Sci.*, **25**, 367.

Akada, Y., Kawano, S., Gotoh, M. and Tanase, Y. (1979) High-speed liquid chromatographic analysis of drugs. IV. Separation analysis of phyllodulcin and hydrangenol in sweet *Hydrangea*. *Yakugaku Zasshi*, **99**, 683–6.

Akisanya, A., Bevan, C. W. L. and Hirst, J. (1959) West African timbers. Part II. Heartwood constituents of the genus *Pterocarpus*. *J. Chem. Soc.*, 2679.

Alcubilla-Martin, M. (1970) Fungus inhibitors in spruce bark. *Landwirt. Forsch. Sonderh.*, **25**, 96–101.

Alvarez-Novoa, J. C., Erdtman, H. and Lindstedt, G. (1950) Consituents of pine heartwood. XIX. Heartwood of *Pinus pinea* L., *Pinus pinaster* Aiton, *Pinus halepensis* Mill. and *Pinus nigra* var. *calabrica* (Loudon) Schneider. *Acta Chem. Scand.*, **4**, 444–7.

Anderson, A. B. (1956) Increasing extractive content in trees for rosin production. Extract stimulation of Jeffrey pine. *Tappi*, **39**, 55–9.

Andres, C. J., Bernardo, J. E., Yan, Q. *et al.* (1993) Combretatropones – hybrids of combretastatin and colchicine – synthesis and biochemical evaluation. *Bioorg. Med. Chem. Lett.* **3**, 565–70.

Andrews, D. H., Hoffman, J. C., Purves, C. B. *et al.* (1968) Isolation, structure and synthesis of a stilbene glucoside from the bark of *Picea glauca* (Moench) Voss. *Can. J. Chem.*, **46**, 2524–9.

Anjaneyulu, A. S. R., Reddy, A. V. R., Reddy, D. S. K. *et al.* (1984) Pacharin:

A new dibenzo(2,3–6,7)oxepin derivative from *Bauhinia racemosa* Lamk. *Tetrahedron*, **40**, 4245–52.

Aquino, R., Behar, I., Desimone, F. and Pizza, C. (1985a) Natural dihydrophenanthrene derivatives from *Tamus communis*. *J. Nat. Prod. (Lloydia)*, **48**, 811–13.

Aquino, R., Behar I., Desimone, F. *et al.* (1985b) Phenanthrene derivatives from *Tamus communis*. *Biochem. Syst. Ecol.*, **13**, 251–2.

Arai, Y., Kamikawa, T. and Kubota, T. (1972) A convenient synthesis of lunularic acid. *Tetrahedron Lett.*, **16**, 1615–17.

Arai, Y., Kamikawa, T., Kubota, T. *et al.* (1973) Synthesis and properties of lunularic acid. *Phytochemistry*, **12**, 2279–82.

Arakawa, H. (1960) Die absolute Konfiguration des Phyllodulcins. *Bull. Chem. Soc. Jap.*, **33**, 200–2.

Arakawa, H. and Nagazaki, M. (1959) Absolute configuration of phyllodulcin. *Chem. Ind.*, 671.

Arditti, J. (1979) Aspects of the physiology of orchids. *Adv. Bot. Res.*, **7**, 422–655.

Arichi, H., Kimura, Y., Okuda, H. *et al.* (1982) Effects of stilbene components of the roots of *Polygonum cuspidatum* Sieb. et Zucc. on lipid metabolism. *Chem. Pharm. Bull.*, **30** 1766–70.

Aritomi, M. and Donnelly, M. X. (1976) Stilbene glucosides in the bark of *Picea sitchensis*. *Phytochemistry*, **15**, 2006–8.

Arriaga-Giner, F. J., Wollenweber, E. and Dörr, M. (1993) Bibenzyls from crowberry leaves. *Phytochemistry*, **33**, 725–6.

Asahina, Y. (1909) *J. Pharm. Soc. Jap.*, 330.

Asahina, Y. and Asano, J. (1929) Uber die Konstitution von Hydrangenol und Phyllodulcin. *Ber. dtsch. chem. Ges.*, **62**, 171–7.

Asahina, Y. and Asano, J. (1930a) Uber die Konstitution von Hydrangenol und Phyllodulcin (II Mitteil.). *Ber. dtsch. chem. Ges.*, **63**, 429–37.

Asahina, Y. and Asano, J. (1930b) Uber die Konstitution von Hydrangenol und Phyllodulcin. III Mitteil. Synthese des Hydrangenols. *Ber. dtsch. chem. Ges.*, **63**, 2059–63.

Asahina, Y. and Asano, J. (1931) Uber die Konstitution von Hydrangenol und Phyllodulcin. IV Mitteil. Synthese des Phyllodulcin-dimethylaethers. *Ber. dtsch. chem. Ges.*, **64**, 1252–6.

Asahina, Y. and Miyake, K. (1916) Hydrangenol, a chemical constituent of *Hydrangea hortensia*. *J. Pharm. Soc. Jap.*, **408**, 121.

Asahina, Y. and Ueno, S. (1916) Phyllodulcin, a chemical constituent of Amacha (*Hydrangea thunbergii* Sieb.). *J. Pharm. Soc. Jap.*, **408**, 146.

Asakawa, Y. (1970) Chemical constituents of *Alnus firma*. I. Phenylpropane derivatives isolated from *Alnus firma*. *Bull. Chem. Soc. Jap.*, **43**, 2223–9.

Asakawa, Y. (1971) Chemical constituents of *Alnus sieboldiana* (Betulaceae) II. The isolation and structure of flavonoids and stilbenes. *Bull. Chem. Soc. Jap.*, **44**, 2761–6.

Asakawa, Y. (1982) Chemical constituents of Hepaticae, in *Progress in the Chemistry of Organic Natural Products*, (eds W. Herz, H. Grisebach and G. W. Kirby), Springer, Vienna, pp. 1–285.

Asakawa, Y. (1984) Phytochemistry of Hepaticae: isolation of biologically active aromatic compounds and terpenoids. *Rev. Latinoam. Quim.*, **14**(3), 109–15.

Asakawa, Y. (1988) Biologically active substances found in Hepaticae, in *Studies in Natural Products Chemistry* (ed. Atta-ur-Rahaman), Elsevier, Amsterdam, pp. 277–92.

Asakawa, Y. (1990a) Terpenoids and aromatic compounds with pharmacological. activity from bryophytes, in *Bryophytes, Their Chemistry and Chemical*

Taxonomy (eds H. D. Zinsmeister and R. Mues), Clarendon Press, Oxford, pp. 369–410.

Asakawa, Y. (1990b) Biologically active substances from Bryophytes, in *Bryophyte Development: Physiology and Biochemistry*, (eds R. N. Chopra and S. C. Bhatla), CRC Press, Florida, p. 259.

Asakawa, Y. and Campbell, E. O. (1982) Terpenoids and bibenzyls from some New Zealand liverworts. *Phytochemistry*, **21**(11), 2663–7.

Asakawa, Y., Hashimoto, T., Takikawa, K. *et al.* (1991a) Prenyl bibenzyls from the liverworts *Radula perrottetii* and *Radula complanata*. *Phytochemistry*, **30**, 235–51.

Asakawa, Y. and Inoue, H. (1984a) Chemical constituents of selected Chilean liverworts, in *Studies on Cryptogams in Southern Chile*, (ed. H. Inoue), Tokai University Press, Tokyo, pp. 109–15.

Asakawa, Y. and Inoue, H. (1984b) Chemical constituents of Chilean *Plagiochila* species, in *Studies on Cryptogams in Southern Chile*, (ed. H. Inoue), Tokai University Press, Tokyo, pp. 117–24.

Asakawa, Y. and Inoue, H. (1987) Chemical constituents of selected Peruvian liverworts, in *Studies on Cryptogams in Southern Peru*, (ed. H. Inoue), Tokai University Press, Tokyo, pp. 107–17.

Asakawa, Y., Kondo, K. and Tori, M. (1991) Cyclopropanochroman derivatives from the liverwort *Radula javanica*. *Phytochemistry*, **30**, 325–8.

Asakawa, Y., Kondo, K., Tori, M. *et al.* (1991b) Prenyl bibenzyls from the liverwort *Radula kojana*. *Phytochemistry*, **30**, 219–34.

Asakawa, Y., Kusube, E., Takemoto, T. and Suire, C. (1978) New bibenzyls from *Radula complanata*. *Phytochemistry*, **17**, 2115–17.

Asakawa, Y., Lin, X., Kondo, K. and Fukuyama, Y. (1991c) Terpenoids and aromatic compounds from selected east Malaysian liverworts. *Phytochemistry*, **30**, 4019–24.

Asakawa, Y. and Matsuda, R. (1982) Riccardin C, a novel cyclic bibenzyl derivative from *Reboulia hemispaerica*. *Phytochemistry*, **21**, 2143–4.

Asakawa, Y., Matsuda, R. and Cheminat, A. (1987) Bibenzyl derivatives from *Frullania* species. *Phytochemistry*, **26**, 1117–22.

Asakawa, Y., Matsuda, R., Toyota, M. *et al.* (1981a) Terpenoids and bibenzyls of 25 liverwort *Frullania* species. *Phytochemistry*, **20**, 2187–94.

Asakawa, Y., Okada, K. and Perold, G. W. (1988) Distribution of cyclic bis(bibenzyl)s in the south African liverwort *Marchantia polymorpha*. *Phytochemistry*, **21**, 161–3.

Asakawa, Y., Takeda, R., Toyota, M. and Takemoto, T. (1981b) Bibenzyls from *Radula tokiensis* and *R. japonica*. *Phytochemistry*, **20**, 858–9.

Asakawa, Y., Takikawa, K. and Aratani, T. (1976) New substituted bibenzyls of *Frullania brittoniae* subsp. *truncatifolia*. *Phytochemistry*, **15**, 1057–9.

Asakawa, Y., Takikawa, K. and Tori, M. (1987) Bibenzyl derivatives from the Australian liverwort *Frullania falciloba*. *Phytochemistry*, **26**, 1023–5.

Asakawa, Y., Takikawa, K., Tori, M. and Campbell, E. O. (1986) Isotachin C and balantiolide, two aromatic compounds from the New Zealand liverwort *Balantiopsis rosea*. *Phytochemistry*, **25**, 2543–6.

Asakawa, Y., Takikawa, K., Toyota, M. and Takemoto, T. (1982) Novel bibenzyl derivatives and ent-cuparane-type sesquiterpenoids from *Radula* species. *Phytochemistry*, **21**, 2481–90.

Asakawa, Y., Tori., M., Masuya, T. and Frahm, J.-P. (1990) Ent-sesquiterpenoids and cyclic bis(bibenzyl)s from the German liverwort *Marchantia polymorpha*. *Phytochemistry*, **29**, 1577–84.

Asakawa, Y., Tori, M., Takikawa, K. *et al.* (1987) Cyclic bis(bibenzyl)s and related compounds from the liverwort *Marchantia polymorpha* and *Marchantia palmata*. *Phytochemistry*, **26**, 1811–16.

Asakawa, Y., Toyota, M., Bischler, H. *et al.* (1984) Comparative study of chemical constituents of *Marchantia* species. *J. Hattori Bot. Lab.*, **57**, 383–9.

Asakawa, Y., Toyota, M., Matsuda, R. *et al.* (1983a) Distribution of novel cyclic bisbibenzyls in *Marchantia* and *Riccardia* species. *Phytochemistry*, **22**, 1413–15.

Asakawa, Y., Toyota, M., Taira, Z. and Takemoto, T. (1983b) Riccardin A and riccardin B, two novel cyclic bis(bibenzyl)s possessing cytotoxicity from the liverwort *Riccardia multifida* (L.) S. Gray. *J. Org. Chem.*, **48**, 2164–7.

Asakawa, Y., Toyota, M. and Takemoto, T. (1978) Seven new bibenzyls and a dihydrochalcone from *Radula variabilis*. *Phytochemistry*, **17**, 2005–10.

Asen, S., Cathey, H. M. and Stuart, N. W. (1960) Enhancement of gibberellin growth-promoting activity by hydrangenol isolated from leaves of *Hydrangea macrophylla*. *Plant Physiol.*, **35**, 816–19.

Askari, A., Worthen, L. R. and Schimiza, Y. (1972) Gaylussacin, a new stilbene derivative from species of *Gaylussacia*. *Lloydia*, **35**, 49–54.

Baba, K., Kido, T., Maeda, K. *et al.*, (1992) Two stilbenoids from *Cassia garretiana*. *Phytochemistry*, **31**, 3215–18.

Baba, K., Maeda, K., Tubota, Y. *et al.* (1988) Chemical studies on the heartwood of *Cassia garrettiana* Craib. 3. Structure of 2 new polyphenolic compounds. *Chem. Pharm. Bull.*, **36**, 2977–83.

Bachelor, F. W., Loman, A. A. and Snowdon, L. R. (1970) Synthesis of pinosylvin and related heartwood stilbenes. *Can. J. Chem.*, **48**, 1554–7.

Bai, L., Kato, T., Inoue, K. *et al.* (1991) Blestrianol A,B and C, biphenanthrenes from *Bletilla striata*. *Phytochemistry*, **30**, 2733–5.

Bai, L., Kato, T., Inoue, K. *et al.* (1993) Stilbenoids from *Bletilla striata*. *Phytochemistry*, **33**, 1481–3.

Bai, L., Yamaki, M., Inoue, K. and Takagi, S. (1990) Blestrin A and B, bis(dihydrophenanthrene)ethers from *Bletilla striata*. *Phytochemistry*, **29**, 1259–60.

Bailey, J. A. and Burden, R. S. (1973) Biochemical changes and phytoalexin accumulation in *Phaseolus vulgaris* following cellular browning caused by tobacco mosaic virus. *Physiol. Plant Pathol.*, **3**, 171–7.

Balke, N. E. and Hodges, T. K. (1979) Comparison of reductions in adenosine triphosphate content, plasma membrane-associated adenosine triphosphate activity, and potassium absorption in oat roots by diethylstilbestrol. *Plant Physiol.* **63**, 53–6.

Banks, H. J. and Cameron, D. W. (1971) A new natural stilbene glucoside from *Rheum rhaponticum* (Polygonaceae). *Aust. J. Chem.*, **24**, 2427–30.

Barlass, M., Miller, R. M. and Douglas, T. J. (1987) Development of methods for screening grapevines for resistance to infection by downy mildew. II. Resveratrol production. *Am. J. Enol. Viticult.*, **38**, 65–8.

Barnes, R. A. and Gerber, N. N. (1955) The antifungal agent of Osage orange wood. *J. Am. Chem. Soc.*, **77**, 3259–62.

Barua, A. K., Ghosh, B. B., Ray, S. and Patra, A. (1990) Cymbinodin-A, a phenanthraquinone, from *Cymbidium aloifolium*. *Phytochemistry*, **29**, 3046–7.

Bassett, C., Sherwood, R. T., Kepler, J. A. and Hamilton, P. B. (1967) Production and biological activity of fomannosin, a toxic sesquiterpene metabolite of *Fomes annosus*. *Phytopathology*, **57**, 1046–52.

Bavaresco, L. and Eibach, R. (1987) Investigations on the influence of N fertilizer on resistance to powdery mildew (*Oidium tuckeri*), downy mildew (*Plasmopara viticola*) and on phytoalexin synthesis in different grapevine varieties. *Vitis*, **26**, 192–200.

Benesová, V. and Herout, V. (1970) Isolation and the structure of pellepiphylin, 2-hydroxy-3-methoxy-4′-methoxydihydrostilbene from liverwort *Pellia epiphylla* (L.) DUM. *Coll. Czech. Chem. Commun.*, **35**, 1926–9.

Bercht, C. A. L., van Dongen, J. P. C. M., Heerma, W. *et al.* (1976) Cannabispirone and cannabispirenone, two naturally occurring spiro-compounds. *Tetrahedron*, **32**, 2939–43.

Bevan, C. W. L. (1958) Wood extractives. *W. Afr. J. Biol. Chem.*, **2**, 36–41.

Bhandari, S. R. and Kapadi, A. H. (1983) A 9,10-dihydrophenanthrene from tubers of *Eulophia nuda. Phytochemistry*, **22**, 747–8.

Bhandari, S. R., Kapadi, A. H., Majumder, P. L. *et al.* (1985) Nudol, a phenanthrene of the orchids *Eulophia nuda, Eria carinata* and *Eria stricta. Phytochemistry*, **24**, 801–4.

Bhaskar, M. U., Jagan Mohan Rao, L., Prakasa Rao, N. S. and Mohana Rao, P. R. (1989) Ochrolide, a phenanthropyrone from *Coelogyne ochracea. Phytochemistry*, **28**, 3545–6.

Bhaskar, M. U., Rao, L. J. M., Rao, N. S. P. and Rao, P. R. M. (1991) Ochrone A, a novel 9,10-dihydro-1,4-phenanthraquinone from *Coelogyne ochracea. J. Nat. Prod.*, **54**, 386–9.

Bhattacharyya, J. (1980) Structure of effusol: a new phenolic constituent from *Juncus effusus. Experientia*, **36**, 27–8.

Bhattacharyya, J. and Miles, D. H. (1977) Structure of juncunol, a novel 9,10-dihydrophenanthrene from *Juncus roemerianus. Tetrahedron Lett.*, 2749–50.

Billek, G. (1964) Stilbene im Pflanzenreich. *Fortschr. org. chem. Naturstoffe*, **22**, 115–57.

Billek, G. and Kindl, H. (1961) Zur Biosynthese pflanzlicher Stilbene. I. Mitt. Zur Biosynthese des Hydrangenols. *Mon. Chem.*, **92**, 493–7.

Billek, G. and Kindl, H. (1962a) Die Pflanzlichen Stilbene. *Osterreichische Chemiker-Zeitung*, **63**, 273–81.

Billek, G. and Kindl, H. (1962b) Uber die phenolischen Inhaltstoffe der Familie Saxifragaceae. *Mon. Chem.*, **93**, 85–98.

Billek, G. and Kindl, H. (1962c) Zur Biosynthese pflanzlicher Stilbene. 2 Mitt. Die Bildung von A ring des Hydrangenols. *Mon. Chem.*, **93**, 814–21.

Billek, G. and Schimpl, A. (1962) Zur Biosynthese pflanzlicher Stilbene. 4 Mitt. Die Bildung von Ring A des Oxyresveratrols. *Mon. Chem.*, **93**, 1457–9.

Billek, G. and Schimpl, A. (1966) Biosynthesis of plant stilbenes, in *Biosynthesis of Aromatic Compounds*. Proceedings of the 2nd meeting F.E.B.S., Vienna, 1965, vol. 3, Pergamon Press, Oxford, pp. 37–44.

Billek, G. and Ziegler, W. (1962) Zur Biosynthese pflanzlicher Stilbene. 3 Mitt. Die Biosynthese des Pinosylvin-monomethylaethers. *Mon. Chem.*, **93**, 1431–40.

Birch, A. J. (1966) Some natural antifungal agents. *Chem. Ind.*, 1173–6.

Birch, A. J., Cameron, D. W. and Rickards, R. W. (1960) Studies in relation to biosynthesis. Part XXIII. The formation of aromatic compounds from β-polyketones. *J. Chem. Soc.*, 4395–400.

Birch, A. J. and Donovan, F. W. (1953) Studies in relation to biosynthesis. I. Some possible routes to derivatives of orcinol and phloroglucinol. *Aust. J. Chem.*, **6** 360–8.

Blaich, R. and Bachmann, O. (1980) Die Resveratrolsynthese bei Vitaceen. Induktion und zytologische Beobachtungen. *Vitis*, **19**, 230–40.

Blair, G. E., Cassady, J. M., Robbers, J. E. *et al.* (1969) Isolation of 3,4′,5-trimethoxy-trans-stilbene, otobaene and hydroxyotobaene from *Virola cuspidata. Phytochemistry*, **8**, 497–500.

Bocskei, Z. and Keseru, G. M. (1994) Crystal structure and conformation of pakyonol, a macrocyclic bis(bibenzyl) constituent of *Mannia fragrans*. *J. Chem. Soc. Perkin Trans.*, **2**, 595–7.

Boeren, E. G., Elsohly, M. A., Turner, C. E. and Salemink, C. A. (1977) Beta-cannabispiranol: a new non-cannabinoid phenol from *Cannabis sativa* L. *Experientia*, **33**, 848.

Bohlmann, F. and Hoffmann, E. (1979) Cannabigerol-ahnliche Verbindungen aus *Helichrysum umbraculigerum*. *Phytochemistry*, **18**, 1371–4.

Bokel, M., Diyasena M. N. C., Gunatilaka, A. A. L. *et al.* (1988) Canaliculatol, an antifungal resveratrol trimer from *Stemonoporous canaliculatus*. *Phytochemistry*, **27**, 377–80.

Bombardelli, E., Martinelli, E. M. and Mustich, G. (1975) Plants of Mozambique. IX. A new hydroxystilbene glycoside from *Terminalia sericea*. *Fitoterapia*, **46**, 199–200.

Boralle, N., Gottlieb, H. E., Gottlieb, O. R. *et al.* (1993) Oligostilbenoids from *Gnetum venosum*. *Phytochemistry*, **34**, 1403–7.

Braun, R. (1963) Orchinol, in *Modern Methods of Plant Analysis*, vol. 6 (eds H. F. Linskens and M. V. Tracey), Springer Verlag, Berlin, pp. 130–4.

Braz Filho, R., Gottlieb, O. R. and Mourao, A. P. (1975a) A stilbene and two flavanones from *Derris rariflora*. *Phytochemistry*, **14**, 261–4.

Braz Filho, R., Gottlieb, O. R. and Mourao, A. P. (1975b) Quoted in: The chemical composition of Amazonian plants. *Acta Amazonica*, **5**, 77.

Braz Filho, R., Gottlieb, O. R., Mourao, A. P. *et al.* (1975) Flavonoids from *Derris* species. *Phytochemistry*, **14**, 1454–6.

Brinker, A. M. and Seigler, D. S. (1991) Isolation and identification of piceatannol as a phytoalexin from sugarcane. *Phytochemistry*, **30**, 3229–32.

Brinker, A. M. and Seigler, D. S. (1993) Time course of piceatannol accumulation in resistant and susceptible sugarcane stalks after inoculation with *Colletotrichum falcatum*. *Physiol. Mol. Plant Pathol.*, **42**, 169–76.

Bryant, J. P., Wieland, G. D., Reichardt, P. B. *et al.* (1983) Pinosylvin methyl ether deters snowshoe hare feeding on green alder. *Science*, **222**, 1023–5.

Calderón, A. A., Zapata, J. M., Munoz, R. *et al.* (1993) Resveratrol production as a part of the hypersensitive-like response of grapevine cells to an elicitor from *Trichoderma viride*. *New Phytologist*, **124**, 455–63.

Cardona, M. L., Fernandez, M. I., Garcia, M. B. and Pedro, J. R. (1986) Synthesis of natural polyhydroxystilbenes. *Tetrahedron*, **42**, 2725–30.

Carvalho, C. F. and Sargent, M. V. (1984) Phenanthrene synthesis: the synthesis of juncunone and the micrandrols – A,B,E, and F. *J. Chem. Soc. Perkin Trans.*, **1**, 1913–18.

Casabuono, A. C. and Pomilio, A. B. (1994) Lignans and a stilbene from *Festuca argentina*. *Phytochemistry*, **35**, 479–83.

Chappell, C. G., Creaser, C. S. and Shepherd, M. J. (1993) Online high-performance liquid-chromatography – multidimensional gas-chromatography and its application to the determination of stilbene hormones in corned beef. *J. High Resol. Chromatogr.*, **16**, 479–82.

Charpentier, B. A. and Cowles, J. R. (1981) Rapid method of analyzing phenolic compounds in *Pinus elliotti* using high-performance liquid chromatography. *J. Chromatogr.*, **208**, 132–6.

Christensen, L. P. and Lam, J. (1989) Excelsaoctaphenol, a stilbene dimer from *Chlorophora excelsa*. *Phytochemistry*, **28**, 917–18.

Christensen, L. P., Lam, J. and Sigsgaard, T. (1988) A novel stilbene from the wood of *Chlorophora excelsa*. *Phytochemistry*, **27**, 3014–16.

Clausen, T. P., Reichardt, P. B. and Bryant, J. P. (1986) Pinosylvin and pinosylvin methyl ether as feeding deterrents in green alder. *J. Chem. Ecol.*, **12**, 2117–31.

Cline, E. I., Adesanya, S. A., Ogundana, S. K. and Roberts, M. F. (1989) Induction of PAL activity and dihydrostilbene phytoalexins in *Dioscorea alata* and their plant growth inhibitory properties. *Phytochemistry*, **28**, 2621–5.

Coggan, P., Janes, N. F., King, F. E. *et al.* (1965) Hopeaphenol, an extractive of the heartwood of *Hopea odorata* and *Balanocarpus heimii*. *J. Chem. Soc.*, 406–8.

Coggan, P., King, T. J. and Wallwork, S. C. (1966) The structure of hopeaphenol. *Chem. Commun.*, 439–40.

Connolly, J. D. (1985) Recent advances in the chemistry of the Hepaticae, in *Natural Products Chemistry* (ed. R. I. Zalewski and J. J. Skolic), Elsevier, Amsterdam, pp. 3–14.

Cooksey, C. J., Dahiya, J. S., Garratt, P. J. and Strange, R. N. (1982) Two novel stilbene-2-carboxylic acid phytoalexins from *Cajanus cajan*. *Phytochemistry*, **21**, 2935–8.

Cooksey, C. J., Garratt, P. J., Richards, S. E. and Strange R. N. (1988) A dienyl stilbene phytoalexin from *Arachis hypogaea*. *Phytochemistry*, **27**, 1015–16.

Cossey, A. L., Gunter, M. J. and Mander, L. N. (1980) Juncunol: revision of structure and synthesis. *Tetrahedron Lett.*, **21**, 3309–12.

Coutts, M. P. (1970) The influence of phenolic compounds in *Pinus radiata* on the growth of *Amylostereum areolatum*. *Aust. J. For. Res.*, **4**, 15–18.

Coxon, D. T., Ogundana, S. K. and Dennis, C. (1982) Antifungal phenanthrenes in yam tubers. *Phytochemistry*, **21**, 1389–92.

Craviero, A. A. and Gottlieb, O. R. (1968) Chemistry of the Brazilian Leguminosae. XVI. Significance of piceatannol in the biosynthesis of centrolobine. *An. Acad. Brasil Cienc.*, **40**, 39–40.

Craveiro, A. A., Prado, A. da C., Gottlieb, O. R. and Welerson de Albuquerque, P. C. (1970) Diarylheptanoids of *Centrolobium* species. *Phytochemistry*, **9**, 1869–75.

Criodain, T. O., O'Sullivan, M., Meegan, M. J. and Donnelly, D. M. X. (1981) Latinone, a phenanthrene-1,4-quinone from *Dalbergia latifolia*. *Phytochemistry*, **20**, 1089–92.

Crombie, L. and Crombie, W. M. L. (1978) Dihydrostilbenes of Thailand cannabis. *Tetrahedron Lett.*, 4711–14.

Crombie, L. and Crombie, W. M. L. (1982) Natural products of Thailand high deltal-THC-strain *Cannabis*. The bibenzyl-spiran-dihydrophenanthrene group: relations with cannabinoids and canniflavones. *J. Chem. Soc. Perkin Trans.*, **1**, 1455–65.

Crombie, L., Crombie, W. M. L. and Jamieson, S. V. (1979) Isolation of cannabispiradienone and cannabidihydrophenanthrene. Biosynthetic relationships between the spirans and dihydrostilbenes of Thailand cannabis. *Tetrahedron Lett.*, 661–4.

Crombie, L., Crombie, W. M. L. and Jamieson, S. V. (1980) Extractives of Thailand *Cannabis*: synthesis of canniprene and isolation of new geranylated and prenylated chrysoeriols. *Tetrahedron Lett.*, 3607–10.

Crombie, L. and Jamieson, S. V. (1982) Dihydrostilbenes of *Cannabis*. Synthesis of canniprene. *J. Chem. Soc. Perkin Trans.*, **1**, 1467–75.

Crombie, L. W., Crombie, W. M. and Firth, D. F. (1988) Synthesis of bibenzyl cannabinoids, hybrids of two biogenetic series found in *Cannabis sativa*. *J. Chem. Soc. Perkin Trans.*, **1**, 1263–70.

Csupor, L. (1970) Desoxy-rhaponticin, ein neues naturliches Stilbenderivat in Rhizoma *Rhei rhapontici*. *L. Arch. Pharm.*, **303**, 681–7.

Cullmann, F., Adam, K.-P. and Becker, H. (1993) Bisbibenzyls and lignans from *Pellia epiphylla*. *Phytochemistry*, **34**, 831–4.

Cunningham, J., Haslam, E. and Haworth, R. D. (1963) The constitution of piceatannol. *J. Chem. Soc.*, 2875–83.

De Alvarenga, M. A., Da Silva, J. J., Gottlieb, H. E. and Gottlieb, O. R. (1981) Diterpenoids from *Micrandropsis scleroxylon*. *Phytochemistry*, **20**, 1159–61.

De Alvarenga, M. A. and Gottlieb, O. R. (1974) Methyl and methylthiophenanthrenes from *Micrandropsis scleroxylon*. *Phytochemistry*, **13**, 1283–8.

De Alvarenga, M. A., Gottlieb, O. R. and Magalhues, M. T. (1976) Methylphenanthrenes from *Sagotia racemosa*. *Phytochemistry*, **15**, 844–5.

De Oliveira, A. B., Fernandes, M. de L. M., Shaat, V. T. *et al.* (1977) The chemistry of Brazilian Leguminosae. LIV. Constituents of *Cassia* species. *Rev. Latinoam. Quim.*, **8**, 82–5.

Deanin, G. G., Oliver, J. M., Burg, D. L. *et al.* (1993) Piceatannol is a selective inhibitor of Fc-epsilon-R1-induced protein tyrosine phosphorylation and a strong inhibitor of receptor-mediated signal transduction in RBL-2H3 rat-tumor mast-cells. *J. Immunol.*, **150**, 221.

Della Greca, M., Fiorentino, A., Molinaro, A. *et al.* (1993) A bioactive dihydrodibenzoxepin from *Juncus effusus*. *Phytochemistry*, **34**, 1182–4.

Delle Monache, G., De Rosa, M. C., Scurria, R. *et al.* (1991) Metabolites from *in vitro* cultures of *Cassia didymobotrya*. *Phytochemistry*, **30**, 1849–54.

Delle Monache, F., Delle Monache, G., Cavalcanti, J. F. and Anheiro, R. M. (1987) An unexpected dihydrophenanthrene from *Clusia paralycola*. *Tetrahedron Lett.*, **28**, 563–6.

Delle Monache, F., Marletti, F., Marini Bettolo, G. B. *et al.* (1977) Isolation and structure of longistylins A, B, C and D, new prenylated stilbenes from *Lonchocarpus violaceus*. *Lloydia*, **40**, 201–8.

Dercks, W. and Creasy, L. L. (1989) The significance of stilbene phytoalexins in the *Plasmopara viticola*-grapevine interaction. *Physiol. Mol. Plant Pathol.*, **34**, 189–202.

Deshpande, V. H., Srinivasan, R. and Rao, A. V. R. (1975) Wood phenolics of *Morus* species. IV. Phenolics of the heartwood of five *Morus* species. *Ind. J. Chem.*, **13**, 453–7.

Dienes, Z., Nógrádi, M., Vermes, B. and Kajtár-Peredy, M. (1989) Synthese von Marchantin I, einem macrocyclischen Bis(bibenzylether) aus *Riccardia multifida*. *Justus Liebigs Ann. Chem.*, 1141–3.

Dietrich, S. M. C. and Valio, I. F. M. (1973) Effect of coumarin and its derivatives on the growth of *Pythium* and other fungi. *Trans. Br. Mycol. Soc.*, **61**, 461–9.

Diyasena, M. N. C., Sotheeswaran, S., Surendrakumar, S. *et al.* (1985) Balanocarpol, a new polyphenol from *Balanocarpus zeylandicus* (Trimen) and *Hopea jucunda* (Thw.) (Dipterocarpaceae). *J. Chem. Soc. Perkin Trans.*, **1**, 1807–9.

Drewes, S. E. (1971) Isolation of 3,3′,4′,5,5′-pentahydroxystilbene from *Schotia* species. *Phytochemistry*, **10**, 2837–8.

Drewes, S. E. and Fletcher, P. (1974) Polyhydroxystilbenes from heartwood of *Schotia brachypetala*. *J. Chem. Soc. Perkin Trans.*, **9**, 961–2.

Dubrovinskaya, G. A., Yakavleva, Z. M. and Tokin, B. (1974) Changes in the ultrastructure of hyphae of *Phytophthora infestans* caused by the monomethyl ether of pinosylvin. *Tr. Biol. Inst. Akad. Nauk. SSSR, Sib. Otd.*, **27**, 195–200.

Eicher, T. (1990) Recent results from the synthesis of chemical constituents of bryophytes, in *Bryophytes, their Chemistry and Chemical Taxonomy*, (eds H. D. Zinsmeister and R. Mues), Oxford University Press, Oxford, pp. 209–23.

El-Feraly, F. S., Elsohly, M. A., Boeren, E. G. and Turner, C. E. (1977) Crystal and molecular structure of cannabispiran and its correlation to dehydrocannbispiran. *Tetrahedron*, **33**, 2373–8.

El-Olemy, M. M. and Reisch, J. (1979a) Isolation of batatasin I from *Dioscorea dumetorum*. *Z. Naturforsch. C.*, **34**, 288–9.

El-Olemy, M. M. and Reisch, J. (1979b) Isolation of batatasin I from non-dormant bulbils of *Dioscorea opposita* – An improved method for isolation of batatasin I. *Planta medica*, **37**, 67–9.

El-Zayat, A. A. E., Degen, D., Drabek, S. *et al.* (1993) *In vitro* evaluation of the antineoplastic activity of combretastatin A-4, a natural product from *Combretum caffrum* (arid shrub). *Anti-Cancer Drugs*, **4**, 19–25.

Elferaly, F. S. (1984) Isolation, characterization and synthesis of 3,5,4′-trihydroxybibenzyl from *Cannabis sativa*. *J. Nat. Prod.*, **47**, 89–92.

ElSohly, H. and Turner, C. E. (1982) Iso-cannabispiran, a new spiro compound isolated from Panamenian variant of *Cannabis sativa* L. *Experientia*, **38**, 229.

Endres, H. (1957) Chemistry of the tannins of spruce bark. IV. *Das Leder*, **8**, 222–9.

Endres, H. and Leppmeier, F. (1961) Uber die Gerbstoffe der Fichtenrinde. XII. Isolierung und Konstitutionsermittlung eines Piceatannol-monoglucosides. *Chem. Ber.*, **94**, 419.

Erdtman, H. (1939a) Die phenolischen Inhaltstoffe des Kiefernkernholzes, ihre physiologische Bedeutung und hemmende Einwirkung auf die normale Aufschliessbarkeit des Kiefernkernholzes nach dem sulfitvarfahre. *Ann. Chem.*, **539**, 116–27.

Erdtman, H. (1939b) The extract materials from pine wood. *Naturwissenschaften*, **27**, 130–1.

Erdtman, H. (1944) U.V.-Absorption und sterischer Bau des Pinosylvins. *Svensk Kem. Tidskr.*, **56**, 134–42.

Erdtman, H. (1949a) Compounds inhibiting the sulphite cook. *Tappi*, **32**, 303–5.

Erdtman, H. (1949b) Heartwood extractives of conifers. Their fungicidal and insect-repellent properties and taxonomic interest. *Tappi*, **32**, 305–10.

Erdtman, H., Kimland, B. and Nordin, T. (1966) Wood constituents of *Ducampopinus krempfii* (Le Comte) Chevalier (*Pinus krempfii* Le Comte). *Phytochemistry*, **5**, 927–31.

Erdtman, H. and Norin, T. (1963) Heartwood constituents of *Laburnum alpinum* Bercht. and Presl. *Acta Chem. Scand.*, **17**, 1781–2.

Erdtman, H. and Rennerfelt, E. (1944) Der Gehalt des Kiefernkernholzes an Pinosylvin-phenolen. *Svensk Papperstidn.*, **47**, 45–56.

Fagboun, D. E., Ogundana, S. K., Adesanya, S. A. and Roberts, M .F. (1987) Dihydrostilbene phytoalexins from *Dioscorea rotundata*. *Phytochemistry*, **26**, 3187–9.

Fang, J.-M., Su, W.-C. and Cheng, Y.-S. (1988) Flavonoids and stilbenes from Armand pine. *Phytochemistry*, **27**, 1395–7.

Favre-Bonvin, J., Jay, M. and Wollenweber, E. (1978) A novel stilbene from bud excretion of *Alnus viridis*. *Phytochemistry*, **17**, 821–2.

Fernandez, M. A., Pedro, J. R. and Seoane, E. (1983) Two polyhydroxystilbenes from stems of *Phoenix dactylifera*. *Phytochemistry*, **22**, 2819–21.

Ferrigni, N. R., McLaughlin, J. L., Powell, R. G. and Smith, C. R. Jr (1984) Use of potato disk and brine shrimp bioassays to detect activity and isolate piceatannol as the antileukemic principle from the seeds of *Euphorbia lagescae*. *J. Nat. Prod.*, **47**, 347–52.

Fisch, M. H. and Arditti, J. (1972) Orchid phytoalexins. *Am. J. Bot.*, **59**, 672.

Fisch, M. H., Flick, B. H. and Arditti, J. (1973) Structure and antifungal activity of hircinol, loroglossol and orchinol. *Phytochemistry*, **12**, 437–41.

Fitzgerald, M. A., Gunning, P. J. M. and Donnelly, D. M. X. (1976) Phytochemical examination of *Pericopsis* species. *J. Chem. Soc. Perkin Trans.*, **1**, 186–91.

Fletcher, T. L., Namkung, M. J., Pan, H.-L. and Cole, C.-A. (1971) New thio derivatives of carcinogenic arylamines. 5. Ring substituted methylthio-4-acetamidostilbenes. *J. Med. Chem.*, **14**, 1113–15.

Fliegmann, J., Schröder, G., Schanz, S. *et al.* (1992) Molecular analysis of chalcone and dihydropinosylvin synthase from Scots pine (*Pinus sylvestris*), and differential regulation of these and related enzymes in stressed plants. *Plant Mol. Biol.*, **18**, 489–503.

Forrest, G. I. (1975a) Polyphenol variation in Sitka spruce. *Can. J. For. Res.*, **5**, 26–37.

Forrest, G. I. (1975b) Within-crown variation in Sitka spruce polyphenols. *Can. J. For. Res.*, **5**, 38–45.

Forrest, G. I. (1975c) Variation in polyphenol content within and between Sitka spruce provenances at different sites. *Can. J. For. Res.*, **5**, 46–54.

Franc, J., Kovar, V. and Mikes, F. (1966) Identification of stilbene derivatives by paper and gas-chromatography and paper electrophoresis. *Mikrochim. Acta*, 133–46.

Frankel, E. N., Waterhouse, A. L. and Kinsella, J. E. (1993) Inhibition of human LDL oxidation by resveratrol. *Lancet*, **341**, 1103–4.

Fries, K. (1964) Über Einen genuinen Keimungs- und Streckungswachstums-aktiven Hemmstoff bei *Marchantia polymorpha*. *Beitr. Biol. Pflanz.*, **40**, 177–235.

Fritzemeier, K.-H. and Kindl, H. (1981) Coordinate induction by U.V. light of stilbene synthase, phenylalanine ammonia lyase and cinnamate-4-hydroxylase in leaves of Vitaceae. *Planta*, **181**, 48–53.

Fritzemeier, K.-H. and Kindl, H. (1983) 9,10-dihydrophenanthrenes as phytoalexins of Orchidaceae. Biosynthetic studies *in vitro* and *in vivo* proving the route from L-phenylalanine, dihydro-*m*-coumaric acid, dihydrostilbene and dihydrophenanthrenes. *Eur. J. Biochem.*, **133**, 543–50.

Fritzemeier, K.-H., Kindl, H. and Schlosser, E. (1984) Two different pathways leading to phenanthrenes and 9,10-dihydrophenanthrenes of the genus *Dioscorea*. *Z. Naturforsch. C.*, **39**, 217–21.

Fritzemeier, K.-H., Rolfs, C.-H., Pfau, J. and Kindl, H. (1983) Action of ultraviolet-C on stilbene formation in callus of *Arachis hypogaea*. *Planta*, **159**, 25–9.

Frykholm, K. O. (1945) Bacteriological studies of pinosylvine, its monomethyl and dimethyl ethers, and toxicological studies of pinosylvine. *Nature*, **155**, 454–5.

Fukai, T., Wang, Q.-H. and Nomura, T. (1991) Six prenylated phenols from *Glycyrrhiza uralensis*. *Phytochemistry*, **30**, 1245–50.

Gardner, J. A. F. and Hillis, W. E. (1962) The influence of extractives on the pulping of wood, in *Wood Extractives*, (ed W. E. Hillis), Academic Press, London.

Gäumann, E. (1963) Weitere Untersuchungen uber die chemische Infektabwehr der Orchideen. *Phytopathol. Z.*, **49**, 211–32.

Gäumann, E. and Hohl, H. R. (1960) Protective reaction of the orchids. *Phytopathol. Z.*, **38**, 93–104.

Gäumann, E. and Kern, H. (1959a) Uber die Isolierung und den chemischen Nachweis des Orchinols. *Phytopathol. Z.*, **35**, 347–56.

Gäumann, E. and Kern, H. (1959b) Uber chemische abwehrreaktionen bei Orchideen. *Phytopathol. Z.*, **36**, 1–26.

Gäumann, E., Nuesch, J. and Rumpau, R. H. (1960) Weitere Untersuchungen uber die chemischen Abwehrreaktion der Orchideen. *Phytopathol. Z.*, **38**, 274–308.

Geewanada, Y. A., Gunawardena, P., Sotheeswaran, S. *et al.* (1986) Another antibacterial polyphenol, copalliferol B, from *Vateria copallifera* (Dipterocarpaceae). *Phytochemistry*, **25**, 1498–500.

Gehlert, R. and Kindl, H. (1991) Induced formation of dihydrophenanthrenes and bibenzyl synthase upon destruction of orchid mycorrhiza. *Phytochemistry*, **30**, 457–60.

Gehlert, R., Schoeppner, A. and Kindl, H. (1990) Stilbene synthase from seedlings of *Pinus sylvestris*: purification and induction in response to fungal infection. *Mol. Plant–Microbe Interact.*, **3**, 444–9.

Gehlert, R., Schwekendiek, A. and Kindl, H. (1991) Gene activation and stilbene formation in pine plants. *Biol. Chem.*, **372**, 660–1.

Gellerstedt, G. and Agnemo, R. (1980) The reactions of lignin with alkaline hydrogen peroxide. V. The formation of stilbenes. *Acta Chem. Scand. ser B.*, **34**, 461–2.

Gellerstedt, G., Gierer, J., Pettersson, E. L. and Szabo-Lin, I. (1976) The reactions of lignin during sulphite pulping. VI. The reactions of intermediates of the stilbene type. *Justus Liebigs Ann. Chem.*, 2021–36.

Gerber, N. N. (1986) Phenolics from osage orange wood. Cleavage of oxyresveratrol. *Phytochemistry*, **25**, 1697–9.

Ghisalberti, E. L., Jefferies, P. R. and McAdam, D. (1981) Isoprenylated resorcinol derivatives from *Glycyrrhiza acanthocarpa*. *Phytochemistry*, **20**, 1959–61.

Ghosh, B. B., Ray, S., Bhattacharrya *et al.* (1992) Cymbinodin-B, a phenanthraquinone from *Cymbidium aloifolium*. *Indian J. Chem.*, **31B**, 557–8.

Gibbs, J. N. (1972) Tolerance of *Fomes annosus* isolates to pine oleoresin and pinosylvins. *Eur. J. For. Pathol.*, **12**, 147–51.

Gierer, J., Lenic, J., Noren, I. and Szabo-Lin, I. (1974) Lignin chromophores. I. Synthesis of chromophores of the 2,4′- and 4,4′-dihydroxystilbene types. *Acta Chem. Scand.*, **28B**, 717–29.

Gierer, J. and Pettersson, I. (1977) Studies on the condensation of lignins in alkaline media. II. The formation of stilbene and arylcoumarin structures through neighbouring group participation reactions. *Can. J. Chem.*, **55**, 593–9.

Gierer, J., Pettersson, I. and Smedman, L. A. (1972) The reactions of lignin during sulphate pulping. XII. Reactions of intermediary *o,p*′-dihydroxy-stilbene structures. *Acta Chem. Scand.*, **26**, 3366–76.

Gierer, J., Pettersson, I. and Szabo-Lin, I. (1974) Lignin chromophores. II. The behaviour of 2,4′- and 4,4′-dihydroxystilbene structures towards oxygen-alkali. *Acta Chem. Scand.*, **28B**, 1129–35.

Gilson, E. (1903) *Bl. Acad. Med. Belg.*, **17**, 161–8.

Goda, Y. and Sankawa, U. (1985) Symposium Papers, 105th Annual Meeting of the Pharmacological Society of Japan, Kanazawa, Japan, p. 468.

Gollapudi, S. R., Telikepalli, H., Keshavarz-Shokri, A. *et al.* (1989) Glepidotin C: a minor antimicrobial bibenzyl from *Glycyrrhiza lepidota*. *Phytochemistry*, **28**, 3556–7.

Gombos, M., Szendrei, K., Novak, I. and Reisch, J. (1974) Isolation of nitrogen-free phenanthrene from opium. *Herba Hungarica*, **13**, 63–8.

Gonzalez, M. J. T. G., Pinto, M. M. M., Kijjoa, A. *et al.* (1993) Stilbenes and other constituents of *Knema austrosiamensis*. *Phytochemistry*, **32**, 433–8.

Gorham, J. (1975) Some aspects of the distribution, metabolism and physiological role of lunularic acid in liverworts. PhD thesis, University of London.

Gorham, J. (1977a) Lunularic acid and related compounds in liverworts, algae and *Hydrangea*. *Phytochemistry*, **16**, 249–53.

Gorham, J. (1977b) Metabolism of lunularic acid in liverworts. *Phytochemistry*, **16**, 915–18.

Gorham, J. (1978) Effect of lunularic acid analogues on liverwort growth and IAA oxidation. *Phytochemistry*, **17**, 99–105.

Gorham, J. (1980) The stilbenoids, in *Progress in Phytochemistry*, vol. 6, (eds L. Reinhold, J. B. Harborne and T. Swain), Pergamon Press, Oxford, pp. 203–52.

Gorham, J. (1989) Stilbenes and phenanthrenes, in *Methods in Plant Biochemistry*, vol. 1, (ed. J. B. Harborne), Academic Press, London, pp. 159–96.

Gorham, J. (1990) Phenolic compounds, other than flavonoids, from bryophytes, in *Bryophytes, Their Chemistry and Chemical Taxonomy*, (eds H. D. Zinsmeister and R. Mues), Oxford University Press, Oxford, pp. 171–200.

Gorham, J. and Coughlan, S. J. (1980) Inhibition of photosynthesis by stilbenoids. *Phytochemistry*, **19**, 2059–64.

Gottlieb, O. R., Alves de Lima, R., Mendes, P. H. and Magalhaes, M. T. (1975) Constituents of Brazilian Moraceae. *Phytochemistry*, **14**, 1674–5.

Gottsegen, A., Nógrádi, M., Vermes, B. *et al.* (1988) The synthesis of riccardin C. *Tetrahedron Lett.*, **29**(39), 5039–40.

Gottsegen, A., Nógrádi, M., Vermes, B. *et al.* (1990) Total syntheses of riccardins A, B, and C, cytotoxic macrocyclic bis(bibenzyl)s from liverworts. *J. Chem. Soc., Perkin Trans.*, **1**, 315–20.

Gracza, L. and Ruff, P. (1984a) Hochleistungs Flussigkeitschromatographische Trennung und Quantitative Bestimmung von pflanzlichen Stilbenderivaten. *J. Chromatogr.*, **287**, 462–5.

Gracza, L. and Ruff, P. (1984b) Stilbene derivatives from plants. II. 3,5,3′,4′-tetrahydroxystilbene-3-D-glucoside, a third stilbene glycoside in *Rhei rhapontici* Radix. *Arch. Pharm.*, **317**, 374–7.

Gradstein, S. R. (1990) Morphology and classification of the Hepaticae: an introduction, in *Bryophytes, Their Chemistry and Chemical Taxonomy*, (eds H. D. Zinsmeister and R. Mues), Oxford University Press, Oxford, pp. 3–17.

Gradstein, S. R., Matsuda, R. and Asakawa, Y. (1985) A chemotaxonomic survey of terpenoids and aromatic compounds in the Lejeuneaceae (Hepaticae), in *Contribution to a Monograph of the Lejeuneaceae subfamily Ptychanthoidae*, (ed. S. R. Gradstein), J. Cramer, Vaduz, pp. 63–86.

Grassmann, W., Deffner, G., Schuster, E. and Pauckner, W. (1956) Uber den Gerbstoff der Fichtenrinde. *Chem. Ber.*, **89**, 2523–35.

Grassmann, W., and Endres, H. (1959) Uber die Gerbstoffe der Fichtenrinde. XI Mitt. Uber Polyhydroxystilbene, eine neue Klasse kondensierbarer Gerbstoffe. *Das Leder*, **10**, 237–44.

Grassmann, W., Endres, H., Brockhaus, R. and Merkle, K. (1957) Oxydative Spaltung des Glucosides E. *Chem. Ber.*, **90**, 2416–22.

Grassmann, W., Endres, H. and Pauckner, W. (1958) Uber die Gerbstoffe der Fichtenrinde. VI. Die Konstitution des Piceatannols. *Chem. Ber.*, **91**, 134.

Gromova, A. S., Lutskii, V. I. and Tyukavkina, N. A. (1974) Hydroxystilbenes of *Picea koraensis*. *Khim. Prir. Soedin.*, **10**, 778–9.

Gromova, A. S., Lutskii, V. I. and Tyukavkina, N. A. (1975) Hydroxystilbenes of the inner and outer bark of *Picea ojanensis*. *Khim. Prir. Soedin.*, **11**, 83–4.

Gromova, A. S., Lutskii, V. I. and Tyukavkina, N. A. (1977) Stilbenes of the outer bark of *Pinus sibirica* and *Picea koraiensis*. *Khim. Prir. Soedin.*, **13**, 275–6.

Gromova, A. S., Tyukavkina, N. A. and Lutskii, V. I. (1975) Hydroxystilbenes of the inner bark of *Pinus sibirica*. *Khim. Prir. Soedin.*, **11**, 677–82.

Grotha, R. and Schwabe, W. W. (1978) On the occurrence of lunularic acid in the liverwort *Riella* and its effect on the regeneration of isolated mature cells of *Riella helicophylla* (Bory et Mont.) Mont. *Biochem. Physiol. Pflanzen*, **172**, 167–71.

Grundon, M. F. and King, F. E. (1949) Chlorophorin, a constituent of Iroko, the wood of *Chlorophora excelsa*. *Nature*, **163**, 564–5.

Gstirner, F. and Holtzem, H. (1949) The medicinal utility of *Rheum undulatum*. *Pharmazie*, **4**, 333–9.

Hachihama, Y. and Jodai, S. (1951) Significance of cationoid groups in lignin. *Jap. Sci. Rev.*, **2**, 21–6.

Hachihama, Y., Sakae, J., Eiichi, F. and Hideik, N. (1952) Sulfite pulping of red pine. III. The phenolic constituents in the Japanese red pine and their inhibiting effect on pulping. *J. Chem. Soc. Japan, Ind. Chem. Sect.*, **55**, 151–2.

Hägglund, E. (1927) Uber den Aufschluss von Kiefernholz nach dem Sulphitverfahren. I. *Cellulosechemie*, **8**, 25–31.

Hain, R., Reif, H.-J., Krause, E. *et al.* (1993) Disease resistance results from foreign phytoalexin expression in a novel plant. *Nature*, **361**, 153–6.

Hamel, E. and Lin, C. M. (1983) Interactions of combretastatin, a new plant-derived antimitotic agent, with tubulin. *Biochem. Pharmacol.*, **32**, 3864–7.

Hanawa, F., Tahara, S. and Mizutani, J. (1992) Antifungal stress compounds from *Veratrum grandiflorum* leaves treated with cupric chloride. *Phytochemistry*, **31**, 3005–7.

Hardegger, E., Biland, H. R. and Carrodi, H. (1963) Welkstoffe und Antibiotika. Synthese von 2,4-dimethoxy-6-hydroxyphenanthren und Konstitution des Orchinols. *Helv. Chim. Acta.*, **46**, 1354–60.

Hardegger, E., Rigassi, N., Seres, J. *et al.* (1963) Synthese von 2,4-Dimethoxy-6-hydroxy-9,10-dihydrophenanthren. *Helv. Chim. Acta*, **46**, 2543–57.

Hardegger, E., Schellenbaum, M. and Carrodi, H. (1963) Welkstoffe und Antibiotika. XXVII. Uber induzierte Abwehrstoffe bei Orchideen II. *Helv. Chim. Acta*, **46**, 1171–80.

Harris, T. M. and Carney, R. L. (1966) Biogenetically modelled synthesis of β-resorcylic acids. *J. Am. Chem. Soc.*, **88**, 2053–4.

Harris, T. M. and Carney, R. L. (1967) Synthesis of 3,5,7-triketo acids and esters and their cyclizations to resorcinol and phloroglucinol derivatives. Models of biosynthesis of phenolic compounds. *J. Am. Chem. Soc.*, **89**, 6734–41.

Hart, J. H. (1981) Role of phytostilbenes in decay and disease resistance. *Ann. Rev. Phytopathol.*, **19**, 437–58.

Hart, J. H. and Hillis, W. E. (1974) Inhibition of wood rotting fungi by stilbenes and other polyphenols in *Eucalyptus sideroxylon*. *Phytopathology*, **64**, 939–48.

Hart, J. H. and Shrimpton, D. M. (1979) Role of stilbenes in resistance of wood to decay. *Phytopathology*, **69**, 1138–43.

Hasegawa, K. and Hashimoto, T. (1973) Quantitative changes of batatasins and abscisic acid in relation to the development of dormancy in yam bulbils. *Plant Cell Physiol.*, **14**, 369–77.

Hasegawa, K. and Hashimoto, T. (1974) Gibberellin-induced dormancy and batatasin content in yam bulbils. *Plant Cell Physiol.*, **15**, 1–6.

Hasegawa, K. and Hashimoto, T. (1975) Variation of abscisic acid and batatasin content in yam bulbils – effect of stratification and light exposure. *J. Exp. Bot.*, **26**, 757–64.

Hashimoto, T. and Hasegawa, K. (1974) Batatasin I: its structure and possible involvement in yam bulbil dormancy, in *Plant Growth Substances*, Proceedings of the International Conference, 1973, Hirokawa Publishing Co. Inc., Tokyo, pp. 150–6.

Hashimoto, T., Hasegawa, K. and Kawarada, A. (1972) Batatasins: new dormancy inducing substances of yam bulbils. *Planta*, **108**, 369–74.

Hashimoto, T., Hasegawa, K., Yamaguchi, H. *et al.* (1974) Structure and synthesis of batatasins, dormancy-inducing substances of yam bulbils. *Phytochemistry*, **13**, 2849–52.

Hashimoto, T., Suzuki, H., Tori, M. and Asakawa, Y. (1991a) Bis(bibenzyl) ethers from *Pellia endiviifolia*. *Phytochemistry*, **30**, 1523–30.

Hashimoto, T., Taira, Z., Tori, M. *et al.* (1991b) Macrocyclic bisbibenzyl compounds, marchantins and biological activity of their related compounds, Abstract, *8th Symposium on the Development and Application of Naturally Occurring Drug Materials*, Tokyo, p. 21.

Hashimoto, T. and Tajima, M. (1978) Structures and synthesis of the growth inhibitors batatasins IV and V, and their physiological activities. *Phytochemistry*, **17**, 1179–84.

Hashimoto, T., Tamura, S. and Kawarada, A. (1973) Batatasin I to regulate the germination of yams. *Japan Kokai* (patent), **73 58**, 724.

Hashimoto, T., Tori, M. and Asakawa, Y. (1987) Three dihydroisocoumarin glucosides from *Hydrangea macrophylla* subsp. *serrata*. *Phytochemistry*, **26**, 3323–30.

Hashimoto, T., Tori, M. and Asakawa, Y. (1988) A highly efficient preparation of lunularic acid and some biological activities of stilbene and dihydrostilbene derivatives. *Phytochemistry*, **27**, 109–13.

Hashimoto, T., Tori, M., Asakawa, Y. and Fukazawa, Y. (1987) Plagiochins A, B, C, and D, new type of macrocyclic bis(bibenzyl)s having a biphenyl linkage between the ortho positions to the benzyl methylenes, from the liverwort *Plagiochila acanthophylla* subsp. *japonica*. *Tetrahedron Lett.*, **28**, 6295–8.

Hashimoto, T., Yoshida, T., Kan, Y. *et al.* (1994) Structures of 4 novel macrocyclic bis(bibenzyl) dimers, pusilatins A–D, from the liverwort *Blasia pusilla*. *Tetrahedron Lett.*, **35**, 909–10.

Hashimoto, Y., Moriyasu, M, Kanai, Y. and Hirase, M. (1979) High-performance liquid chromatographic determination of the sweet component in sweet *Hydrangea*. *J. Chromatogr.*, **171**, 494–6.

Hata, K. (1952a) The phenolic substances of the Japanese red pine (*P. densiflora*) and their antipulping effect in the sulfite cook. *J. Jap. Tech. Assoc. Pulp Paper Ind.*, **6**, 277–82.

Hata, K. (1952b) Pulp of *Pinus densiflora* wood. XVI. The difficulty of sulfite digestion of the pine heartwood 5. *Tech. Bull. Kagawa Agr. Coll.*, **4**, 8–13.

Hata, K. (1955) Chemical properties of *Pinus pinaster* wood from the Ehime district. *J. Jap. Forest Soc.* **Special Issue**, 335–7.

Hata, K., Baba, K. and Kozawa, M. (1979) Chemical studies on the heartwood of *Cassia garrettiana* Craib. II. Nonanthraquinone constituents. *Chem. Pharm. Bull.*, **27**, 984–9.

Hata, K., Kozawa, M. and Baba, K. (1975) A new stilbene glucoside from Chinese crude drug 'heshouwu', the roots of *Polygonum multiflorum* Thunb. *Yakugaku Zasshi*, **95**, 211–13.

Hata, K. and Sogo, M. (1952) Pulp of *Pinus densiflora* wood. XIX. The difficulty of sulphite digestion of the pine heartwood. *J. Jap. Forest Soc.*, **34**, 386–90.

Hata, K. and Sogo, M. (1954) Detection of the phenolic substances of four species of Japanese pine trees by paper partition chromatography. *J. Jap. Forest Soc.*, **36**, 8–11.

Hathaway, D. E. (1962) The use of hydroxystilbene compounds as taxonomic tracers in the genus *Eucalyptus*. *Biochem. J.*, **83**, 80–4.

Hathaway, D. E. and Seakins, J. W. T. (1959) Hydroxystilbenes of *Eucalyptus wandoo*. *Biochem. J.*, **72**, 369–74.

Hayashi, Y., Sakurai, K., Takahashi, T. and Kitao, K. (1974) Chemistry of wood extractives. XX. Heartwood constituents of *Haplormosia monaphylla*. *Mokuzai Gakkaishi*, **20**, 595–9.

Hemingway, R. W., Hillis, W. E. and Bruerton, K. (1970) A gas–liquid chromatographic examination of stilbene derivatives. *J. Chromatogr.*, **50**, 391–9.

Hemingway, R. W., McGraw, G. W. and Barras, S. J. (1977) Polyphenols in *Ceratocystis minor* infected *Pinus taeda*: fungal metabolites, phloem and xylem phenols. *Agric. Food Chem.*, **25**, 717–22.

Hergert, H. L. (1960) Chemical composition of tannins and polyphenols. *Forest. Prod. J.*, **10**, 610–17.

Hesse, O. (1899) Uber Rhubarberstoffe und damit verwandte Korper. *Liebigs Ann. Chem.*, **309**, 32.

Higuchi, T., Shimada, M. and Watanabe, K. (1967) Mechanism of heartwood formation. VI. Artificial heartwood of *Cryptomeria japonica* and *Pinus densiflora*. *Nippon Makuzai Gakkaisi*, **13**, 274–9.

Hillis, W. E. (1966) Variation in polyphenol composition within species of *Eucalyptus* L'Herit. *Phytochemistry*, 35, 541–56.
Hillis, W. E. and Carle, A. (1962) The origin of the wood and bark polyphenols of *Eucalyptus* species. *Biochem. J.*, **82**, 435–9.
Hillis, W. E., Hart, J. H. and Yazaki, Y. (1974) Polyphenols of *Eucalyptus sideroxylon* wood. *Phytochemistry*, **13**, 1591–5.
Hillis, W. E. and Hasegawa, M. (1962a) Biosynthesis of hydroxystilbenes. *Chem. Ind.*, 1330–1.
Hillis, W. E. and Hasegawa, M. (1962b) The polyphenols in leaves of *Eucalyptus sideroxylon*. *Biochem. J.*, **83**, 503–26.
Hillis, W. E. and Inoue, T. (1967) The polyphenols of *Nothofagus* species. II. The heartwood of *Nothofagus fusca*. *Phytochemistry*, **6**, 59–67.
Hillis, W. E. and Inoue, T. (1968) The formation of polyphenols in trees. IV. The polyphenols formed in *Pinus radiata* after *Sirex* attack. *Phytochemistry*, **7**, 13–22.
Hillis, W. E. and Ishikura, N. (1968) The chromatographic and spectral properties of stilbene derivatives. *J. Chromatogr.*, **32**, 323–36.
Hillis, W. E. and Ishikura, N. (1969a) An enzyme from *Eucalyptus* which converts cinnamoyl triacetic acid into pinosylvin. *Phytochemistry*, **8**, 1079–88.
Hillis, W. E. and Ishikura, N. (1969b) The extractives of the mycorrhizas and roots of *Pinus radiata* and *Pseudotsuga menziesii*. *Aust. J. Biol. Sci.*, **22**, 1425–36.
Hillis, W. E. and Ishikura, N. (1970) The biosynthesis of polyphenols in tissues with low phenylalanine ammonia lyase activity. *Phytochemistry*, **9**, 1517–28.
Hillis, W. E. and Isoi, K. (1965a) The biosynthesis of polyphenols in *Eucalyptus* species. *Phytochemistry*, **4**, 905–18.
Hillis, W. E. and Isoi, K. (1965b) Variation in the chemical composition of *Eucalyptus sideroxylon*. *Phytochemistry*, **4**, 541–50.
Hillis, W. E., Nelson, P., and Zadow, G. (1966) The cause of discoloration of *Pinus radiata* bisulphite pulp. *Appita*, **19**, 111–14.
Hillis, W. E. and Orman, H. R. (1962) The extractives of New Zealand *Nothofagus* species. *J. Linn. Soc. Bot.*, **58**, 175–84.
Hillis, W. E. and Yazaki, Y. (1971) The biosynthesis of stilbenes in eucalypt leaves. *Phytochemistry*, **10**, 1051–4.
Hillis, W. E. and Yazaki, Y. (1973) Polyphenols of *Intsia* heartwoods. *Phytochemistry*, **12**, 2491–5.
Hills, K. A., Stoessl, A., Oliva, A. P. and Arditti, J. (1984) Effects of orchinol, loroglossol, dehydroorchinol, batatasin III, and 3,4′-dihydroxy-5-methoxydihydrostilbene on orchid seedlings. *Bot. Gaz.*, **145**, 298–301.
Homans, A. L. and Fuchs, A. (1970) Direct bioautography on thin-layer chromatograms as a method for detecting fungitoxic substances. *J. Chromatogr.*, **51**, 327–9.
Hoos, G. and Blaich, R. (1988) Metabolism of stilbene phytoalexins in grapevines: oxidation of resveratrol in single-cell cultures. *Vitis*, **27**, 1–12.
Hoos, G. and Blaich, R. (1990) Influence of resveratrol on germination of conidia and mycelial growth of *Botrytis cinerea* and *Phomopsis viticola*. *J. Phytopathol.*, **129**, 102–10.
Hopkins, B. J. and Perold, G. W. (1974) (*S*)-2-Hydroxycuparene (*p*-(1,2,2-trimethylcyclopentyl)-*o*-cresol) and 3,4′-ethylenebisphenol from a liverwort, *Marchantia polymorpha* L. *J. Chem. Soc. Perkin Trans.*, **1**, 32–6.
Hornemann, A. (1822) *Berl. Jahrbuch für die Pharmazie*, 252.
Huneck, S., Connolly, J. D., Rycroft, D. S. and Woods, N. (1988) Pakyonol, a macrocyclic bisbibenzyl diether from the liverwort *Mannia fragrans*. *J. Chem. Res.*, (**Suppl.**), 78–9.

Huneck, S. and Pryce, R. J. (1971) Inhaltstoffe der Moose. II. Lunularsäure aus *Marchantia alpestris*. *Z. Naturforsch.*, **26b**, 738.

Huneck, S. and Schreiber, K. (1972) Wachstumregulatorische Eigenschaften von Flechten- und Moos-Inhaltstoffen. *Phytochemistry*, **11**, 2429–34.

Huneck, S. and Schreiber, K. (1977) Synthese der Lunularsäure. *Phytochemistry*, **16**, 1013–16.

Ibrahim, R. K. and Towers, G. H. N. (1960) Studies of hydrangenol in *Hydrangea macrophylla* Ser. I. Isolation, identification and biosynthesis from C14-labelled components. *Can. J. Biochem. Physiol.*, **38**, 627–34.

Ibrahim, R. K. and Towers, G. H. N. (1962) Studies of hydrangenol in *Hydrangea macrophylla* Ser. II. Biosynthesis of hydrangenol from C14-labelled compounds. *Can. J. Biochem. Physiol.*, **40**, 449.

Iino, M., Hashimoto, T. and Heber, U. (1978) Inhibition of photosynthesis and respiration by batatasins. *Planta*, **138**, 167–72.

Iinuma, M., Tanaka, T., Kawai, M. *et al.* (1991) Two complex flavanones from the roots of *Sophora leachiana*. *Phytochemistry*, **30**, 3773–5.

Imoto, S. A. and Ohta, Y. (1985) Intracellular localization of lunularic acid and prelunularic acid in suspension cultured cells of *Marchantia polymorpha*. *Plant Physiol.* **79**, 751–5.

Inamori, Y., Kato, Y., Kubo, M. *et al.* (1984a) Physiological activities of 3,3′,4,5′-tetrahydroxystilbene isolated from the heartwood of *Cassia garrettiana* Craib. *Chem. Pharm. Bull.*, **32**, 213–18.

Inamori, Y., Kobayashi, T., Ogawa, M. and Tsujibo, H. (1988a) Biological activities of hexestrol. *Chem. Pharm. Bull.*, **36**, 815–18.

Inamori, Y., Kubo, M., Kato, Y. *et al.* (1984b) The coronary vasodilatory and hypotensive effects of 3,3′,4,5′-tetrahydroxystilbene and its derivatives. *Yakugaku Zasshi (J. Pharm. Soc. Japan)*, **104**, 819–21.

Inamori, Y., Kubo, M., Kato, Y. *et al.* (1984c) The antifungal activity of stilbene derivatives. *Chem. Pharm. Bull.*, **32**, 801–4.

Inamori, Y., Kubo, M., Kato, Y. *et al.* (1985) Biological activities of stilbene derivatives. *J. Pharmacobio-dynamics*, **8**, 3.

Inamori, Y., Kubo, M., Ogawa, M. *et al.* (1987a) Biological activities of 3,3′-dihydroxy-α,β-diethylstilbene and its isomer, diethylstilbestrol. *Chem. Pharm. Bull.*, **35**, 3502–6.

Inamori, Y., Kubo, M., Tsujibo, H., *et al.* (1987b) The ichthyotoxicity and coronary vasodilator action of 3,3′-dihydroxy-α,β-diethylstilbene. *Chem. Pharm. Bull.*, **35**, 887–90.

Inamori, Y., Ogawa, M., Amino, H. *et al.* (1990) Hypotensive effects on spontaneously hypertensive rats and antifungal activity on various species of *Fusarium oxysporum* of diethylstilbestrol-related compounds. *Chem. Pharm. Bull.*, **38**, 2045–6.

Inamori, Y., Ogawa, M., Nishiguchi, K. *et al.* (1991a) Phytogrowth inhibitory and antimicrobial activities of 3,4′-dihydroxy-α,β-diethylstilbene, the isomer of diethylstilbestrol. *Chem. Pharm. Bull.*, **39**, 218–20.

Inamori, Y., Ogawa, M., Ohno, Y. *et al.* (1989) The hypotensive effect and antifungal activity of 3,3′-dihydroxy-α,β-diethyldiphenylethane. *Chem. Pharm. Bull.*, **37**, 3137–8.

Inamori, Y., Ogawa, M., Tsujibo, H. *et al.* (1991b) Inhibitory effects of 3,3′,4,5′-tetrahydroxystilbene and 3,3′,4,5′-tetrahydroxybibenzyl, constituents of *Cassia garretiana* on antigen-induced histamine release *in vitro*. *Chem. Pharm. Bull.*, **39**, 3353–4.

Inamori, Y., Ogawa, M., Tsujibo, H. *et al.* (1991c) The inhibitory effect of

3,3′,4,5′-tetrahydroxystilbene, a constituent of *Cassia garretiana*, on antigen-IgE-induced histamine release from human basophils *in vitro*. *Chem. Pharm. Bull.*, **39**, 805–7.

Inamori, Y., Tsujibo, H., Hachiken, H. *et al.* (1988a) Effect of 3,3′-dihydroxy-α,β-diethyldiphenylethane on coronary and mesenteric arteries isolated from guinea pigs. *Chem. Pharm. Bull.*, **36**, 1252–5.

Inamori, Y., Tsujibo, H., Ogawa, M. *et al.* (1988b) Antifungal activity against plant-pathogenic fungi and phytogrowth inhibitory activity of 3,3′-dihydroxy-α,β-diethyldiphenylethane. *Chem. Pharm. Bull.*, **36**, 3173–5.

Ingham, J. (1978) Isoflavonoid and stilbene phytoalexins of the genus *Trifolium*. *Biochem. Syst. Ecol.*, **6**, 217–23.

Ingham, J. L. (1972) Phytoalexins and other natural products are factors in plant disease resistance. *Bot. Rev.*, **38**, 343–424.

Ingham, J. L. (1976) 3,5,4′-trihydroxystilbene as a phytoalexin from groundnuts (*Arachis hypogaea*). *Phytochemistry*, **15**, 1791–3.

Ireland, C. R. and Passam, H. C. (1985) Effect of exogenous batatasin, batatasin analogues and gibberellins on the dormancy of stored yam tubers. *Trop. Agricult.*, **62**, 41–6.

Ireland, C. R., Schwabe, W. W. and Coursey, D. G. (1981) The occurrence of batatasins in the Dioscoreaceae. *Phytochemistry*, **20**, 1569–71.

Ito, T. (1961) Chemical constituents of Coniferae and allied orders. XLVII. Relation between the distribution of the components and taxonomical position of the leaves of *Picea bicolor* and other *Picea* plants. *Yakugaku Zasshi*, **81**, 236–8.

Iyoda, M., Sakaitani, M., Tsuka, H. and Oda, M. (1985) Synthesis of riccardin B by nickel-catalyzed intramolecular cyclization. *Tetrahedron Lett.*, **26**, 4777–80.

Janthasoot, W., Ratanabanangkoon, K. and Chudapongse, P. (1977) Inhibition of mitochondrial ATPase by 2,4,3′,5′-tetrahydroxystilbene. *Chem.–Biol. Interact.*, **18**, 355–64.

Jayatilake, G. S., Jayasuriya, H., Lee, E. S. *et al.* (1993) Kinase inhibitors from *Polygonum cuspidatum*. *J. Nat. Prod.*, **56**, 1805–10.

Jeandet, P., Bessis, R. and Gautheron, B. (1992) The production of resveratrol [(3,5,4″)-trihydroxystilbene] by grape berries in different developmental stages. *Am. J. Enol. Viticult.*, **42**, 41–6.

Jeandet, P., Sbaghi, M. and Bessis, R. (1992) The production of resveratrol (3,4,5′-trihydroxystilbene) by grapevine *in vitro* cultures, and its application to screening for grey mould resistance. *J. Wine Res.*, **3**, 47–57.

Jorgensen, E. (1961) The formation of pinosylvin and its monomethyl ether in the sapwood of *Pinus resinosa* Act. *Can. J. Bot.*, **39**, 1765–72.

Jorgensen, E. and Balsillie, D. (1969) Formation of heartwood phenols in callus tissue cultures of red pine (*Pinus resinosa*). *Can. J. Bot.*, **47**, 1015–16.

Juneja, R. K., Sharma, S. C. and Tandon, J. S. (1985) A substituted 1,2-diarylethane from *Cymbidium giganteum*. *Phytochemistry*, **24**, 321–4.

Juneja, R. K., Sharma, S. C. and Tandon, J. S. (1987) Two substituted bibenzyls and a dihydrophenanthrene from *Cymbidium aloifolium*. *Phytochemistry*, **26**, 1123–5.

Kajtar-Peredy, M., Keseru, G. M. and Nogradi, M. (1993) Determination of the preferred conformation of a macrocyclic bis(bibenzyl) by nuclear magnetic resonance spectroscopy and molecular mechanics calculations. *J. Chem. Soc., Perkin Trans.*, **2**, 563–6.

Kametler, L., Keseru, G. M., Nogradi, M. *et al.* (1992) Synthesis of plagiochins A and B, macrocyclic bis(bibenzyl) ethers from *Plagiochila acantophylla*. *Justus Liebigs Ann. Chem.*, 1239–42.

Kaouadji, M., Agban, A., Mariotte, A. M. and Tissut, M. (1986) Lonchocarpene, a stilbene, and lonchocarpusone, an isoflavone, two new pyranopolyphenols from *Lonchocarpus nicou* roots. *J. Nat. Prod. (Lloydia)*, **49**, 281–5.

Kariyone, T., Takahashi, M., Ito, T. and Masutani, K. (1958) Chemical constituents of conifers and related plants. XXI. Components of the leaves of *Picea glehnii* 1. *Yakugaku Zasshi*, **78**, 935–6.

Kariyone, T., Takahashi, M., Ito, T. and Masutani, K. (1959a) Chemical constituents of the plants of Coniferae and allied orders. XXX. Constituents of the leaves of *Picea glehnii*. 3. Occurrence of picein and resveratrol. *Yakugaku Zasshi*, **79**, 394–5.

Kariyone, T., Takahashi, M., Ito, T. and Masutani, K. (1959b) Chemical constituents of the plants of the Coniferae and allied orders. XXVIII. Component of the leaves of *Picea glehnii*. 2. Chemical structure of a new glucoside, piceid. *Yakugaku Zasshi*, **79**, 265–7.

Kariyone, T., Takahashi, M., Ito, T. and Masutani, K. (1959c) Chemical constituents of the plants of Coniferae and allied orders. XXXII. Existence of resveratrol and 3,4-dihydroxyacetophenone in the leaves of the plants of *Picea*. *Yakugaku Zasshi*, **79**, 1326–7.

Kashiwada, Y., Nonaka, G.-I. and Nishioka, I. (1984) Studies on rhubarb (*Rhei rhizoma*). VI. Isolation and characterization of stilbenes. *Chem. Pharm. Bull.*, **32**, 3501–17.

Katekar, G. F., Venis, M. A. and Geissler, A. E. (1993) Binding of lunularic acid, hydrangeic acid and related compounds to the receptor for 1-*N*-naphthylphthalamic acid. *Phytochemistry*, **32**, 527–32.

Kawabata, J., Ichikawa, S., Kurihara, H. and Mizutani, J. (1989) Kobophenol A, a unique tetrastilbene from *Carex kobomugi* Ohwi (Cyperaceae). *Tetrahedron Lett.* **30**, 3785–8.

Kawabata, J., Mishima, M., Kurihara, H. and Mizutani, J. (1991) Kobophenol B, a tetrastilbene from *Carex pumila*. *Phytochemistry*, **30**, 645–7.

Kawamura, S. (1938) Ueber die Konstitution des Rhapontins. *J. Pharm. Soc. Jap.*, **58**, 83–5.

Keen, N. T. and Ingham, J. L. (1976) New stilbene phytoalexins from American cultivars of *Arachis hypogaea*. *Phytochemistry*, **15**, 1794–5.

Kemal, M., Khalil, S. K. W., Rao, N. G. S. and Woolsey, N. F. (1979) Isolation and identification of a cannabinoid-like compound from *Amorpha* species. *J. Nat. Prod.*, **42**, 463–8.

Kemp, M. S. and Burden, R. S. (1986) Phytoalexins and stress metabolites in the sapwood of trees. *Phytochemistry*, **25**, 1261–9.

Kende, A. S. and Curran, D. P. (1979) Regiochemical control in dihydrophenanthrene synthesis. A photochemical total synthesis of juncusol. *J. Am. Chem. Soc.*, **101**, 1857–64.

Keserú, G. M. and Nógrádi, M. (1992) Biosynthesis and molecular strain. A computational study on the conformation of cyclic bis(bibenzyl) constituents of liverwort species. *Phytochemistry*, **31**, 1573–6.

Keserú, G. M., Mezey-Vándor, G., Nógrádi, M. *et al.* (1992) The total synthesis of plagiochins B, and D, macrocyclic bis(bibenzyl) constituent of *Plagiochila acanthophylla*. *Tetrahedron Lett.*, **48**, 913–22.

Kettanes-Van den Bosch, J. J. and Salemink, C. A. (1978) Cannabis XIX. Oxygenated 1,2-diphenylethanes from marijuana. *Recl. Trav. Chim. Pays-Bas (J. R. Netherlands Chem. Soc.)*, **97**, 221–2.

Khafagy, S. M., Gireis, A. N., Kayyal, S. E. and Olama, H. Y. (1974) Identification and microestimation of rhaponticin in rhubarb and its galenical preparations. *J. Drug Res.*, **6**, 47–54.

Khan, M. A., Nabi, S. G., Prakash, S. and Zaman, A. (1986) Pallidol, a resveratrol dimer from *Cissus pallida*. *Phytochemistry*, **25**, 1945–8.

Kimura, K. and Hato, K. (1952) Purity test of rhubarbs by capillary analysis. *Jap. J. Pharmacol.*, **6**, 57–9.

Kimura, Y., Ohminami, H., Okuda, H. *et al.* (1983) Effects of stilbene components of roots of *Polygonum* spp. on liver injury in peroxidized oil fed rats. *Planta Med.*, **49**, 51–4.

Kimura, Y., Okuda, H. and Arichi, S. (1985) Effects of stilbenes on arachidonate metabolism in leukocytes. *Biochim. Biophys. Acta*, **834**, 275–8.

Kindl, H. (1971a) Formation of a stilbene in isolated chloroplasts. *Hoppe-Seyler's Z. Physiol. Chem.*, **352**, 767–8.

Kindl, H. (1971b) Aromatische Aminosauren im Stoffwechsel hoherer Pflanzen. *Naturwissenschaften*, **58**, 554–63.

Kindl, H. (1985) Biosynthesis of stilbenes, in *Biosynthesis and Biodegradation of Wood Components*, (ed. T. Higuchi), Academic Press, New York, pp. 349–77.

King, F. E., Cotterill, C. B., Godson, D. H. *et al.* (1953) The chemistry of extractives of hardwoods. Part XIII. Colourless constituents of *Pterocarpus* species. *J. Chem. Soc.*, 3693–7.

King, F. E. and Grundon, M. F. (1949) The constitution of chlorophorin, a constituent of Iroko, the timber of *Chlorophora excelsa*. Part I. *J. Chem. Soc.*, 3348–52.

King, F. E. and Grundon, M. F. (1950) The constitution of chlorophorin. Part II. Further oxidation experiments, and the completion of the structural problem. *J. Chem. Soc.*, 3547–52.

King, F. E., King, T. J., Godson, D. H. and Manning, L. C. (1956) The chemistry of extractives from heartwoods. Part XXVIII. The occurrence of 3,4,3′, 5′-tetrahydroxy and 3,4,5,3′,5′-pentahydroxystilbene in *Vouacapoua* sp. *J. Chem. Soc.*, 4477–80.

Klimek, B. (1973) Evaluation of the usefulness of some *Rheum* L. species in medical treatment. Part I. Occurrence of anthraquinone compounds and stilbene derivatives. *Ann. Acad. Med. Lodz*, **14**, 133–48.

Knauf, P. A., Ries, E. A., Romanow, L. A. *et al.* (1993) DNDS (4,4′-dinitrostilbene-2,2′-disulphonate) does not act as a purely competitive inhibitor of red-blood cell band 3-mediated anion exchange. *Biophys. J.*, **64**, 307.

Knorr, K., Lehr, H. and Probst, V. (1956) Hormone therapy with plant estrogens. *Die Medizinische*, 195–7.

Kodama, M., Shiobara, Y., Sumitomo, H. *et al.* (1988) Total synthesis of marchantin A and riccardin B, cytotoxic bis(bibenzyl)s from liverworts. *J. Org. Chem.*, **53**, 72–7.

Kondo, T. and Kitamura, Y. (1955) A quantitative method of pinosylvin phenols. *J. Jap. Forest Soc.*, **37**, 157–60.

Kondo, T. and Kitamura, Y. (1956) The antipulping substances in pine heartwood. V. Effect of air seasoning on pinosylvin phenols. *J. Jap. Tech. Assoc. Pulp. Paper Ind.*, **10**, 69–72.

Kondo, T., Ito, H. and Miyashi, T. (1955) Wood extractives. III. Heartwood of *Pinus pumila*. *Nippon Nagei-Kagaku Kaishi*, **29**, 110–12.

Kondo, T., Ito, H. and Suda, M. (1956) Wood extractives. VII. Wood components of *Morus bombycis*. *Nippon Nogei-Kagaku Kaishi*, **32**, 1–4.

Kondo, T. and Tanaka, J. (1955) Screenings of sulphite pulp. I. Chemical properties of phenol lignin. *J. Jap. Forest Soc.*, **Special Issue**, 361.

Kumar, N., Ravindranath, B. and Seshadri, T. (1974) Terpenoids of *Pterocarpus santolinus* heartwood. *Phytochemistry*, **13**, 633–6.

Kumar, R. J., Jyostna, D., Krupadanam, G. L. D. and Srimannarayana, G. (1988)

Phenanthrene and stilbenes from *Pterolobium hexapetallum*. *Phytochemistry*, **27**, 3625–6.

Kunz, S. and Becker, H. (1992) Bibenzyl glycosides from the liverwort *Ricciocarpos natans*. *Phytochemistry*, **31**, 3981–3.

Küppers, F. J. E. M., Salemink, C. A., Bastart, M. and Paris, M. (1976) Alkaloids of *Papaver bracteatum:* presence of codeine, neopine and alpinine. *Phytochemistry*, **15**, 444–5.

Kurata, K., Amiya, T. and Nakano, N. (1976) 3,3′-Dibromo-4,4′,5,5′-tetrahydroxybibenzyl, a new bromophenol from the red alga *Polysiphonia urceolata*. *Chem. Lett.*, 821–2.

Kurihara, H., Kawabata, J., Ichikawa, S. *et al.* (1991) Oligostilbenes from *Carex kobomugi*. *Phytochemistry*, **30**, 649–53.

La Rue, C. D. and Narayanaswami, S. (1957) Auxin inhibition in the liverwort *Lunularia*. *New Phytologist*, **56**, 61–9.

Laarhoven, W. H., Nivard, R. J. F. and Havinga, E. (1961) Synthesis, infrared spectra and molecular refractions of some sterically hindered *p,p′*-dimethoxystilbenes. *Recuil*, **80**, 775–91.

Laidlaw, R. A. and Smith, G. A. (1959) Heartwood extractives of some timbers of the family Moraceae. *Chem. Ind.*, 1604–5.

Lamuela-Raventós, R. M. and Waterhouse, A. L. (1993) Occurrence of resveratrol in selected California wines by a new HPLC method. *J. Agricult. Food Chem.*, **41**, 521–3.

Lang, W. (1957) Rhaponticin in rhubarb. *Mitt. Deut. Pharm. Ges.*, **27**, 61–3.

Langcake, P. (1981) Disease resistance of *Vitis* spp. and the production of the stress metabolites resveratrol, ϵ-viniferin and pterostilbene. *Physiol. Plant Pathol.*, **18**, 213–26.

Langcake, P., Cornford, C. A. and Pryce, R. J. (1979) Identification of pterostilbene as a phytoalexin from *Vitis vinifera* leaves. *Phytochemistry*, **18**, 1025–7.

Langcake, P. and McCarthy, W. V. (1979) The relationship of resveratrol production to infection of grapevine leaves by *Botrytis cinerea*. *Vitis*, **18**, 244–53.

Langcake, P. and Pryce, R. J. (1976) The production of resveratrol by *Vitis vinifera* and other members of the Vitaceae as a response to infection or injury. *Physiol. Plant Pathol.*, **9**, 77–86.

Langcake, P. and Pryce, R. J. (1977a) A new class of phytoalexins from grapevines. *Experientia*, **33**, 151–2.

Langcake, P. and Pryce, R. J. (1977b) The production of resveratrol and the viniferins by grapevines in response to ultraviolet irradiation. *Phytochemistry*, **16**, 1193–6.

Langcake, P. and Pryce. R. J. (1977c) Oxidative dimerization of 4-hydroxystilbenes *in vitro*: production of a grapevine phytoalexin mimic. *J. Chem. Soc. Chem. Commun.*, 208–10.

Lanz, T., Schroeder, G. and Schroeder, J. (1990) Differential regulation of genes for resveratrol synthase in cell cultures of *Arachis hypogaea*. L. *Planta*, **181**, 169–75.

Lanz, T., Tropf, S., Marner, F.-J. *et al.* (1991) The role of cysteines in polyketide synthases. Site-directed mutagenesis of resveratrol and chalcone synthases, two key enzymes in different plant-specific pathways. *J. Biol. Chem.*, **266**, 9971–6.

Lee, T. T., Rock, G. L. and Stoessl, A. (1978) Effects of orchinol and related phenanthrenes on the enzymic degradation of indole-3-acetic acid. *Phytochemistry*, **17**, 1721–6.

Letcher, R. M. (1973) Structure and synthesis of the growth inhibitor batatasin I from *Dioscorea batatas*. *Phytochemistry*, **12**, 2789–90.

Letcher, R. M. and Nhamo, L. R. M. (1971) Chemical constituents of the Combretaceae. Part I. Substituted phenanthrenes and 9,10-dihydrophenanthrenes from the heartwood of *Combretum apiculatum*. *J. Chem. Soc. C.*, 3070–6.

Letcher, R. M. and Nhamo, L. R. M. (1972a) Chemical constituents of the Combretaceae. Part III. Substituted phenanthrenes, 9,10-dihydrophenanthrenes and bibenzyls from the heartwood of *Combretum psidioides*. *J. Chem. Soc. Perkin Trans.*, **1**, 2941–6.

Letcher, R. M. and Nhamo, L. R. M. (1972b) A revised structure for the tetra-substituted phenanthrene from *Tamus communis*. *Tetrahedron Lett.*, **48**, 4869–70.

Letcher, R. M. and Nhamo, L. R. M. (1973a) Chemical constituents of the Combretaceae. Part IV. Phenanthrene derivatives from the heartwood of *Combretum hereroense*. *J. Chem. Soc. Perkin Trans.* **11**, 1179–81.

Letcher, R. M. and Nhamo, L. R. M. (1973b) Structure of orchinol, loroglossol and hircinol. *J. Chem. Soc. Perkin Trans.*, **1**, 1263–5.

Letcher, R. M., Nhamo, L. R. M. and Gumiro, I. T. (1972) Chemical constituents of the Combretaceae. Part II. Substituted phenanthrenes and 9,10-dihydrophenanthrenes and a substituted bibenzyl from the heartwood of *Combretum molle*. *J. Chem. Soc. Perkin Trans.*, **1**, 206–10.

Letcher, R. M. and Wong, K.-M. (1978) Structure and synthesis of the phenanthrenes TaI and TaV from *Tamus communis*. *J. Chem. Soc. Perkin Trans.*, **1**, 739–42.

Letcher, R. M. and Wong, K.-M. (1979) Structure and synthesis of the phenanthrenes TaIV and TaVIII from *Tamus communis*. *J. Chem. Soc. Perkin Trans.*, **1**, 2449–50.

Li, J. B., Lin, M., Li, S. Z. and Song, W. Z. (1991) Structure of gnetifolin A of *Gnetum parvifolium* (Warb.) C. Y. Cheng. *Yaoxue Xuebao*, **26**, 437–41.

Lin, C. M., Ho, H. H., Pettit, G. R. *et al.*, (1989) Antimitotic natural products combretastatin A-4 and combretastatin A-2; studies on the mechanism of their inhibition of the binding of colchicine to tubulin. *Biochemistry*, **28**, 6984–91.

Lin, C. M., Singh, S. B., Chu, P. S. *et al.*, (1988) Interactions of tubulin with potential natural and synthetic analogs of the antimitotic agent combretastatin: a structure–activity study. *Mol. Pharmacol.*, **34**, 200–8.

Lin, M., Li, J. B., Li, S. Z. *et al.*, (1992) A dimeric stilbene from *Gnetum parvifolium*. *Phytochemistry*, **31**, 633–8.

Lin, M., Li, J. B., Wu, B. and Zheng, Q. T. (1991) A stilbene derivative from *Gnetum parvifolium*. *Phytochemistry*, **30**, 4201–3.

Lindstedt, G. (1949a) Constituents of pine heartwood. IX. The heartwood of *Pinus montana*. *Acta Chem. Scand.*, **3**, 755–8.

Lindstedt, G. (1949b) Constituents of pine heartwood. XI. The heartwood of *Pinus radiata* D.Don. *Acta Chem. Scand.*, **3**, 763–6.

Lindstedt, G. (1949c) Constituents of pine heartwood. X. The heartwood of *Pinus contorta* var. *latifolia* S. Wats. *Acta Chem. Scand.*, **3**, 759–62.

Lindstedt, G. (1949d) Constituents of pine heartwood. XIII. The heartwood of *Pinus jeffreyi* Bolf. *Acta Chem. Scand.*, **3**, 770–2.

Lindstedt, G. (1949e) Constituents of pine heartwood. XII. The heartwood of *Pinus ponderosa* Dougl. *Acta Chem. Scand.*, **3**, 767–9.

Lindstedt, G. (1949f) Constituents of pine heartwood. XIV. Heartwood of *Pinus monticola* Dougl. *Acta Chem. Scand.*, **3**, 1147–52.

Lindstedt, G. (1949g) Constituents of pine heartwood. XV. Heartwood of *Pinus dansa* Wall. *Acta Chem. Scand.*, **3**, 1375–80.

Lindstedt, G. (1949h) Constituents of pine heartwood. XVI. Heartwood of *Pinus virginiana* Mill. *Acta Chem. Scand.*, **3**, 1381–4.

Lindstedt, G. (1950) Constituents of pine heartwood. XXII. Isolation of pinostrobin and 3,5-dihydroxy-7-methoxyflavanone from the heartwood of *Pinus clausa* Vosey. *Acta Chem. Scand.*, **4**, 1042–6.

Lindstedt, G. (1951) Constituents of pine heartwood. XXVI. A general discussion. *Acta Chem. Scand.*, **5**, 129–38.

Lindstedt, G. and Misiorny, A. (1951) Constituents of pine heartwood. XXV. Investigation of forty-eight *Pinus* species by paper partition chromatography. *Acta Chem. Scand.*, **5**, 121–8.

Lindstedt, G. and Misiorny, A. (1952) Constituents of pine heartwood. XXVIII. Investigation of four additional *Pinus* species by paper partition chromatography. *Acta Chem. Scand.*, **6**, 744–6.

Lindstrom, J. O. and Westfelt, L. (1967) Chemistry of the order Pinales. XLIII. Sesqui- and di-terpenes from the wood of *Pinus albicaulis*. *Ark. Chem.*, **26**, 539–43.

Lins, A. P., Felicia J. D'A., Braggio, M. M. and Roque, L. C. (1991) A resveratrol dimer from *Parthenocissus tricuspidata*. *Phytochemistry*, **30**, 3144–6.

Lins, A. P., Nilce, M, Ribeiro, S. *et al.* (1982) Gnetins: resveratrol oligomers from *Gnetum* species. *J. Nat. Prod.*, **45**, 754–61.

Lins, A. P., Yoshida, M., Gottlieb, O. R. *et al.* (1986) Gnetins in *Welwitschia*. *Bull. Soc. Chim. Belg.*, **95**, 737–48.

Lisina, A. I., Kashtonova, N. K., Dzizenko, A. K. and Pentegova, V. A. (1967a) The heartwood extractives of *Pinus sibirica* I. Neutral compounds and acids. *Izv. Sib. Otd. Akad. Nauk SSSR. Ser. Khim. Nauk.*, **1**, 113–17.

Lisina, A. I., Val'Skii, L. N., Kalistratova, E. F. and Pentegova, V. H. (1967b) The composition of the neutral fraction of the oleoresin from *Pinus sibirica*. *Izv. Sib. Otd. Akad. Nauk. SSSR. Ser. Khim. Nauk.*, **1**, 165–6.

Liswidowati, F. M., Hohmann, F., Schwer, B. and Kindl, H. (1991) Induction of stilbene synthase by *Botrytis cinerea* in cultured grapevine cells. *Planta*, **183**, 307–14.

Loman, A. A. (1970a) The effect of heartwood fungi of *Pinus contorta* var. *latifolia* on pinosylvin, pinosylvin monomethyl ether, pinobanksin and pinocembrin. *Can. J. Bot.*, **48**, 737–47.

Loman, A. A. (1970b) Bioassays of fungi isolated from *Pinus contorta* var. *latifolia* with pinosylvin, pinosylvin monomethyl ether, pinobanksin and pinocembrin. *Can. J. Bot.*, **48**, 1303–8.

Lorimer, S. D., Perry, N. B. and Tangney, R. S. (1993) An antifungal bibenzyl from the New Zealand liverwort, *Plagiochila stephensoniana*. Bioactively-directed isolation, synthesis and analysis. *J. Nat. Prod.*, **56**, 1444–50.

Lutskii, V. I., Tyukavkina, N. A., Gromova, A. S. and Borodina, N. M. (1973) Phenolic compounds of *Pinus sibirica* and *P. sylvestris* wood and bark. *Fenol'nye Soedin Ikh. Fiziol. Svoistva Mater. Vses. Sump. Fenol'nym Soedin*, **2**, 150.

Lutskii, V. T., Gromova, A. S. and Tyukavkina, N. A. (1970) Phenolic components from the bark of *Pinus sibirica and Pinus sylvestris*. *Khim. Prir. Soedin*, **6**, 367.

Lyr, H. (1961) Analytical studies with inhibitors of various enzymes of wood-rotting fungi. *Enzymologia*, **23**, 231–48.

Lyr, H. (1962a) Detoxification of heartwood toxins and chlorophenols by higher fungi. *Nature*, **195**, 289–90.

Lyr, H. (1962b) Enzymatische Detoxifikation der Kernholztoxine. *Flora*, **152**, 570–79.

Lyr, H. (1962c) Die Wirkungsweise toxischer Kernholzinhaltstoffe (Thuyaplicine und Pinosylvine) auf den Stoffwechsel von Mikroorganismen. *Flora*, **150**, 227–42.

MacRae, W. D. and Towers, G. H. N. (1985) Non-alkaloidal constituents of *Virola elongata* bark. *Phytochemistry*, **24**, 561–6.

Madhav, R., Seshadri, T. R. and Subramanian, G. B. V. (1967) Identity of the polyphenol of *Shorea* species with hopeaphenol. *Phytochemistry*, **6**, 1155–6.

Mahesh, V. B. and Seshadri, T. R. (1954) Chemical components of commercial woods and related plant materials. II. The heartwood of *Pinus griffithii*. *J. Sci. Ind. Res. (India)*, **13B**, 835–41.

Majumder, P. L., Bandyopadhyay, D. and Joarder, S. (1982) Coelogin and coeloginin: two novel 9,10-dihydrophenanthrene derivatives from the orchid *Coelogyne cristata*. *J. Chem. Soc. Perkin Trans.*, **1**, 1131–6.

Majumder, P. L. and Banerjee, S. (1988a) Structure of flavanthrin, the 1st dimeric 9, 10-dihydrophenanthrene derivative from the orchid *Eria flava*. *Tetrahedron*, **44**, 7303–8.

Majumder, P. L. and Banerjee, S. (1988b) A ring-B oxygenated phenanthrene derivative from the orchid *Bulbophyllum gymnopus*. *Phytochemistry*, **27**, 245–8.

Majumder, P. L. and Banerjee, S. (1990) Two stilbenoids from the orchid *Eria flava*. *Phytochemistry*, **29**, 3052–5.

Majumder, P. L. and Basak, M. (1990) Cirrhopetalin, a phenanthrene derivative from *Cirrhopetalum andersonii*. *Phytochemistry*, **29**, 1002–4.

Majumder, P. L. and Basak, M. (1991a) Two bibenzyl derivatives from the orchid *Cirrhopetalum andersonii*. *Phytochemistry*, **30**, 321–4.

Majumder, P. L. and Basak, M. (1991b) Two stilbenoids from the orchid *Cirrhopetalum andersonii*. *Phytochemistry*, **30**, 3429–32.

Majumder, P. L. and Datta, N. (1981) Abstracts, 5th National Symposium on Organic Chemistry (NASOC-V), Calcutta University, Calcutta, pp. 35 and 41.

Majumder, P. L., Datta, N., Sarkar, A. K. and Chakraborti, J. (1982) Flavidin, a novel 9,10-dihydrophenanthrene derivative of the orchid *Coelogyne flavida*, *Pholidota articulata* and *Otochilus fusca*. *J. Nat. Prod.*, **45**, 730–2.

Majumder, P. L. and Ghosal, S. (1994) Two stilbenoids from the orchid *Arundina bambusifolia*. *Phytochemistry*, **35**, 205–8.

Majumder, P. L. and Ghosal, S. (1993) Two stilbenoids from the orchid *Arundina bambusifolia*. *Phytochemistry*, **32**, 439–44.

Majumder, P. L. and Joardar, M. (1984) Structure of erianin, a new bibenzyl derivative from the orchid *Eria carinata*. *Indian J. Chem. B.*, **23**, 1040–2.

Majumder, P. L. and Joarder, M. (1985) Erianthridin, a new 9,10-dihydrophenanthrene derivative from the orchids *Eria carinata* and *Eria stricta*. *Indian J. Chem. B*, **24**, 1192–4.

Majumder, P. L. and Kar, A. (1987) Confusarin and confusaridin, two phenanthrene derivatives of the orchid *Eria confusa*. *Phytochemistry*, **26**, 1127–9.

Majumder, P. L., Kar, A. and Shoolery, J. N. (1985) Bulbophyllanthrin, a phenanthrene of the orchid *Bulbophyllum leopardium*. *Phytochemistry*, **24**, 2083–7.

Majumder, P. L., Laha, S. and Datta, N. (1982) Coelonin, a 9,10-dihydrophenanthrene from the orchids *Coelogyne ochracea* and *Coelogyne elata*. *Phytochemistry*, **21**, 478–80.

Majumder, P. L. and Lahiri, S. (1989) Chemical constituents of the orchid *Lusia indivisa*. *Indian J. Chem.*, **28B**, 771–4.

Majumder, P. L. and Lahiri, S. (1990a) Volucrin, a new dimeric phenanthrene derivative from the orchid *Lusia volucris*. *Tetrahedron*, **46**, 3621–6.

Majumder, P. L. and Lahiri, S. (1990b) Lusianthrin and lusianthridin, two stilbenoids from the orchid *Lusia indivisa*. *Phytochemistry*, **29**, 621–4.

Majumder, P. L. and Maiti, D. C. (1988) Flaccidin, a 9,10-dihydrophenanthropyran derivative from the orchid *Coelogyne flaccida*. *Phytochemistry*, **27**, 899–901.

Majumder, P. L. and Maiti, D. C. (1989) Flaccidinin and oxoflaccidin, two phenanthrene derivatives of the orchid *Coelogyne flaccida*. *Phytochemistry*, **28**, 887–90.

Majumder, P. L. and Maiti, D. C. (1991) Isoflaccidinin and isooxoflaccidin, stilbenoids from *Coelogyne flaccida*. *Phytochemistry*, **30**, 971–4.

Majumder, P. L. and Pal, S. (1992) Rotundatin, a new 9,10-dihydrophenathrene derivative from *Dendrobium rotundatum*. *Phytochemistry*, **31**, 3225–8.

Majumder, P. L. and Pal, S. (1993) Cumulatin and tristin, two bibenzyl derivatives from the orchids *Dendrobium cumulatum* and *Bulbophyllum triste*. *Phytochemistry*, **32**, 1561–5.

Majumder, P. L., Pal, A. and Joarder, M. (1990) Cirrhopetalanthrin, a dimeric phenanthrene derivative from the orchid *Cirrhopetalum maculosum*. *Phytochemistry*, **29**, 271–4.

Majumder, P. L. and Sabzabadi, E. (1988) Agrostophyllin, a naturally occurring phenanthropyran derivative from *Agrostophyllum khasiyanum*. *Phytochemistry*, **27**, 1899–901.

Majumder, P. L. and Sarkar, A. K. (1982) Imbricatin, a new modified 9,10-dihydrophenanthrene derivative of the orchid *Pholidota imbricata*. *Indian J. Chem. B*, **21**, 829–31.

Majumder, P. L., Sarkar, A. K. and Chakrabarti, J. (1982) Isoflavidinin and isooxoflavidinin, two 9,10-dihydrophenanthrenes from the orchids *Pholidota articulate*, *Otochilus porecta* and *Otochilus fusca*. *Phytochemistry*, **21**, 2713–16.

Majumder, P. L. and Sen, R.C. (1987) Moscatilin, a bibenzyl derivative from the orchid *Dendrobium moscatum*. *Phytochemistry*, **26**, 2121–4.

Majumder, P. L. and Sen, R. C. (1991a) Pendulin, a polyoxygenated phenanthrene derivative from the orchid *Cymbidium pendulum*. *Phytochemistry*, **30**, 2432–4.

Majumder, P. L. and Sen, R. C. (1991b) Bulbophyllanthrone, a phenanthraquinone from *Bulbophyllum odoratissimum*. *Phytochemistry*, **30**, 2092–4.

Maksimov, O. B., Krivoshchekova, O. E., Stepanenko, L. S. and Boguslavskaya, L. V. (1985) Isoflavines and stilbenes of the kernel wood of *Maackia amurensis*. *Khim. Prir. Soedin (Tashk)*, **6**, 775–80.

Malan, E. and Swinny, E. (1993) Substituted bibenzyls, phenanthrenes and 9,10-dihydrophenanthrenes from the heartwood of *Combretum apiculatum*. *Phytochemistry*, **33**, 1139–42.

Malan, E., Swinny, E., Ferreira, D. and Hall, A. J. (1988) Metabolites from *Chlorophora excelsa*: possible intermediates in the biogenesis of a pentasubstituted stilbene. *Phytochemistry*, **27**, 2309–12.

Maniwa, H. (1915) On the chemical constituents of 'Amacha' (*Hydrangea thunbergii*). *J. Pharm. Soc. Jap.*, p. 1399.

Maniwa, H. (1924) Phyllodulcin, a sweet principle of the leaves of *Hydrangea thunbergii* Sieb. *J. Pharm. Soc. Jap.*, 348–98.

Manners, G. D. and Swan, E. P. (1971) Stilbenes in the barks of five Canadian *Picea* species. *Phytochemistry*, **10**, 607–10.

Mannila, E. and Talvitie, A. (1992) Stilbenes from *Picea abies* bark. *Phytochemistry*, **31**, 3288–9.

Mannila, E. and Talvitie, A. (1993) Combretastatin analogs via hydration of stilbene derivatives. *Justus Liebigs Ann. Chem.*, **1993**, 1037–9.

Mannila, E., Talvite, A. and Kolehmainen, E. (1993) Anti-leukaemic compounds derived from stilbenes in *Picea abies* bark. *Phytochemistry*, **33**, 813–16.

Manson, D. W. (1954) The leucoanthocyanin from black spruce inner bark. *Tappi*, **43**, 59–64.

Markham, K. R. and Porter, L. J. (1973) Extractives of *Pinus radiata* bark. I. Phenolic components. *NZ J. Sci.*, **16**, 751–61.

Marta, M., Lupi, A., Di Chiara, G. *et al.* (1979) Synthesis of longistylin A, B, C, D and other new prenylated stilbenes. *Gazz. Chim. Ital.*, **109**, 323–4.

Marten, I., Busch, H., Raschke, K. and Hedrich, R. (1993) Modulation and block of the plasma-membrane anoin channel of guard cells by stilbene derivatives. *Eur. Biophys. J.*, **21**, 403–8.

Martin, B. and Bassham, J. A. (1980) Photosynthetic CO_2 fixation in spinach chloroplasts inhibited by pine chloroplasts or extracts of pine chloroplasts. *Physiol. Plant.*, **48**, 213–20.

Mathew, J., Subha Rao, A. V. and Subha Rao, N. V. (1977) Photooxidation of pterostilbene from *Pterocarpus marsupium* Roxb. *Curr. Sci.*, **46**, 337–8.

Mato, M. C. and Calvo, R. (1977) Effect of lunularic acid on auxin-oxidase activity. *Biologia Plantarum*, **19**, 394–6.

Matsuo, A., Nozaki, H., Suzuki, M. and Nakayama, M. (1985) A new antifungal 9,10-dihydrophenanthrene from the liverwort *Riccardia jakii*. *J. Chem. Res. Synop.*, 174–5.

Mattivi, F. (1993) Solid-phase extraction of *trans*-resveratrol from wines for HPLC analysis. *Zeitschrift Lebensmittel-untersuchungen -Forschung*, **196**, 522–5.

McGown, A. T. and Fox, B. W. (1989) Structural and biochemical comparison of the anti-mitotic agents colchicine, combretastatin A-4 and amphethinile. *Anti-cancer Drug Design*, **3**, 249–54.

McGown, A. T. and Fox, B. W. (1990) Differential cytotoxicity of combretastatins A1 and A4 in two daunorubicin-resistant P388 cell lines. *Cancer Chemother. Pharmacol.*, **26**, 79–81.

Mezey-Vándor, G., Nógrádi, M., Novikov, V. P. *et al.* (1989) Die Synthese von Perrottetin F und G, zwei cyclische Bis(bibenzyl)ether aus *Radula perrottetii*. *Justus Liebigs Ann. Chem.*, 401–3.

Miles, D. H., Bhattacharya, J., Mody, N. V. *et al.* (1977) Antineoplastic agents. 3. The structure of juncusol. A novel cytotoxic dihydrophenanthrene from the estuarine marine plant *Juncus roemerianus*. *J. Am. Chem. Soc.*, **99**, 618–20.

Miles, D. H. and De la Cruz, A. A. (1976) Pharmacological potential of marsh plants, in *Estuarine Processes*, Proceedings of the International Estuarine Research Conference, (ed. M. Wiley), Academic Press, London.

Miles, D. H., Pelletier, S. W., Bhattacharyya, J. *et al.* (1978) Structural studies on juncusol. A novel cytotoxic 9,10-dihydrophenanthrene derivative from the marsh plant *Juncus roemerianus*. *J. Org. Chem.*, **43**, 4371–3.

Miles, D. H., Randle, S., Shakir, R. and Atwood, J. L. (1981) Structure of juncunone: a biogenetically intriguing molecule from the marsh plant *Juncus roemerianus*. *J. Org. Chem.*, **46**, 2813–15.

Mitscher, L. A., Gollapudi, S. R., Drake, S. and Oburn, D. S. (1985) Amorphastilbol, an antimicrobial agent from *Amorpha nana*. *Phytochemistry*, **24**, 1481–3.

Mitscher, L. A., Park, Y. H., Al-Shamma, A. *et al.* (1981) Antimicrobial agents from higher plants. Amorfrutin A and B. New bibenzyls from *Amorpha fruticosa*. *Phytochemistry*, **20**, 781–5.

Mitscher, L. A., Rao, G. S. R., Khanna, I. *et al.* (1983) Antimicrobial agents from higher plants: prenylated flavonoids and other phenols from *Glycyrrhiza lepidota*. *Phytochemistry*, **22**, 573–6.

Miyase, T. and Ueno, A. (1991) Ionone and bibenzyl glycosides from *Epimedium grandiflorum* var. *thunbergianum*. *Phytochemistry*, **30**, 1727–8.

Miyase, T., Ueno, A., Takizawa, N. *et al.* (1987) Studies on the glycosides of

Epimedium grandiflorum Morr. var. *thunbergianum* (Miq.) *Nakai I. Chem. Pharm. Bull.*, **35**, 1109–17.

Miyase, T., Ueno, A., Takizawa, N. *et al.* (1988) Studies on the glycosides of *Epimedium grandiflorum* Morr. var. *thunbergianum* (Miq.) *Nakai III. Chem. Pharm. Bull.*, **36**, 2475–84.

Mody, N. V., Mahmoud, I. I., Finer-Moore J. and Pelletier, S. W. (1982) Constituents of *Juncus effusus*: the X-ray analysis of effusol diacetate. *J. Nat. Prod.*, **45**, 733–7.

Mongulsuk, S., Robertson, A. and Towers, R. (1957) 2,4,3′,5′-tetrahydroxystilbene from *Artocarpus lakoocha*. *J. Chem. Soc.*, 2231–3.

Morgan, J. W. W. and Orsler, R. J. (1968) The chemistry of colour changes in wood. V. The significance of stilbenes. *Holzforschung*, **22**, 11–16.

Morgante, A. (1976) Variation of the content of polycyclic aromatic hydrocarbons during the maturation process of wheat. *Quad. Merceol.*, **15**, 173–87.

Mues, R. and Zinsmeister, H. D. (1988) The chemotaxonomy of phenolic compounds in Bryophytes. J. Hattori Bot. Lab., **64**, 109–41.

Nagashima, F., Toyota, M. and Asakawa, Y. (1990) Terpenoids from some Japanese liverworts. *Phytochemistry*, **29**, 2169–74.

Nagashima, F., Nishioka, E., Kameo, K. *et al.* (1991) Terpenoids and aromatic compounds from selected Ecuadorian liverworts. *Phytochemistry*, **30**, 215–17.

Nakajima, K., Taguchi, H., Endo, T. and Yasioka, I. (1978) The constituents of *Scirpus fluviatilis* (Torr.) A. Gray. I. The structure of two new hydroxystilbene dimers, scirpusin A and B. *Chem. Pharm. Bull.*, **26**, 3050–7.

Neuhoff, E. W. and Auterhoff, H. (1954) Analysis of rhubarb drugs. VI. Anthraquinones. *Deut. Apoth. Ztg. ver. Suddeut. Apoth. Ztg.*, **94**, 541–4.

Nimmanpisut, S., Chudapongse, P. and Ratanabanangkoon, K. (1976) Effects of 2,4,3′,5′-tetrahydroxystilbene on oxidative phosphorylation by rat liver mitochondria. *Biochem. Pharmacol.*, **25**, 1245–8.

Nógrádi, M., Vermes, B. and Kajtár-Peredy, M. (1987) The constitution of riccardin B. *Tetrahedron Lett.*, **28**, 2899–900.

Nógrádi, M., Vermes, B., Kajtár-Peredy, M. and Novikov, V. P. (1990) Synthese von Pakyonol, einem macrocyclischen Bis(bibenzylether) aus *Mannia fragrans*. *Liebigs Ann. Chem.*, 299–301.

Nonaka, G., Minami, M. and Nishioka, I. (1977) Studies on rhubarb (*Rhei rhizoma*). III. Stilbene glycosides. *Chem. Pharm. Bull.*, **25**, 2300–5.

Nonaka, G. I., Miwa, N. and Nishioka, I. (1982) Stilbene glycoside gallates and proanthocyanidins from *Polygonum multiflorum*. *Phytochemistry*, **21**, 429–32.

Nuesch, J. (1963) Defense reactions in orchid bulbs. Symposium of the Society of General Microbiology, No. 13, Symbiotic Associations, pp. 335–43.

Nunn, J. R. and Rapson, W. S. (1949) The phenolic resin of the wood of *Chlorophora excelsa* (Benth. and Hook.). *J. Chem. Soc.*, 3151–5.

Ohta, Y., Abe, S., Komura, H. and Kobayashi, M. (1983) Prelunularic acid, a probable immediate precursor of lunularic acid. First example of a 'prearomatic' intermediate in the phenylpropanoid-polymalonate pathway. *J. Am. Chem. Soc.*, **105**, 4480–1.

Ohta, Y., Abe, S., Komura, H. and Kobayashi, M. (1984a) Prelunularic acid. *Phytochemistry*, **23**, 1607–9.

Ohta, Y., Abe, S., Komura, H. and Kobayashi, M. (1984b) Prelunularic acid, a probable immediate precursor of lunularic acid, in suspension-cultured cells of *Marchantia polymorphs*. *J. Hattori Bot. Lab.*, **56**, 249–54.

Ohta, Y., Katoh, K. and Takeda, R. (1990) Growth and secondary metabolite production in cultured cells of liverworts, in *Bryophyte Development: Physiology and Biochemistry*, (ed. R. N. Chopra and S. C. Bhatla), CRC Press, Florida, p. 209.

Ohwaki, Y., Ogino, J. and Shibano, K. (1993) 3-Hydroxy-5-methoxystilbene-2-carboxylic acid, a phytotoxic compound isolated from methanolic extracts of pigeonpea (*Cajanus cajan* Millsp.) leaves. *Soil Sci. Plant Nutr.*, **39**, 55–61.

Oshima, Y., Ueno, Y. and Hikino, H. (1993) Ampelopsins D, E, H, and *cis*-ampelopsin E, oligostilbenes from *Ampelopsis brevipedunculata* var. *hancei* roots. *Phytochemistry*, **33**, 179–82.

Oshima, Y., Ueno, Y., Hikino, H. *et al.* (1990) Ampelopsin-A, ampelopsin-B and ampelopsin-C, new oliogostilbenes of *Ampelopsis brevipedunculata* var. *hancei*. *Tetrahedron*, **46**, 5121–6.

Ottersen, T., Aasen, A., El-Feraly, F. S. and Turner, C. E. (1976) X-ray structure of cannabispiran: a novel *Cannabis* constituent. *J. Chem. Soc. Chem. Commun.*, 580–1.

Pearson, T. W., Kriz, G. S. and Taylor, R. J. (1977) Absolute identification of hydroxystilbenes, chemical markers in Engelmann spruce. *Wood Sci.*, **10**, 93–8.

Pettit, G. R., Cragg, G. M., Herald, D. L. *et al.* (1982) Isolation and structure of combretastatin. *Can J. Chem.*, **60**, 1374–6.

Pettit, G. R., Cragg, G. M. and Singh, S. B. (1987) Antineoplastic agents, 122. Constituents of *Combretum caffrum*. *J. Nat. Prod.*, **50**, 386–91.

Pettit, G. R. and Singh, S. B. (1987) Isolation, structure and synthesis of combretastin A-2, A-3 and B-2. *Can. J. Chem.*, **65**, 2390–6.

Pettit, G. R., Sing, S. B. and Cragg, G. M. (1985) Synthesis of (−)-combretastatin. *J. Org. Chem.*, **50**, 3404–6.

Pettit, G. R., Singh, S. B., Hamel, E. *et al.* (1989) Antineoplastic agents 145. Isolation and structure of the strong cell growth and tubulin inhibitor combretastatin A-4. *Experientia*, **45**, 209–11.

Pettit, G. R., Singh, S. B., Niven, M. L. and Schmidt, J. M. (1988a) Cell growth inhibitory dihydrophenanthrene and phenanthrene constituents of the African tree *Combretum caffrum*. *Can. J. Chem.*, **66**, 406–13.

Pettit, G. R., Singh, S. B., Niven, M. L. *et al.* (1987) Isolation, structure, and synthesis of combretastatins A-1 and B-1, potent new inhibitors of microtubule assembly, derived from *Combretum caffrum*. *J. Nat. Prod.*, **50**, 119–31.

Pettit, G. R., Singh, S. B., Schmidt, J. M. *et al.* (1988b) Isolation, structure, synthesis and antimitotic properties of combretastatins B-3 and B-4 from *Combretum caffrum*. *J. Nat. Prod.*, **51**, 517–27.

Pettit, G. R., Smith, C. R. and Singh, S. B. (1987) Recent advances in the chemistry of plant antineoplastic constituents, in *Biologically Active Natural Products*, Annual Proceedings of the Phytochemical Society of Europe, vol. 27, (eds K. Hostettmann and P. J. Lea), Oxford University Press, Oxford, pp. 105–16.

Pezet, R. and Pont, V. (1988a) Mise en évidence de pterostilbene dans les grappes de *Vitis vinifera*. *Plant Physiol. Biochem.*, **26**, 603–7.

Pezet, R. and Pont, V. (1988b) Activite antifongique dans les baies de *Vitis vinifera*: effects d'acides organiques et du pterostilbene. *Revue suisse Viticult.*, d'Arboricult. d'Horticult., **20**, 303–9.

Pezet, R. and Pont, V. (1990) Relation between the chemical structure and the biological activity of hydroxystilbenes against *Botrytis cinerea*. *J. Phytopathol.*, **130**, 1–8.

Pezet, R., Pont, V. and Hoang-Van, K. (1991) Evidence for oxidative detoxication of pterostilbene and resveratrol by a laccase-like stilbene oxidase produced by *Botrytis cinerea*. *Physiol. Mol. Plant Pathol.*, **39**, 441–50.

Pont, V. and Pezet, R. (1990) Relation between the chemical structure and the biological activity of hydroxystilbenes against *Botrytis cinerea*. *J. Phytopathol.*, **130**, 1–8.

Pool, R. M., Creasy, L. L. and Frackleton, A. S. (1981) Resveratrol and the viniferins, their application to screening for disease resistance in grape breeding programs. *Vitis*, **20**, 136–45.

Powell, R. G., Bajaj, R. and McLauchlin, J. L. (1987) Bioactive stilbenes of *Scirpus maritimus*. *J. Nat. Prod. (Lloydia)*, **50**, 293–6.

Powell, R. G., TePaske, M. R., Plattner, R. D. *et al.* (1994) Isolation of resveratrol from *Festuca versuta* and evidence for the widespread occurrence of this stilbene in the Poaceae. *Phytochemistry*, **35**, 335–8.

Prakash, S., Ahmed, J. and Zaman, A. (1981) Acetophenone and stilbene derivatives from *Gnetum ula*. *Phytochemistry*, **20**, 1455–6.

Prakash, S., Khan, M. A., Khan, K. Z. and Zaman, A. (1985) Stilbenes of *Gnetum ula*. *Phytochemistry*, **24**, 622–4.

Prakash, S., Wizarat, K., Zaman, A. *et al.* (1983) Isolation and structure of gnetol, a novel stilbene from *Gnetum ula*. *Indian J. Chem.*, **22B**, 101–4.

Prior, C. (1976) Resistance by Corsican pine to attack by *Heterobasidion annosum*. *Ann. Bot.*, **40**, 261–79.

Pryce, R. J. (1971a) Lunularic acid, a common endogenous growth inhibitor of liverworts. *Planta*, **97**, 354–7.

Pryce, R. J. (1971b) Biosynthesis of lunularic acid – a dihydrostilbene endogenous growth inhibitor of liverworts. *Phytochemistry*, **10**, 2679–85.

Pryce, R. J. (1972a) Metabolism of lunularic acid to a new plant stilbene by *Lunularia cruciata*. *Phytochemistry*, **11**, 1355–64.

Pryce, R. J. (1972b) The occurrence of lunularic acid and abscisic acid in plants. *Phytochemistry*, **11**, 1759–61.

Pryce, R. J. and Langcake, P. (1977) Alpha-viniferin: an antifungal resveratrol trimer from grapevines. *Phytochemistry*, **16**, 1452–4.

Rajaraman, K. and Rangaswami, S. (1975) Structures of two new 9,10-dihydrophenanthrenes from *Dioscorea prazeri*. *Indian J. Chem.*, **13**, 1137–8.

Rao, V. S. S. and Rajadurai, S. (1968) Isolation of 3,4,3′,5′-tetrahydroxystilbene (piceatannol) from *Cassia marginata* heartwood. *Aust. J. Chem.*, **21**, 1921–2.

Rasamoelisendra, R., Voirin, B., Favre-Bonvin, J. *et al.* (1991) (+)- and (−)-androyol and 6-C-methylandroyol, dihydrostilbeno-flavonoids from *Alluaudia dumosa*. *Phytochemistry*, **30**, 1665–7.

Ratanabanangkoon, K., Poopyruchpong, N. and Charoensiri, K. (1976) A preliminary study on the antifungal activity of 2,4,3′,5′-tetrahydroxystilbene on dermatophytes. *J. Sci. Soc. Thailand*, **2**, 202–5.

Rathi, S. S. (1970) PhD thesis, University of Poona.

Reimann, E. (1969) Natural stilbenes. Synthesis of polyhydroxystilbene ethers by the Wittig reaction. *Chem. Ber.*, **102**, 2881–8.

Reimann, E. (1970) Naturliche Polyhydroxystilbene. Die Synthese von oxyresveratrol, piceatannol and rhapontigenin. *Tetrahedron Lett.*, **47**, 4051–3.

Reimann, E. (1971) Natural stilbenes. II. Synthesis of polyhydroxystilbenes. *Justus Liebigs Ann. Chem.*, **750**, 109–27.

Reinecke, T. and Kindl, H. (1994) Characterization of bibenzyl synthase catalysing the biosynthesis of phytoalexins of orchids. *Phytochemistry*, **35**, 63–6.

Reisch, J. *et al.* (1969) Chemistry of natural substances. XXII. Nitrogen-free phenanthrene derivatives from *Tamus communis* rhizomes. *Tetrahedron Lett.*, 67–8.

Reisch, J., Bathory, M., Novak, I. and Szendrei, K. (1970) Structure and biochemistry of natural phenanthrene derivatives. *Herba Hungarica*, **9**, 43–8.

Reisch, J., Bathory, M., Szendrei, K. *et al.* (1973) Weitere Phenanthrene aus dem Rhizom von *Tamus communis*. *Phytochemistry*, **12**, 228–9.

Reisch, J., Gamboo, M., Szendrei, K. and Novak, I. (1974) Chemistry of natural substances. 47. Thebaol from opium. *Arch. Pharm.*, **307**, 814–6.

Reitz, D. B. and Massey, S. M. (1990) A directed metalation of *N-tert*-butyl-*N*-methyl-2-methoxybenzamide. Short syntheses of 2-methoxy-6-methylbenzoic acid and lunularic acid. *J. Org. Chem.*, **55**, 1375–9.

Rennerfeldt, E. (1949) The effect of some antibiotic substances on the germination of the conidia of *Polyporus annosus*. *Acta Chem. Scand.*, **3**, 1343–9.

Richtzenhain, H. and von Hofe, C. (1939) Ein Stilbenederivat aus Sulfitablauge. *Chem. Ber.*, **72**, 1890–2.

Robinson, R. (1955) *The Structural Relations of Natural Products*, Clarendon Press, Oxford.

Rolfs, C. H., Fritzemeier, K.-H. and Kindl, H. (1981) Cultured cells of *Arachis hypogaea* susceptible to induction of stilbene synthase (resveratrol forming). *Plant Cell Rep.*, **1**, 83–5.

Rolfs, C.-H. and Kindl, H. (1984) Stilbene synthase and chalcone synthase. Two different constitutive enzymes in cultured cells of *Picea excelsa*. *Plant Physiol.*, **75**, 489–92.

Rosemann, D. (1990) Die Induktion der Stilben-Biosynthese als fruher Indikator einer Ozonbelstung bei der Kiefer (*Pinus sylvestris* L.). PhD thesis, University of Munich, Germany.

Rosemann, D., Heller, W. and Sandermann, H. (1991) Biochemical plant responses to ozone. II. Induction of stilbene biosynthesis in Scots pine (*Pinus sylvestris* L.) seedings. *Plant Physiol.*, **97**, 1280–6.

Rowe, J. W., Bower, C. L. and Wagner, E. R. (1969) Extractives of jack pine bark. Occurrence of *cis*- and *trans*-pinosylvin dimethylether and ferulic acid esters. *Phytochemistry*, **8**, 235–41.

Rudolff, E. von and Jorgensen, E. (1963) The biosynthesis of pinosylvin in the sapwood of *Pinus resinosa*. *Phytochemistry*, **2**, 297–304.

Rudolff, E. von and Sato, A. (1963) The heartwood extractives of *Pinus banksiana* Lamb. *Can. J. Chem.*, **41**, 2165–74.

Rupprich, N., Hildebrandt, H. and Kindl, H. (1980) Substrate specificity *in vivo* and *in vitro* in the formation of stilbenes. Biosynthesis of rhaponticin. *Arch. Biochem. Biophys.*, **200**, 72–8.

Rupprich, N. and Kindl, H. (1977) Stilbene synthases and stilbenecarboxylate synthases. I. Enzymatic synthesis of 3,5,4′-trihydroxystilbene from *p*-coumaryl Coenzyme A and malonyl Coenzyme A. *Hoppe Seyler's Z. Physiol. Chem.*, **359**, 165–72.

Rycroft, D.(1990) Some recent NMR studies of diterpenoids from the Hepaticas, in *Bryophytes, Their Chemistry and Chemical Taxonomy*, (eds H. D. Zinsmeister and R. Mues), Oxford University Press, Oxford, pp. 109–19.

Sachdev, K. and Kulshreshtha, D. K. (1986) Phenolic constituents of *Coelogyne ovalis*. *Phytochemistry*, **25**, 499–502.

Sackett, D. L. (1993) Podophyllotoxin, steganacin and combretastatin – natural products that bind at the colchicine site of tubulin. *Pharmacol. Therapeut.*, **59**, 163–228.

Saito, M., Kondo, N., Yamaguchi, H. and Hashimoto, T. (1976) Plant growth-regulating activities of batatasin III analogs. *Plant Cell Physiol.*, **17**, 411–16.

Sakakibara, Y., Saito, I., Oda, T. *et al.* (1992) Effects of 3,3′-dihydroxy-α, β-diethystilbene and 3,3′,4,5′-tetrahydroxystilbene on microtubule assembly *in vitro*, aneuploidy induction, and cellular microtubule and actin networks. *Chem. Pharm. Bull.*, **40**, 2115–19.

Samaraweera, U., Sotheeswaran, S. and Sultanbawa, M. U. S. (1982) Polyphenols from *Stemonoporus* species. *Phytochemistry*, **21**, 2585–7.

Sambhandkaraksa, C., Thantivantana, P. and Ratanachai, T. (1962) Phytochemical studies of *Artocarpus lakoocha*. *J. Natl. Res. Council Thailand*, **3**, 245–55.

Sandermann, W. and Dietrichs, H. H. (1957a) Untersuchungen uber Termiten-resistante Holzer. *Holz als Roh-Werkstoff*, **15**, 281.

Sandermann, W. and Dietrichs, H. H. (1957b) Accessory substances contributing to the properties of woods. Results of paper chromatography. *Umschau Fortschr. Wiss. Tech.*, **57**, 197–200.

Sano, Y. and Sakakibara, A. (1977) Studies on the chemical components of Akaezomatsu (*Picea glehnii*) bark. 2. Isolation of stilbenes. *Hokkaido Daigaku Nogakubu Enshurin Kenkyu Hokoku*, **34**, 287–304.

Saraswathy, A., Purushothaman, K. K., Patra, A. *et al.* (1992) Shoreaphenol, a polyphenol from *Shorea robusta*. *Phytochemistry*, **31**, 2561–2.

Sargent, M. V. and Stanojevic, E. (1984) Phenanthrene synthesis: coeloginin, a novel 9,10-dihydrophenanthrene from the orchid *Coëlogyne cristata*. *J. Chem. Soc. Perkin Trans.*, **1**, 1919–21.

Sarkar, H., Zerezghi, M. and Bhattacharyya, J. (1988) Dehydrojuncusol, a constituent of the roots of *Juncus roemerianus*. *Phytochemistry*, **27**, 3006–8.

Sawhney, P. L. and Seshadri, T. R. (1956) Special chemical components of commercial woods and related plant materials. IV. Phenolic components of *Pterocarpus* species. *J. Sci. Ind. Res.*, **15c**, 154–5.

Schanz, S., Schroder, G. and Schroder, J. (1992) Stilbene synthase from Scots pine (*Pinus sylvestris*). *FEBS Lett.*, **313**, 71–4.

Scheffer, T. C. and Cowling, E. B. (1966) Natural resistance of wood to microbial deterioration. *Ann. Rev. Phytopathol.*, **4**, 147–70.

Schmalle, H. and Hausen, B. M. (1979) New sensitizing quinone from Lady Slipper (*Cypripedium calceolus*). *Naturwissenschaften*, **66**, 527–8.

Schöppner, A. and Kindl, H. (1979) Stilbene synthase (pinosylvin synthase) and its induction by U. V. light. *FEBS Lett.*, **108**, 349–352.

Schöppner, A. and Kindl, H. (1984) Purification and properties of a stilbene synthase from induced cell suspension cultures of peanut. *J. Biol. Chem.*, **259**, 6806–11.

Schröder, J. and Schröder, G. (1990) Stilbene and chalcone synthases: related enzymes with key functions in plant-specific pathways. *Z. Naturforschung*, **45c**, 1–8.

Schröder, G., Brown, J. W. S. and Schröder, J. (1988) Molecular analysis of resveratrol synthase cDNA, genomic clones and relationship with chalcone synthase. *Eur. J. Biochem.*, **172**, 161–9.

Schultz, T. P., Cheng, Q., Boldin, W. D. *et al.* (1991) Comparison of the fungicidal activities of (*E*)-4-hydroxylated stilbenes and related bibenzyls. *Phytochemistry*, **30**, 2939–45.

Schultz, T. P., Hubbard, T. F., LeHong *et al.* (1990) Role of stilbenes in the natural durability of wood: fungicidal structure–activity relationships. *Phytochemistry*, **29**, 1501–7.

Schwabe, W. W. and Nachmony-Bascomb, S. (1963) Growth and dormancy in *Lunularia cruciata* (L.) Dum. II. The response to daylength and temperature. *J. Exp. Bot.*, **14**, 353–78.

Schwabe, W. W. and Valio I. F. M. (1970) Growth and dormancy in *Lunularia cruciata* (L.) Dum. V. The control of growth by a natural endogenous inhibitor. *J. Exp. Botany*, **21**, 112–21.

Schwabe, C. G. and Ekenstam, A. (1929) Heartwood of the pine. III. Heating in sealed tubes. *Cellulosechemie*, **10**, 27–34.

Schwekendiek, A., Pfeffer, G. and Kindl, H. (1992) Pine stilbene synthesis cDNA, a tool for probing environmental stress. *FEBS Lett.*, **301**, 41–4.

Seshadri, T. R. (1957) Occurrence of C8 units in naturally occurring stilbenes and phenylisocoumarins. *Curr. Sci.*, **10**, 310–11.

Sethi, M. L., Taneja S. C., Agarwal, S. G. *et al.* (1980) Isoflavones and stilbenes from *Juniperus macropoda*. *Phytochemistry*, **19**, 1831–2.

Shain, L. (1967) Resistance of sapwood in stems of loblolly pine to infection by *Fomes annosus*. *Phytopathology.*, **57**, 1034–45.

Shain, L. and Hillis, W. E. (1971) Phenolic extractives in Norway spruce and their effects on *Fomes annosus*. *Phytopathology.*, **61**, 841–5.

Shain, L. and Hillis, W. E. (1973) Ethylene production in xylem of *Pinus radiata* in relation to heartwood formation. *Can. J. Bot.*, **51**, 1331–5.

Shan, C. W. (1988) Effects of polydatin on platelet aggregation of rabbits. *Yaoxue Xuebao*, **23**, 394–6.

Sheers, E. H. (1971) U.S. Patent 3,577,230, 4th May 1971, Arizona Chemical Company.

Shibanova, G. I., Gromova, A. S., Kislitsina, L. D. and Tyukavkina, N. A. (1977) Phenolic compounds of the inner and outer bark of *Picea obovata* Ledb. *Izv. Sib. Otd. Akad. Nauk SSSR, Ser. Khim. Nauk*, 153–5.

Shibanova, G. I. and Pentagava, V. A. (1970) Polyphenols in Siberian spruce bark. *Isv. Vyssh. Uch. Zav. Tech. Li. Prom.*, 58–60.

Shibata, S. and Takido, M. (1952) Studies on the evaluation of crude drugs. II. Paper chromatographic studies of rhubarbs. *J. Pharm. Soc. Jap.*, **72**, 1311–14.

Shima, K., Toyota, M. and Asakawa, Y. (1991) Phenanthrene derivatives from the medullae of *Juncus effusus*. *Phytochemistry*, **30**, 3149–51.

Shiobara, Y., Sunitoino, H., Tsukamoto, M. *et al.* (1985) Synthesis and structure confirmation of riccardin B, a macrocyclic bis(bibenzyl) fron the liverwort *Riccardia multifida*. *Chem. Lett.*, 1587–8.

Shirataki, Y., Noguchi, M. , Yokoe, *et al.* (1991) Sophoraflavanones H, I and J, flavonostilbenes from *Sophora moorcroftiana*. *Chem. Pharm. Bull.*, **39**, 1568–72.

Shostakovskii, S. F., Tyukavkina, N. A., Lutskii, V. I. and Borodin, N. M. (1969) Hydroxystilbenes of *Pinus sibirica* and *Pinus sylvestris*. *Khim. Prir. Soedin*, **5**, 48–9.

Shoyama, Y. and Nishioka, I. (1978) Cannabis XIII. Two new spiro-compounds cannabispirol and acetyl cannabispirol. *Chem. Pharm. Bull.*, **26**, 3641–6.

Shrimpton, D. M. (1973) Extractives associated with wound response of lodgepole pine attacked by the mountain pine beetle and associated microorganisms. *Can. J. Bot.*, **51**, 527–34.

Shrimpton, D. M. and Watson, J. A. (1971) Response of lodgepole pine seedlings to inoculation with *Europhium clavigerum*, a blue stain fungus. *Can. J. Bot.*, **49**, 373–5.

Siddiqui, Z. S., Rahman, M., Khan, M. A. *et al.* (1993) Gnetulin, a dimer of 3′,4,5′-trihydroxy-3-methoxystilbene from *Gnetum ula*. *Tetrahedron Lett.*, **49**, 10393–6.

Siegel, U., Mues, R., Donig, R. and Eicher, T. (1991) A cyclohexenone from *Plagiochila longispina*. *Phytochemistry*, **30**, 3643–6.

Siegfried, B. (1943) Rhubarb, rhapontic and rhaponticin. *Hundert Jahre Schweiz. Apoth. -Ver. 1843–1943*, 531–4.

Siesto, A.J. (1958) Rhubarb grown on Sila mountains. *Fitoterapia*, **29**, 857–79.

Sinsheimer, J. E. and Smith, R. V. (1967) Methods for the qualitative and quantitative analysis of some hydroxystilbenes. *J. Pharm. Sci.*, **56**, 1280–3.

Skinnider, L. and Stoessl, A. (1986) The effect of the phytoalexins, lubimin, (−)-maackiain, pinosylvin, and the related compounds dehydroloroglossol and hordatine M on human lymphoblastoid cell lines. *Experientia*, **42**, 568–70.

Sogo, M. and Hata, K. (1953) Detection of the constituents of the wood of *Pinus pentaphylla* var. *himekomatsu* by paper partition chromatography. *Tech. Bull. Kagawa Agr. Coll.*, **5**, 15–18.

Soni, P. L. (1975) The chemistry of extractives of *Dalbergia sissoo*. Part I. The occurrence of 3,5-dihydroxy-trans-stilbene in the heartwood. *J. Indian Acad. Wood Sci.*, **6**, 57–8.

Sotheeswaran, S., Diyasena, M. N. C., Gunatilaka, A. A. L. *et al.* (1987) Further evidence for the structure of vaticaffinol and a revision of its stereochemistry. *Phytochemistry*, **26**, 1505–7.

Sotheeswaran, S. and Pasupathy, V. (1993) Distribution of resveratrol oligomers in plants. *Phytochemistry*, **32**, 1083–92.

Sotheeswaran, S., Sultanbawa, M. U. S., Surendrakumar, S. and Bladon, P. (1983) Polyphenols from Dipterocarp species. Copalliferol A and stemonoporol. *J. Chem. Soc. Perkin Trans.*, **1**, 699–702.

Sotheeswaran, S., Sultanbawa, M.U.S., Surendrakumar, S. *et al.* (1985) Polyphenols from Dipterocarp species. Vaticaffinol and E-viniferin. *J. Chem. Soc. Perkin Trans.*, **1**, 159–62.

Späth, E. and Schläger, J. (1940) Uber die Inhaltstoffe des roten Sandelholzes. II. Die Konstitution des Pterostilbens. *Chem. Ber.*, **73**, 881–4.

Spörle, J. Becker, H., Gupta, M. P. *et al.* (1989) Novel C-35 terpenoids from the Panamanian liverwort *Plagiochila moritziana*. *Tetrahedron Lett.*, **45**, 5003–14.

Stein, U. and Blaich, R. (1985) Untersuchungen über Stilbenenproduktion und Botrytisanfälligkeit bei *Vitis*-Arten. *Vitis*, **24**, 75–87.

Stein, U. and Hoos, G. (1984) Methods for induction and determination of stilbenes in Vitaceae. *Vitis*, **23**, 179–94.

Stermitz, F. R., Suess, T. R., Schauer, C. K. *et al.* (1983) New and old phenanthrene derivatives from *Oncidium cebolleta*, a peyote-replacement plant. *J. Nat. Prod. (Lloydia)*, **46**, 417–23.

Steynberg, J. P., Brandt, E. V., Burger, J. F. W. *et al.* (1988) Stilbene glycosides from *Guibourtia coleosperma*. Determination of glycosidic connectivities by homonuclear nuclear Overhauser effect difference spectroscopy. *J. Chem. Soc. Perkin Trans.*, **1**, 37–41.

Steynberg, J. P., Ferriera, D. and Roux, D. G. (1983) The first condensed tannins based on a stilbene. *Tetrahedron Lett.*, **24**, 4147–50.

Steynberg, J. P., Ferriera, D. and Roux, D. G. (1987) Synthesis of condensed tannins. Part 18. Stilbenes as potent nucleophiles in regio- and stereo-specific condensations: novel guibourtinidol-stilbenes from *Guibourtia coleosperma*. *J. Chem. Soc. Perkin Trans.*, **1**, 1705–12.

Stoessl, A. (1970) Antifungal compounds produced by higher plants, in *Recent Advances in Phytochemistry*, vol. 3, Applelton-Century-Crofts, New York, pp. 158–80.

Stoessl, A. (1982) in *Phytoalexins*, (eds J. A. Bailey and J. W. Mansfield), Blackie, Glasgow, pp. 130–80.

Suga, T., Iwata, N. and Asakawa, Y. (1972) Chemical constituents of the male flower of *Alnus pendula* (Betulaceae). *Bull. Chem. Soc. Jap.*, **45**, 2058–60.

Suga, T., Ohta, S., Munesada, K. *et al.* (1993) Endogenous pine wood nematicidal substances in pines, *Pinus massoniana*, *P. strobus* and *P. palustris*. *Phytochemistry*, **33**, 1395–401.

Sultanbawa, M. U. S., Surendrakumar, S. and Balden, P. (1980) New antibacterial polyphenol, copalliferol A, from *Vateria copallifera* (Retz.) Alston (Dipterocarpaceae). *J. Chem. Soc. Chem. Commun.*, 619–20.

Sultanbawa, M. U. S., Surendrakumar, S. and Bladon, P. (1987) Distichol, an antibacterial polyphenol from *Shorea disticha*. *Phytochemistry*, **26**, 799–801.

Sultanbawa, M. U. S., Surendrakumar, S., Wazeer, M. I. M. and Bladon, P. (1981) Novel resveratrol tetramer, vaticaffinol, from *Vatica affinis* Thw. (Dipterocarpaceae). *J. Chem. Soc. Chem. Commun.*, 1204–6.

Sunder, R., Rangaswami, S. and Reddy, G. C. S. (1978) A new dihydrophenanthrene from *Dioscorea decipiens*. *Phytochemistry*, **17**, 1067.

Suzuki, H., Ikeda, T., Matsumoto, T. and Noguchi, M. (1977) Isolation and identification of hydrangenol and umbelliferone from cultured cells of amacha (*Hydrangea macrophylla* Seringe var. *thunbergii* Makino). *Agricult. Biol. Chem.*, **41**, 205–7.

Suzuki, H., Ikeda, T., Matsumoto, T. amd Noguchi, M. (1978) Polyphenol components in cultivated cells of amacha (*Hydrangea macrophylla* Seringe var. *thunbergii* Nakino). *Agricult. Biol. Chem.*, **42**, 1133–7.

Suzuki, H., Matsumoto, T., Kisaki, T. and Noguchi, M. (1979a) Polyphenol components in the leaves of Amacha (*Hydrangea macrophylla* Seringe var. *thunbergii* Makino). *Agricult. Biol. Chem.*, **43**, 653–4.

Suzuki, H., Matsumoto, T., Kisaki, T. and Noguchi, M. (1979b) Isolation and identification of two novel glycosides from the fresh leaves of amacha (*Hydrangea macrophylla* Seringe var. *thunbergii* Makino). *Agricult. Biol. Chem.*, **43**, 1785–8.

Suzuki, H., Matsumoto, T., Kisaki, T. and Noguchi, (1981) Influences of cultural conditions on polyphenol formation and growth of amacha cells (*Hydrangea macrophylla* Seringe var. *thunbergii* Makino) and changes of polyphenol contents in leaves. *Agricult. Biol. Chem.*, **45**, 1067–78.

Susuki, K., Shimizu, T., Kawabata, J. and Miyutani, J. (1987) New 3,5,4′-trihydroxystilbene (resveratrol) oligomers from *Carex fedia* Nees var. *miyabei* (Franchet) Koyama, T. (Cyperaceae). *Agricult. Biol. Chem.*, **51**, 1003–8.

Suzushino, G. (1954a) Plant pigments. VIII. Coloring matter of *Morus alba*. I. Morin, yellow pigment of the wood. *Misc. Repts. Res. Inst. Nat. Resources Jap.*, **34**, 21–3.

Suzushino, G. (1954b) Plant pigments. IX. Coloring matter of *Morus alba*. II. Some chemical properties of cudranin, a new colouring principle from the wood. *Misc. Repts. Res. Inst. Nat. Resources Jap.*, **35**, 8–13.

Tadros, W. and Ekladuis, L. (1950) Preparation of deoxyanisoin and some stilbenes. Reducing action of sodium glycollate in glycols. *Nature*, **166**, 525.

Tadros, W., Ekladuis, L. and Sakla, A. B. (1954) Reducing action of glycols in alkaline medium. Part I. Mechanism of formation of stilbenes from substituted benzaldehydes. *J. Chem. Soc.*, 2351–3.

Tahara, S., Matsukura, Y., Katasuta, H. and Mizutani, J. (1993) Naturally-occurring antidotes against benzimidazole fungicides. *Z. Naturforschung C – J. Biosci.*, **48**, 757–65.

Taira, Z., Hashimoto, T., Takei, M. *et al.* (1990) Structure–activity relationship of non-nitrogen macrocyclic aromatic compounds, marchantin A and its analogues from some liverworts, possessing d-tubocurarine-like activity. 11th Symposium on Medicinal Chemistry, Tokushima, Japan, p. 124.

Takagi, S., Yamaki, M. and Inoue, K. (1983) Antimicrobial agents from *Bletilla striata*. *Phytochemistry*, **22**, 1011–15.

Takaoka, M. (1939) Constitution of hydroxyresveratrol. *J. Chem. Soc. Jap.*, **61**, 96–8.

Takaoka, M. (1940a) The phenolic substances of white hellebore (*Veratrum grandiflorum* Loes). I. Synthesis of resveratrol (3,5,4′-trihydroxystilbene) and its derivatives. *J. Chem. Soc. Jap.*, **61**, 1067–9.

Takaoka, M. (1940b) Phenolic substances of white hellebore (*Veratrum grandiflorum* Loes. Fil.). *J. Faculty Sci. Hokkaido Imp. Univ. Ser. III*, **3**, 1–16.

Takaoka, M. (1940c) Constitution of rhapontigenin (pontigenin). *Proc. Imp. Acad. (Tokyo)*, **16**, 408–10.

Takaoka, M. (1940d) The synthesis of resveratrol and its derivatives. *Proc. Imp. Acad. (Tokyo)*, **16**, 405–7.

Takasugi, M., Kawashima, S., Monde, K. *et al.* (1987) Antifungal compounds from *Dioscorea batatas* inoculated with *Pseudomonas cichorii*. *Phytochemistry*, **26**, 371–5.

Takasugi, M., Munoz, L., Masamune, T. *et al.* (1978) Studies on phytoalexins of the Moraceae. 3. Stilbene phytoalexins from diseased mulberry. *Chem. Lett.*, 1241–2.

Takeuchi, N., Murase, M., Ochi, K. and Tobinaga, S. (1976) Biogenetic-type synthesis of (+)-phyllodulcin, a sweet principle of *Hydrangea serrata* Seringe var. *thunbergii* Sugimoto. *J. Chem. Soc. Chem. Commun.*, 820–1.

Takeuchi, N., Murase, M., Ochi, K. and Tobinaga, S. (1980) Biogenetic-type synthesis of (+)-phyllodulcin, a sweet principle of *Hydrangea serrata* Seringe var. *thunbergii* Sugimoto. *Chem. Pharm. Bull.*, **28**, 3013–19.

Takikawa, K., Tori, M. and Asakawa, Y. (1989) Chemical constituents and chemosystematics of *Radula* species (liverworts). *J. Hattori Bot. Lab.*, **67**, 365–71.

Talapatra, B., Mukhopadhyay, P., Chaudhury, P. and Talapatra, S.K. (1982) Denbinobin, a new phenathraquinone from *Dendrobium nobile* Lindl. (Orchidaceae). *Indian J. Chem.*, **21B**, 386–7.

Talvitie, A., Mannila, E. and Kolehmainen, E. (1992) Synthesis of some biologically active compounds from stilbenes isolated from the bark of *Picea abies*. *Justus Liebigs Ann. Chem.*, **1992**, 399–401.

Tandon, M. and Shukla, Y. N. (1993) A bibenzyl xyloside from *Chlorophytum arundinaceum*. *Phytochemistry*, **32**, 1624–5.

Taylor, R. J. and Shaw, D. C. (1983) Allelopathic effects of Engelmann spruce bark stilbenes and tannin stilbene combinations on seed germination and seedling growth of selected conifers. *Can. J. Bot.*, **61**, 279–89.

Tezuka, Y., Ueda, M. and Kikuchi, T. (1989) Studies on the constituents of orchidaceous plants. VIII. Constituents of *Spiranthes sinensis* (Pers.) Ames var. *amoena* (M. Bieberson) Hara. (1) Isolation and structure elucidation of spiranthol-A, spiranthol-B and spirasineol-A, new isopentenyldihydrophenanthrenes. *Chem. Pharm. Bull.*, **37**, 3195–6.

Tezuka, Y., Ji, L., Hirano, H., Ueda *et al.* (1990) Studies on the constituents of orchidaceous plants. IX. Constituents of *Spiranthes sinensis* (Pers.) Ames var. *amoena* (M. Bieberson) Hara. (2) Structure of spiranthesol, spiranthoquinone, spiranthol-C, and spirasineol-B, new isopentenyldihydrophenanthrenes. *Chem. Pharm. Bull.*, **38**, 629–35.

Thakkar, K., Geahlen, R. L. and Cushman, M. (1993) Synthesis and protein-tyrosine kinase inhibitory activity of polyhydroxylated stilbene analogs of piceatannol. *J. Med. Chem.*, **36**, 2950–5.

Theuns, H. G., Janssen, R. H. A. M., Biessels, H. W. A. and Salemink, C. A. (1985) Constituents of *Papaver bracteatum*: *O*-methyl-alpha-thebaol and 10-*n*-nonacosanol. Lanthanide-induced chemical shifts in ^{1}H NMR and ^{13}C NMR. *Phytochemistry*, **24**, 163–9.

Thi Thu Ha, N., Nógrádi, M., Brlik, J. *et al.* (1991) Synthesis of marchantin H and B, macrocyclic bis(bibenzyl)s from liverworts. *J. Chem. Res. (S)*, **137**(M), 1240–54.

Thomas, B. D. (1977) Studies of physiological effects of lunularic acid in cryptogams and higher plants. PhD thesis, University of London.

Tori, M. (1988) Application of 2D NMR techniques to structure determination of natural products, in *Natural Products Chemistry*, (ed. Atta-ur-Rahaman), Elsevier, Amsterdam, pp. 81–114.

Tori, M., Aoki, M. and Asakawa, Y. (1994) Chenopodene, marchantin P and riccardin G from the liverwort *Marchantia chenopoda*. *Phytochemistry*, **36**, 73–6.

Tori, M., Masuya., T. and Asakawa, Y. (1990) Neomarchantin A and B, new macrocyclic bisbibenzyls from the liverwort *Schistochila glaucescens*. *J. Chem. Res. (S)*, 36–7.

Tori, M., Toyota, M., Harrison, L. J. *et al.* (1985) Total assignment of ^{1}H and ^{13}C NMR spectra of marchantins isolated from liverworts and its application to structure determination of two new macrocyclic bis(bibenzyl)s from *Plagiochasma intermedium* and *Riccardia multifida*. *Tetrahedron Lett.*, **26**, 4735–8.

Toscano Underwood, C. D. and Pearce, R. B. (1991) Astringin and isorhapontin distribution in Sitka spruce trees. *Phytochemistry*, **30**, 2183–9.

Toscano Underwood, C. D. and Pearce, R. B. (1992) Stilbene glucoside levels and the resistance of Sitka spruce (*Picea sitchensis*) tissues to colonization by root- and butt-rotting fungi. *Plant Pathol.*, **41**, 722–9.

Toyota, M. (1987) Chemical constituents of *Marchantia polymorpha, Riccardia multifida* and *Plagiochila* genus (Hepaticae), Dissertation, Tokushima Bunri University.

Toyota, M. and Asakawa, Y. (1993) Bisbibenzyl and sesquiterpenoids from the liverwort *Jubulu juponicu*. *Phytochemistry*, **34**, 1135–7.

Toyota, M., Nagashima, F. and Asakawa, Y. (1988) Fatty acids and cyclic bis(bibenzyl)s from the New Zealand liverwort *Monoclea forsteri*. *Phytochemistry*, **27**, 2603–8.

Toyota, M., Tori, M., Takikawa, K. *et al.* (1985) Perrottetins E, F, and G from *Radula perrottetii* (liverwort) – isolation, structure determination, and synthesis of perrottetin E. *Tetrahedron Lett.*, **26**, 6097–100.

Tschirch, A. and Christofoletti, U. (1905) Uber die Rhaponticwurzel. *Arch. Pharm.*, **243**, 443.

Tsukida, K. (1954) Stilbene derivatives as plant components. *Yakugaku Kenkyu*, **26**, 668–95.

Tsukida, K. and Yokota, M. (1954) Plant components XXXVI. Constituents of *Polygonum multiflorum* by paper chromatography. *J. Pharm. Soc. Jap.*, **74**, 230–1.

Tsukida, K. and Yoneshigo, M. (1954) Constituents of Japanese rhubarb (*Rheum undulatum*). *J. Pharm. Soc. Jap.*, **74**, 382–5.

Tuchinda, P., Udchachon, J., Khumtaveeporn, K. and Taylor, W. C. (1989) Benzylated phenanthrenes from *Eulophia nuda*. *Phytochemistry*, **28**, 2463–6.

Tuchinda, P., Udchachon, J., Khumtaveeporn, K. *et al.* (1988) Phenanthrenes of *Eulophia nuda*. *Phytochemistry*, **27**, 3267–71.

Tyukavkina N. A., Lutskii, V. T., Dzizenko, A. K. and Pentegova, V. A. (1968) Phenolic components in the extracts of *Pinus sibirica* heartwood. *Khim. Prir. Soedin*, **4**, 249–50.

Tyukavkina, N. A., Govokhova, V. G., Babkin, V. A. *et al.* (1979) Liquid chromatography of plant phenolic compounds. 4. Reverse-phase chromatography of stilbenes. *Khim. Drev.*, 81–3.

Tyukavkina, N. A., Gromova, A. S., Lutskii, V. I. and Voronov, V. K. (1972) Hydroxystilbene from the bark of *Pinus sibirica*. *Khim. Prir. Soedin*, **8**, 600–3.
Ueno, M. (1931) Constitution of phyllodulcin. *J. Pharm. Soc. Jap.*, **51**, 207–27.
Ueno, Y. (1937) Hydrangenol glucoside. *J. Pharm. Soc. Jap.*, **57**, 602–5.
Updhyay, C. M. and Dutta, N. L. (1968) Chemical investigation of the heartwood of *Cassia siamea*. *Ind. J. Appl. Chem.*, **31**, 239.
Urech, J., Fechtig, B., Nuesch, J. and Vischer, E. (1963) Hircinol, eine antifungal wirksame Substanze aus Knollen von *Loroglossum hircinum* (L.) Rich. *Helv. Chim. Acta*, **46**, 2758–66.
Valio, I. F. M., Burdon, R. S. and Schwabe, W. W. (1969) New natural growth inhibitor in the liverwort *Lunularia cruciata* (L.) Dum. *Nature*, **223**, 1176–8.
Valio, I. F. M. and Schwabe, W. W. (1970) Growth and dormancy in *Lunularia cruciata* (L.) Dum. VII. The isolation and bioassay of lunularic acid. *J. Exp. Bot.*, **21**, 138–50.
Veerraju, P., Prakasa Rao, N. S., Jaganmohana Rao, L. *et al.* (1989a) Amoenumin, a 9,10-dihydro-5H-phenanthro-(4,5-b,c,d)-pyran from *Dendrobium amoenum*. *Phytochemistry*, **28**, 950–1.
Veerraju, P., Prakasa Rao, N. S., Jaganmohan Rao, L. *et al.* (1989b) Bibenzyls and phenanthrenoids of some species of Orchidaceae. *Phytochemistry*, **28**, 3031–4.
Venkatamaran, K. (1972) Wood phenolics of the Morceae. *Phytochemistry*, **11**, 1571–86.
Vornam, B., Schön, H. and Kindl, H. (1988) Control of gene expression during induction of cultured peanut cells: mRNA levels, protein synthesis and enzyme activity of stilbene synthase. *Plant Mol. Biol.*, **10**, 235–43.
Ward, E. W. B., Unwin, C. H. and Stössl, A. (1975a) Postinfectional inhibitors from plants. XV. Antifungal activity of the phytoalexin orchinol and related phenanthrenes and stilbenes. *Can. J. Bot.*, **53**, 964–71.
Ward, E. W. B., Unwin, C. H. and Stössl, A. (1975b) Loroglossol, an orchid phytoalexin. *Phytopathology.*, **65**, 632–3.
Waterhouse, A. L. and Lamuela-Raventos, R. M. (1994) The occurrence of piceid, a stillbene glucoside, in grape berries. *Phytochemistry*, **37**, 571–3.
Wei, H.-C. and Wu, C.-L. (1991) A new macrocyclic bisbibenzyl diether, marchantinquinone, from the liverwort *Mannia subpilosa*. *J. Chem. Res. Suppl.*, 230–1.
Weiler, E. W. (1979) Radioimmunoassay for the detection of free and conjugated abscisic acid. *Planta*, **144**, 255–63.
Weissmann, G. (1981) Investigation of *Picea abies* bark extractives. *Holz als Roh und Werkst.*, **39**, 457–62.
Westfeld, L. (1966) High-boiling neutral constituents from the wood of *Pinus sylvestris*. *Acta Chem. Scand.*, **20**, 2829–40.
Wheeler, O. H. and Battle de Pabon, H. N. (1965) Synthesis of stilbenes. A comparative study. *J. Org. Chem.*, **30**, 1473–7.
Wij, M. and Rangaswami, S. (1978) Chemical components of *Dioscorea bulbifera*: isolation and structures of a new dihydrophenanthrene (2,4,6,7-tetrahydroxy-9, 10-dihydrophenanthrene) and a new phenanthrene (2,4,5,6-tetrahydroxyphenanthrene). *Indian J. Chem.*, **16B**, 643–4.
Wolcott, G. N. (1951) Termite resistance of pinosylvin and other new insecticides. *J. Econ. Entomol.*, **44**, 263–4.
Wolcott, G. N. (1953) Stilbenes and comparable materials for dry-wood termite control. *J. Econ. Entomol.*, **46**, 374–5.
Wolfrom, M. L. and Bhat, H. B. (1965) Osage orange pigments. XVII. 1,3,6,7-tetrahydroxyxanthone from the heartwood. *Phytochemistry*, **4**, 765–8.
Wollenweber, E. (1982) in *The Plant Cuticle*, Linn. Soc. Symp. Series no. 10, (eds. D. F. Cutler, K. L. Alvin, K. L. and C. E. Price), Academic Press, London.

Wollenweber, E., Dörr, M., Waton, H. and Favre-Bonvin, J. (1993) Flavonoid aglycones and a novel dihydrostilbene from the frond exudate of *Notholaena nivea*. *Phytochemistry*, **33**, 611–12.

Wollenweber, E., Dörr, M., Stelzer, R. and Arriaga-Giner, J. F. (1992) Lipophilic phenolics from the leaves of *Empetrum nigrum* – chemical structures and exudate localization. *Acta Botanica*, **105**, 300–5.

Wollenweber, E. and Favre-Bonvin, J. (1979) Novel dihydrostilbene from fronds of *Notholaena dealbata* and *Notholaena limitanea*. *Phytochemistry*, **18**, 1243–4.

Woodward, S. and Pearce, R. B. (1988) The role of stilbenes in resistance of Sitka spruce (*Picea sitchensis* (Bong.) (Carr.) to entry of fungal pathogens. *Physiol. Mol. Plant Pathol.*, **33**, 127–49.

Wotton, H. R. and Strange, R. N. (1985) Circumstantial evidence for phytoalexin involvement in the resistance of peanuts to *Aspergillus flavus*. *J. Gen. Microbiol.*, **131**, 487–94.

Wu, C.-L. (1990) Some new findings on the chemistry of Taiwanese liverworts, in *Bryophytes, Their Chemistry and Chemical Taxonomy* (ed. H. D. Zinsmeister and R. Mues), Oxford University Press, Oxford, pp. 71–82.

Wurzel, G. and Becker, H. (1990) Sesquiterpenoids from the liverwort *Ricciocarpos natans*. *Phytochemistry*, **29**, 2565–8.

Yagi, A., Ogata, Y., Yamauchi, T. and Nishioka, I. (1977) Metabolism of phenylpropanoids in *Hydrangea serrata* var. *thunbergii* and the biosynthesis of phyllodulcin. *Phytochemistry*, **16**, 1098–100.

Yagi, A., Washida, Y., Takata, N. and Nishioka, I. (1972) Studies on *Hydrangea* species. I. Phenolic components of *Hydrangea serrata* Seringe var. *thunbergii* Sugimoto. *Chem. Pharm. Bull.*, **20**, 1755–61.

Yahama, . and Kamishiro, (1951) *Jap. Sci. Rev.*, **2**, 121.

Yamaki, M., Bai, L., Inoue, K. and Takagi, S. (1989) Biphenanthrenes from *Bletilla striata*. *Phytochemistry*, **28**, 3503–5.

Yamaki, M., Bai, L., Inoue, K. and Takagi, S. (1990) Benzylphenanthrenes from *Bletilla striata*. *Phytochemistry*, **29**, 2285–7.

Yamaki, M., Bai, L., Kato, T. *et al.* (1993a) Three dihydrophenanthropyrans from *Bletilla striata*. *Phytochemistry*, **32**, 427–30.

Yamaki, M., Bai, L., Kato, T. *et al.* (1993b) Blespirol, a phenanthrene with a spirolactone ring from *Bletilla striata*. *Phytochemistry*, **31**, 1497–8.

Yamaki, M., Kato, T., Bai, L. *et al.* (1991) Methylated stilbenoids from *Bletilla striata*. *Phytochemistry*, **30**, 2759–60.

Yamaki, M., Kato, T., Bai, L. *et al.* (1993c) Phenanthrene glucosides from *Bletilla striata*. *Phytochemistry*, **34**, 535–7.

Yazaki, Y. and Hillis, W. E. (1977) Polyphenolic extractives of *Pinus radiata* bark. *Holzforschung*, **31**, 20–5.

Youngken, H. W. (1946) Studies on Indian Rhubarb. II. *Rheum emodi*. *J. Am. Pharmacol. Assoc.*, **35**, 148–54.

Zaman, A., Prakash, S., Wizarat, K. *et al.* (1983) Isolation and structure of gnetol, a novel stilbene from *Gnetum ula*. *Ind. J. Chem. B*, **22**, 101–4.

Zinsmeister, H. D., Becker, H. and Eicher, T. (1991) Bryophytes, a source of biologically active, naturally occurring material? *Angew. Chem. Int. Ed. Engl.* **30**, 130–47.

Appendix: Stilbenoid structures

Group A

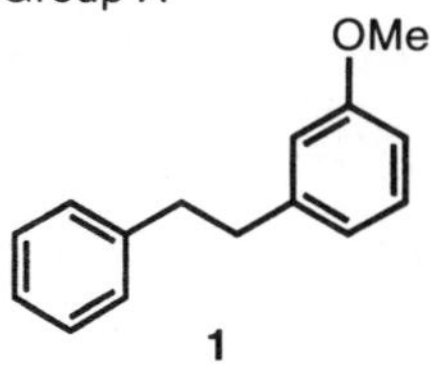

1

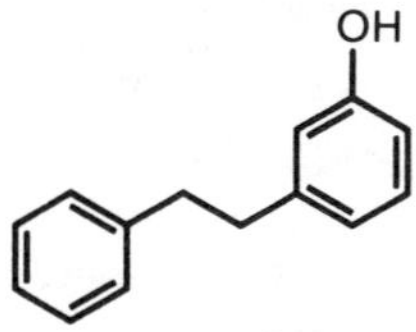

2

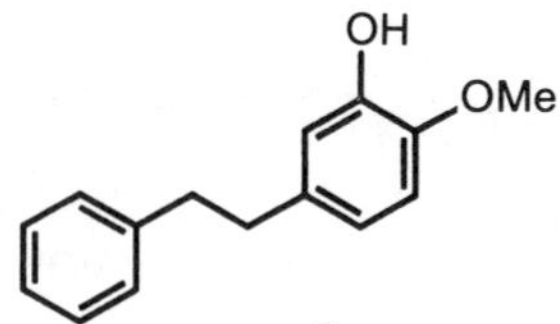

3

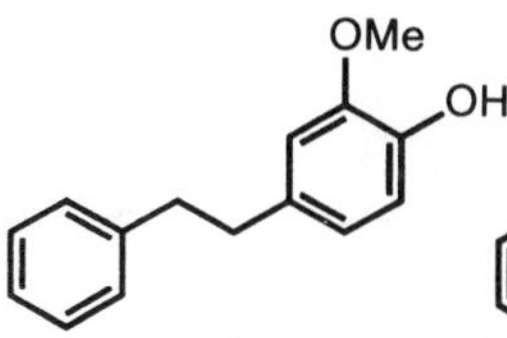

4

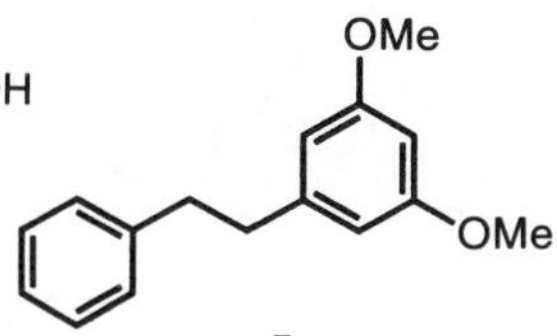

5

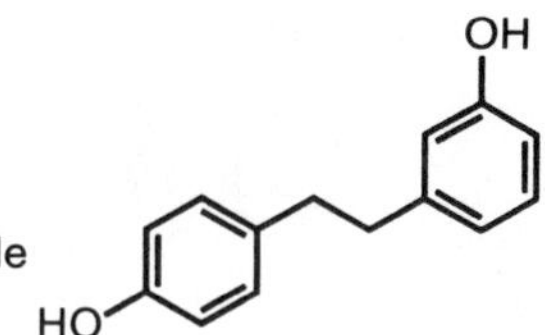

6

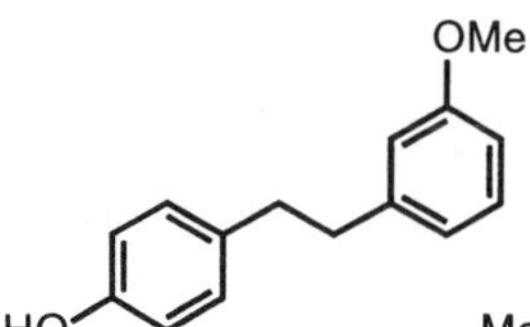

7

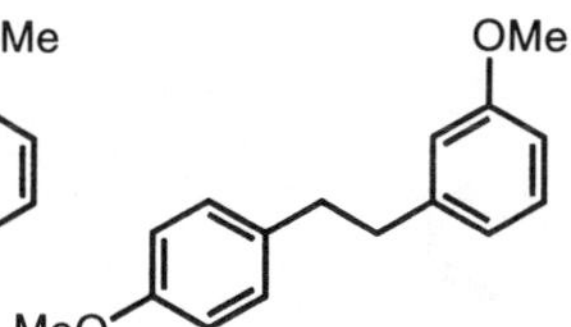

8

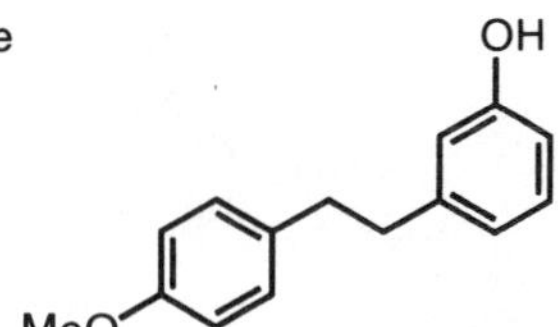

9

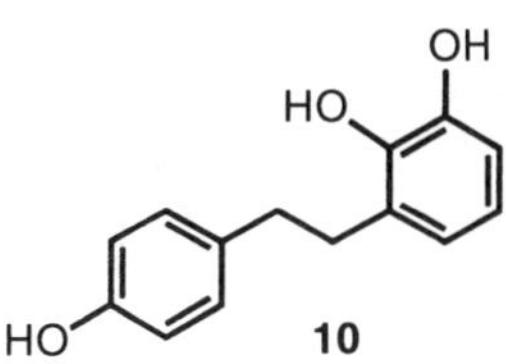

10

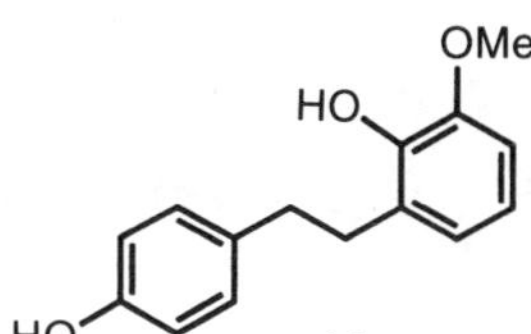

11

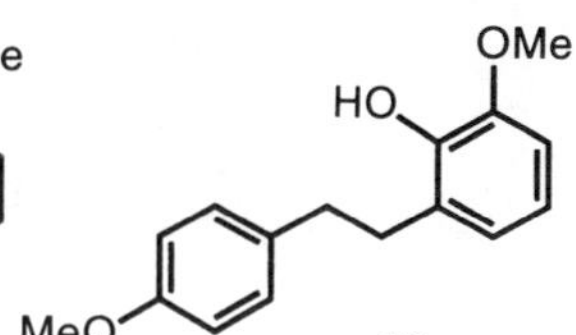

12

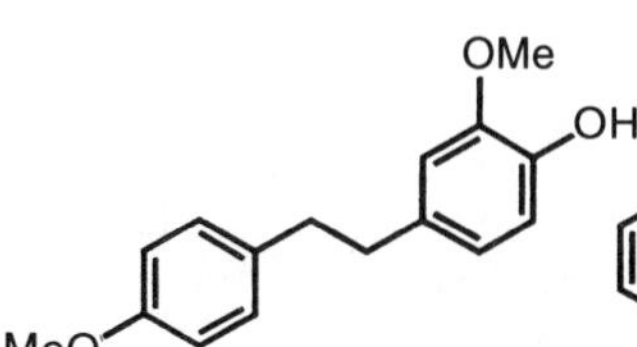

13

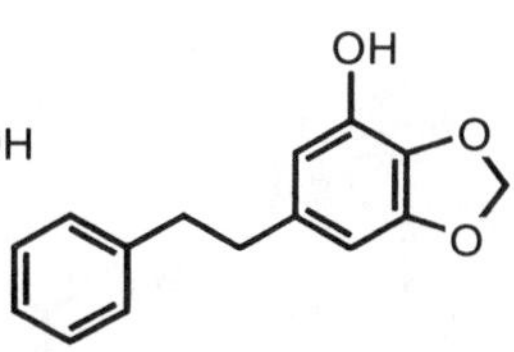

14

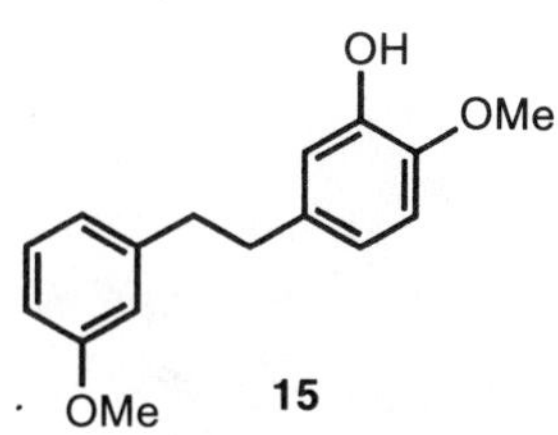

15

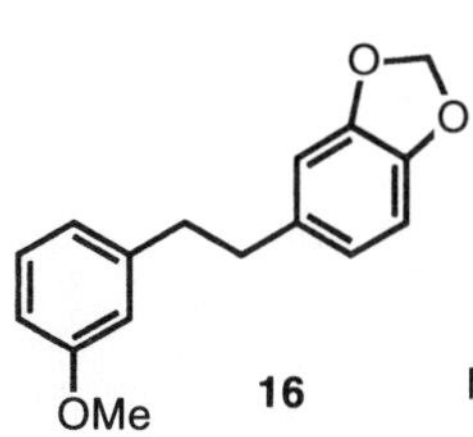

16

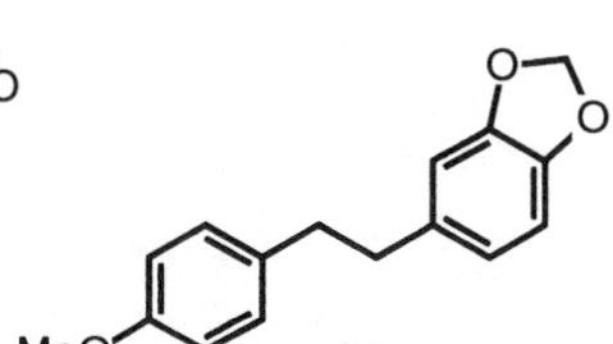

17

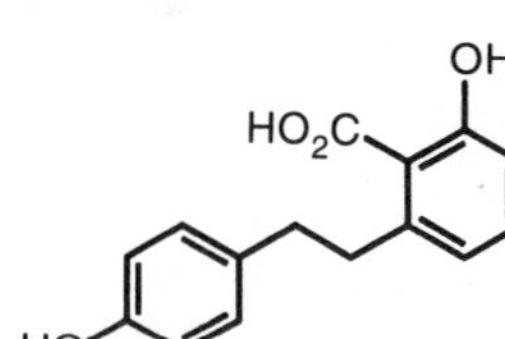

18

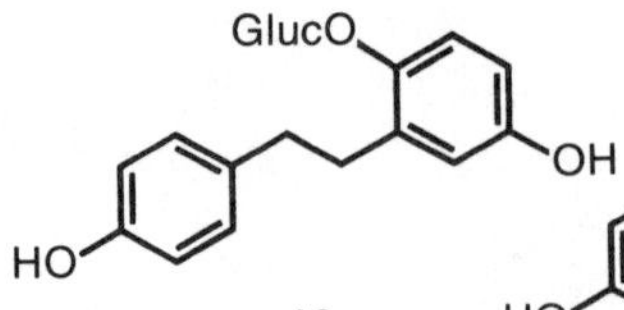

19

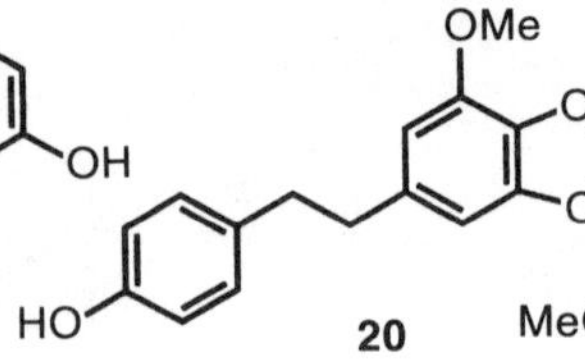

20

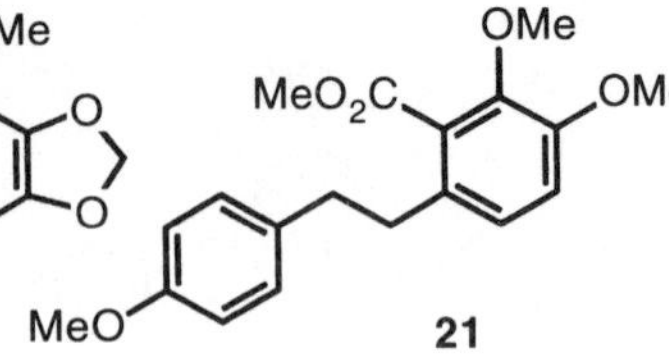

21

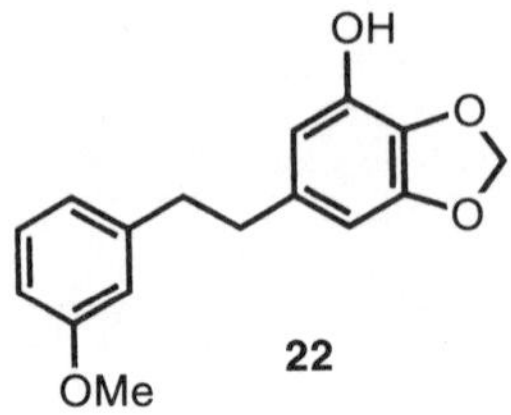

22

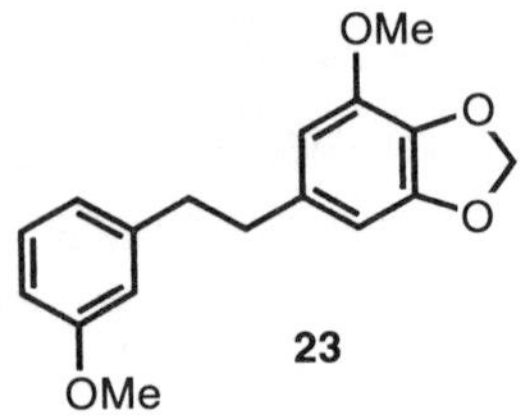

23

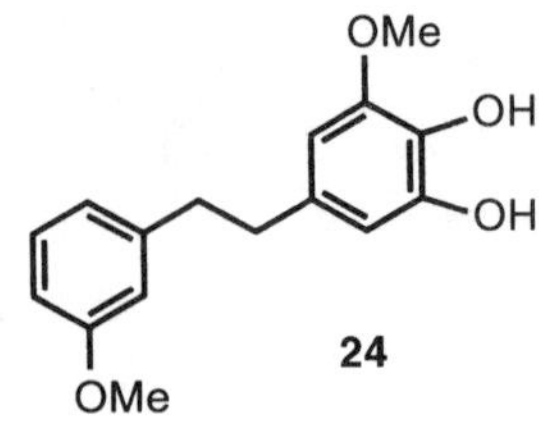

24

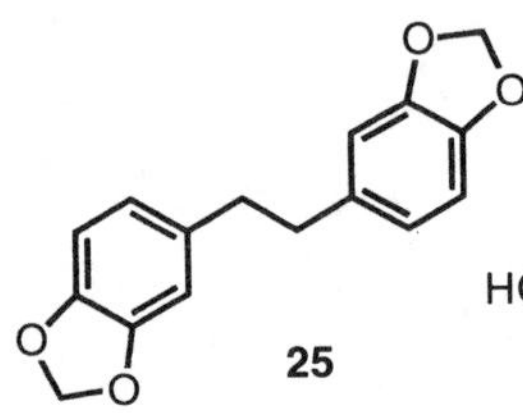

25

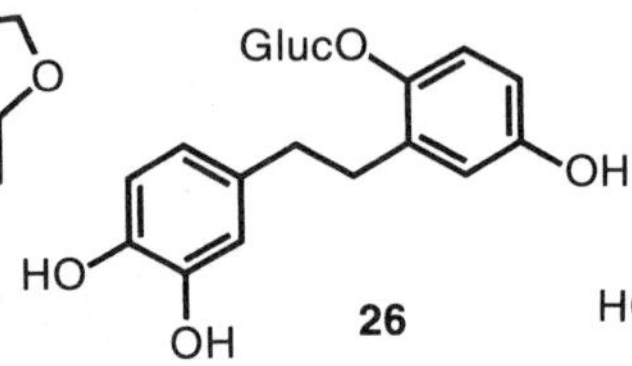

26

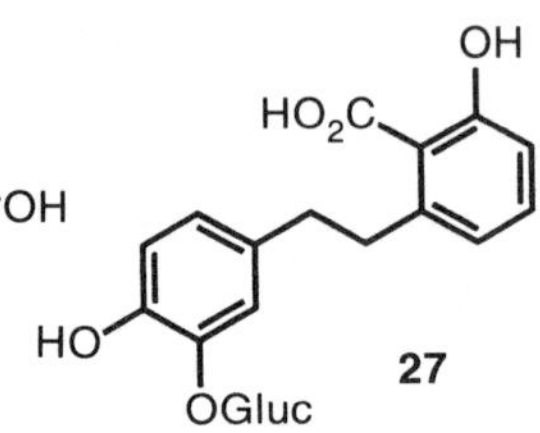

27

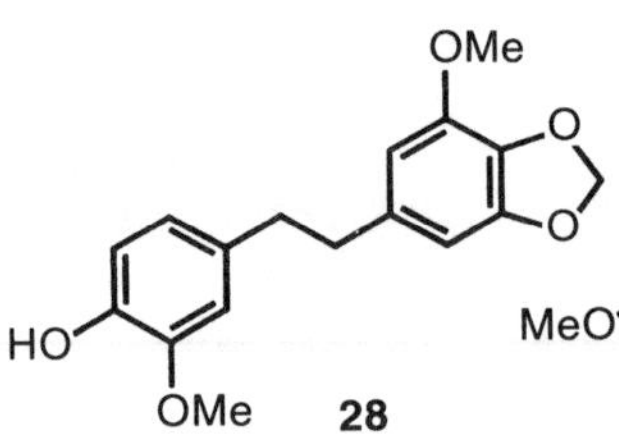

28

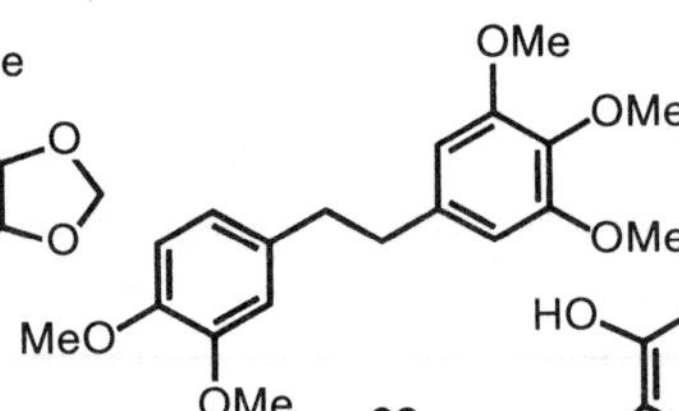

29

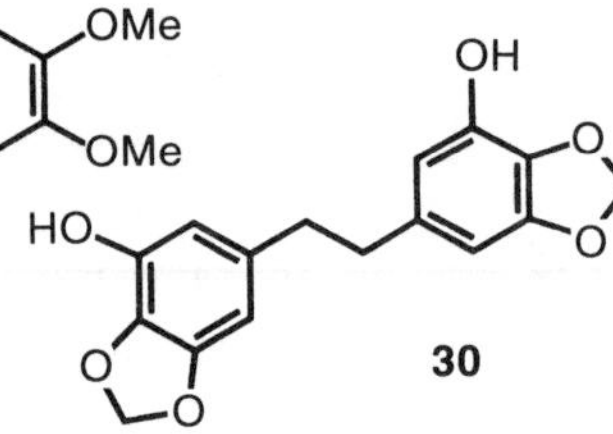

30

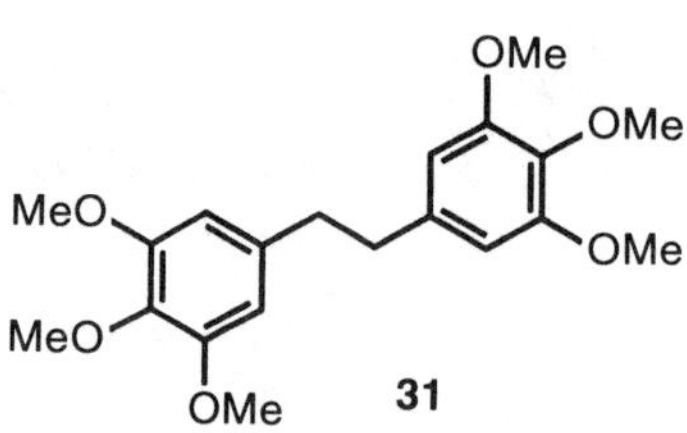

31

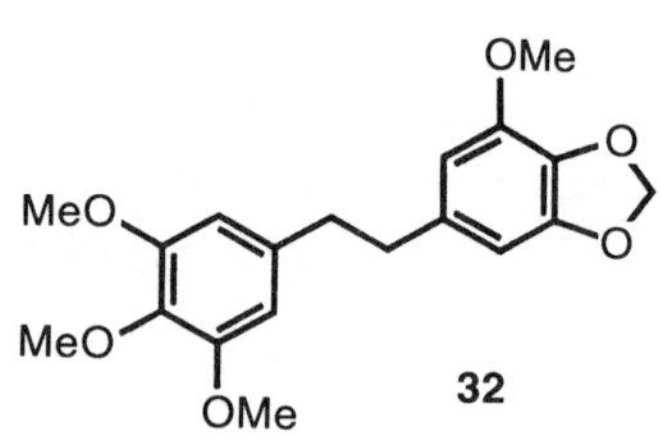

32

Group B (prenyl and geranyl bibenzyls)

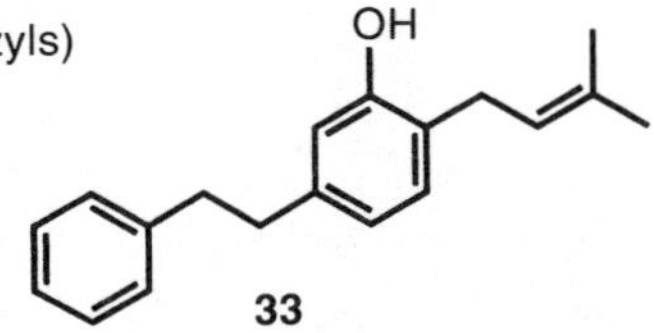

33

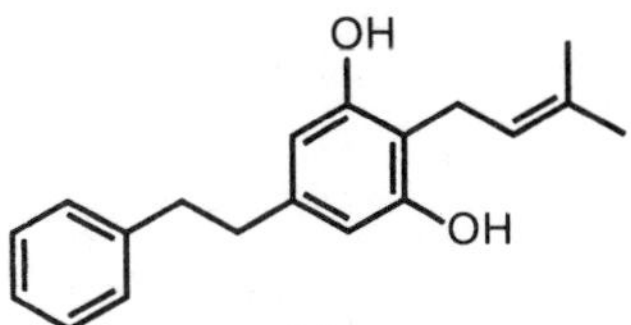

34

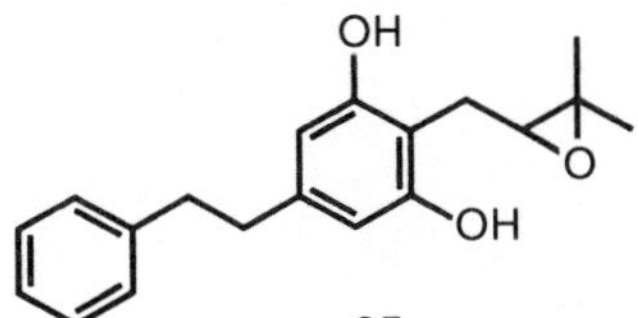

35

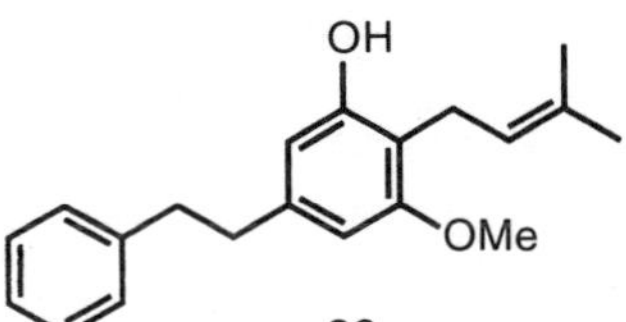

36

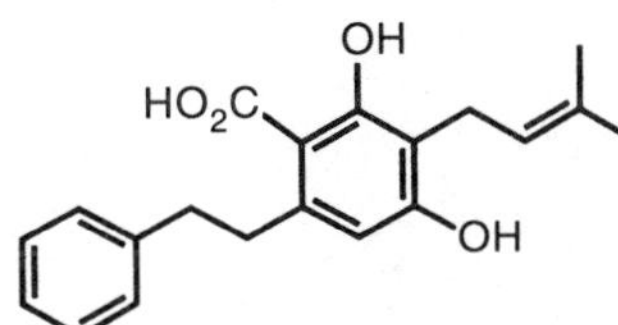

37

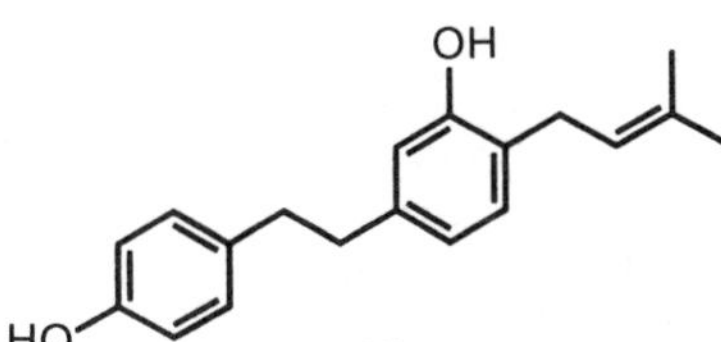

38

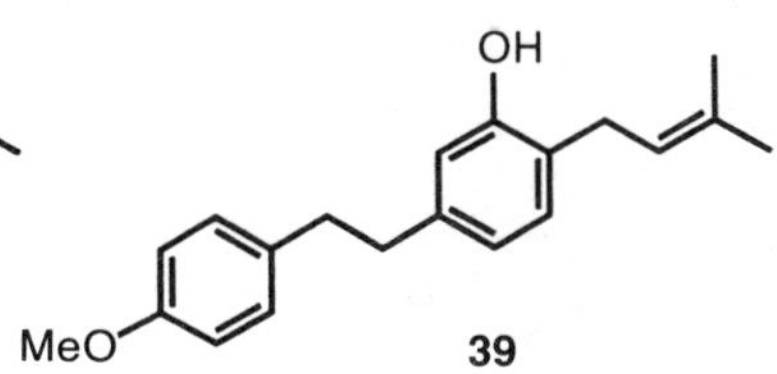

39

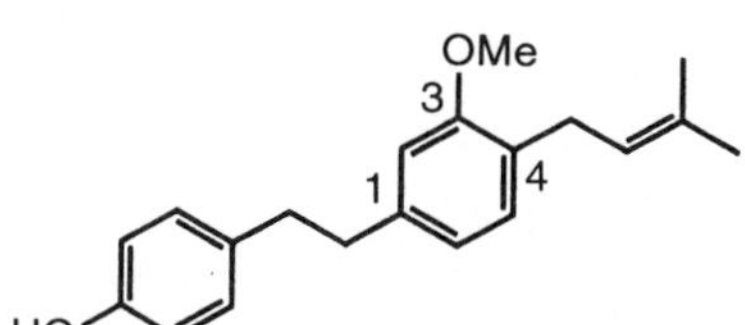

40

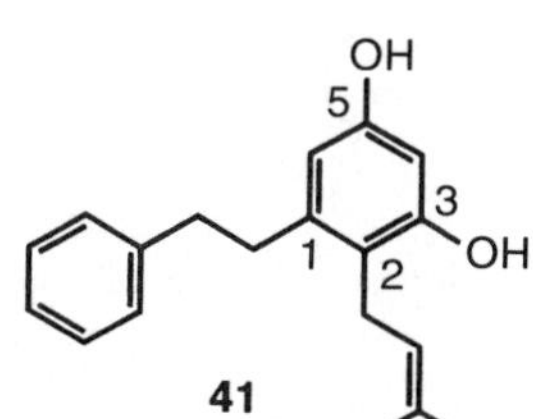

41

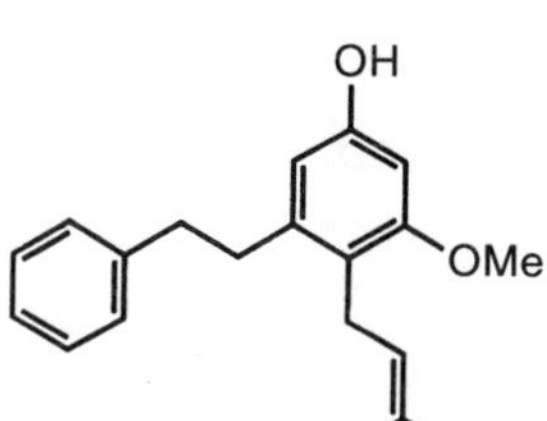

42

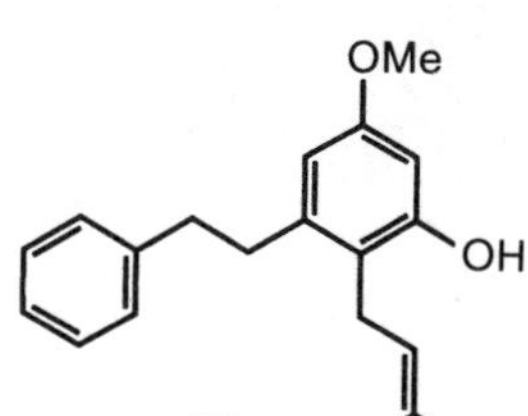

43

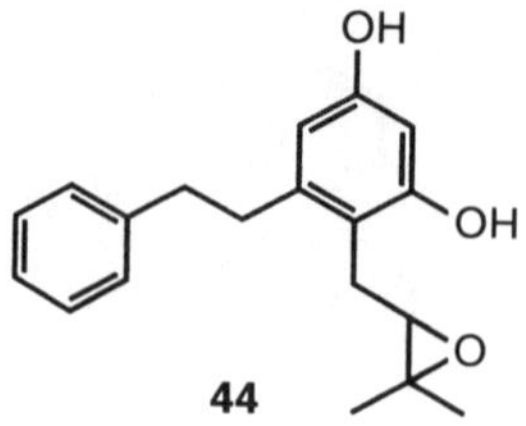

44

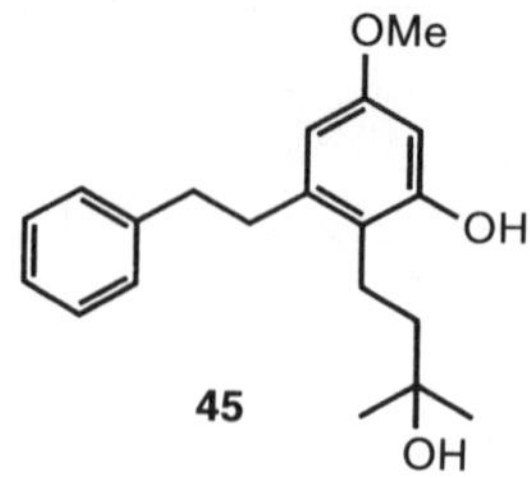

45

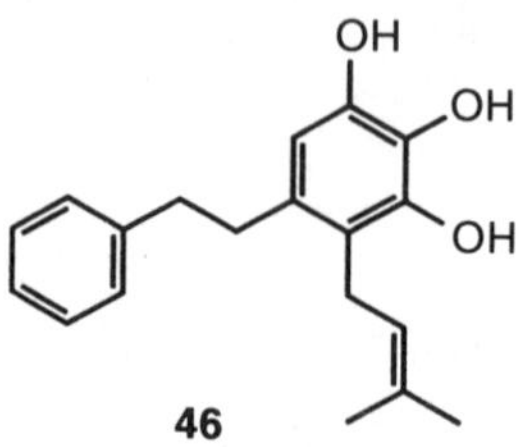

46

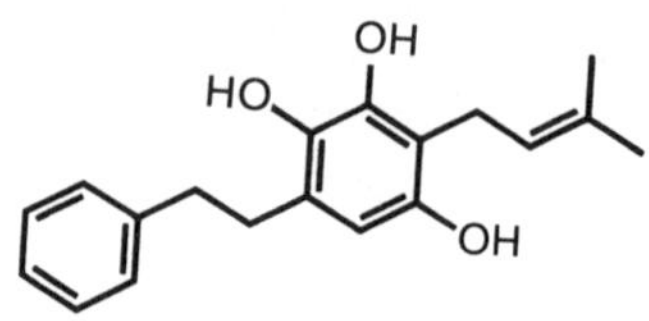

46'

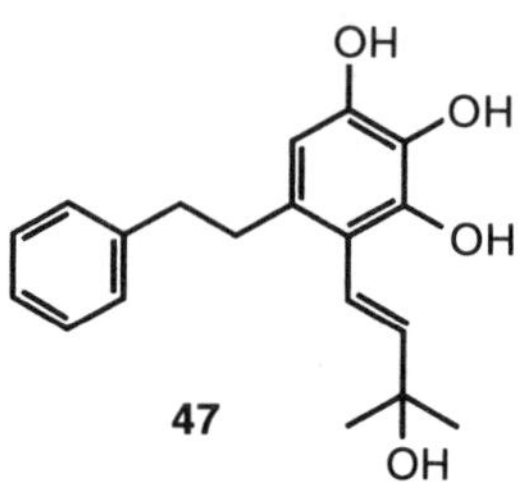

47

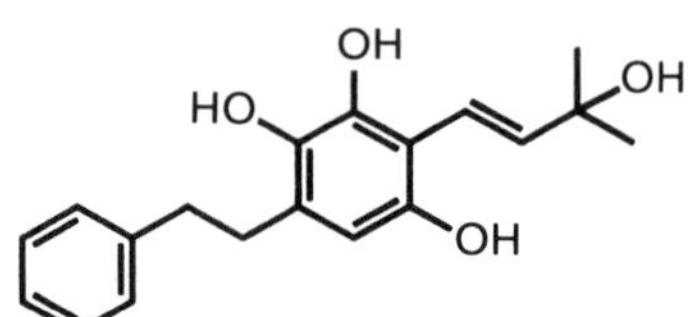

47'

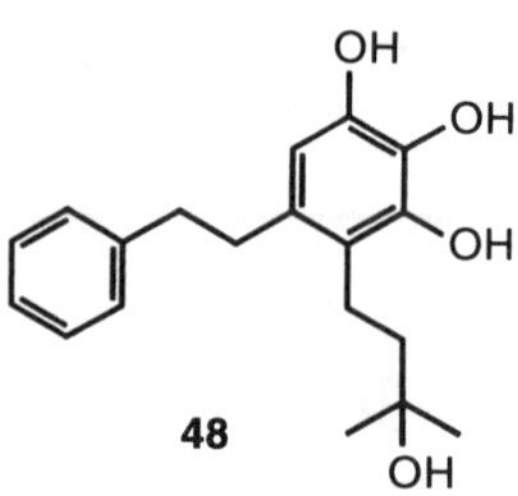

48

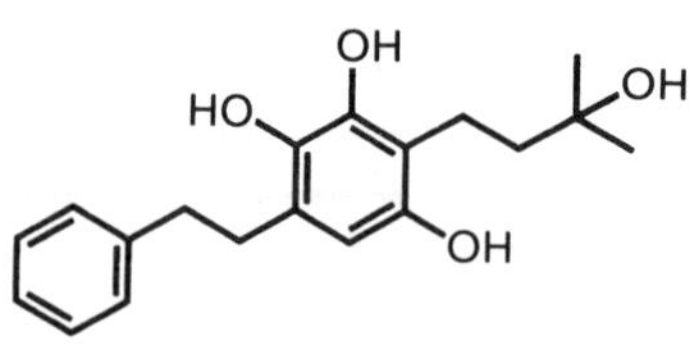

48'

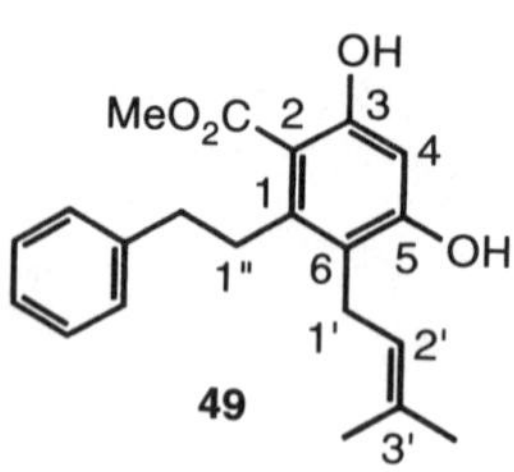

49

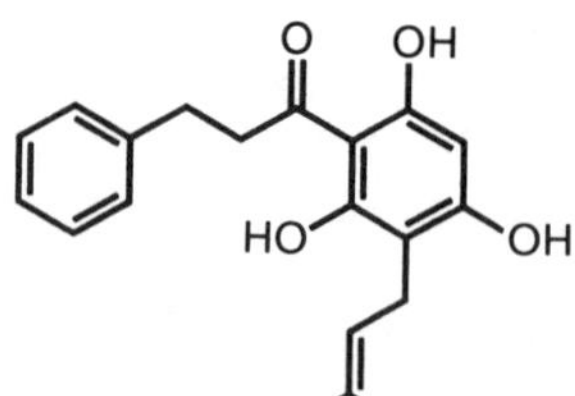

49'

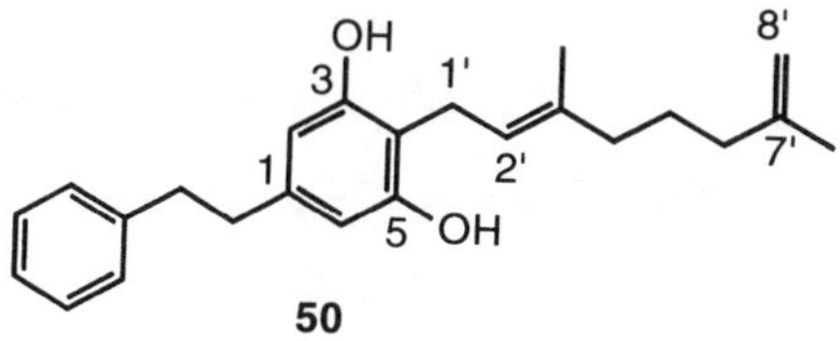

50

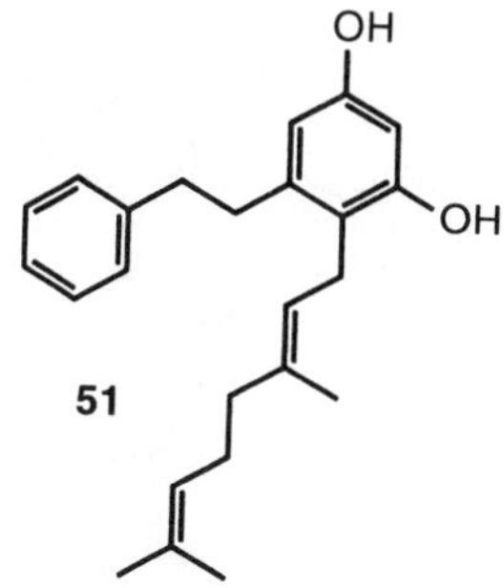

51

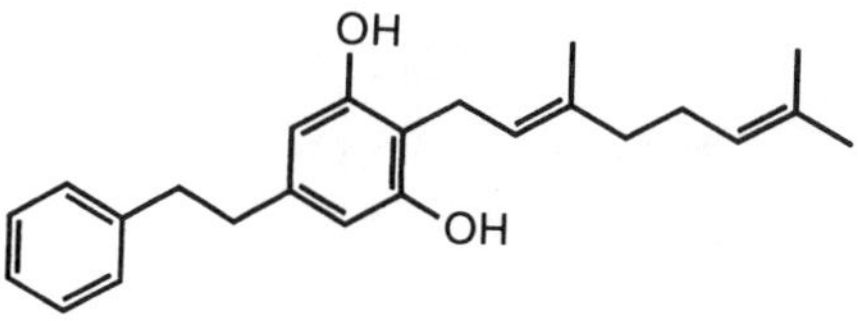

52

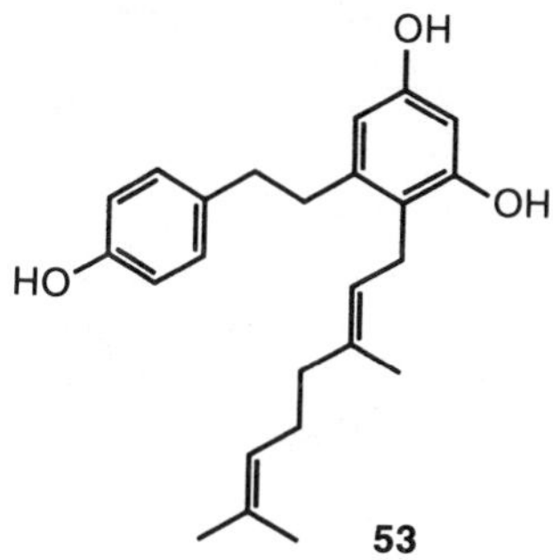

53

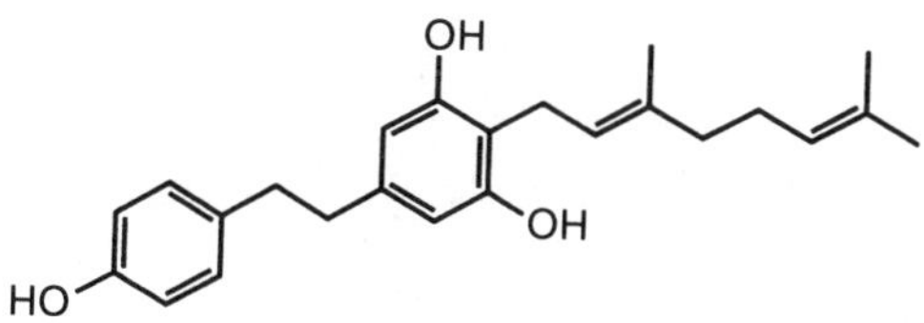

54

Group C (with one more carbocycle)

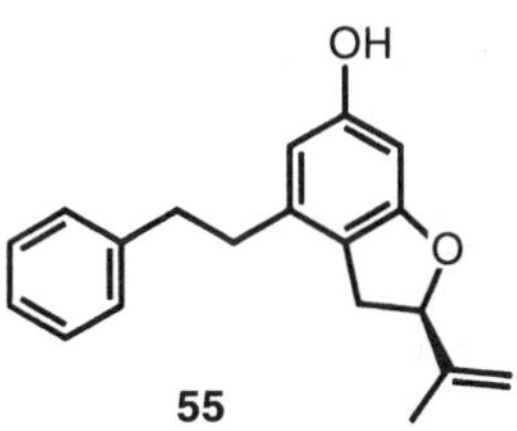

55

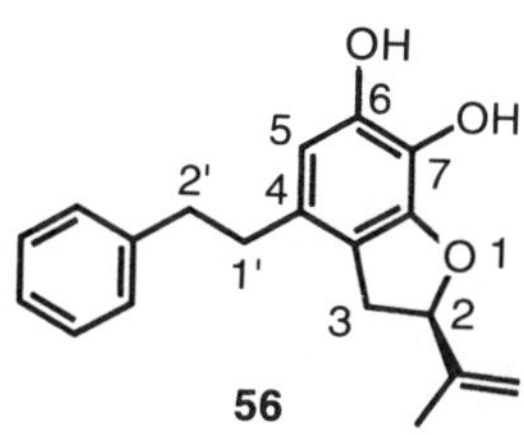

56

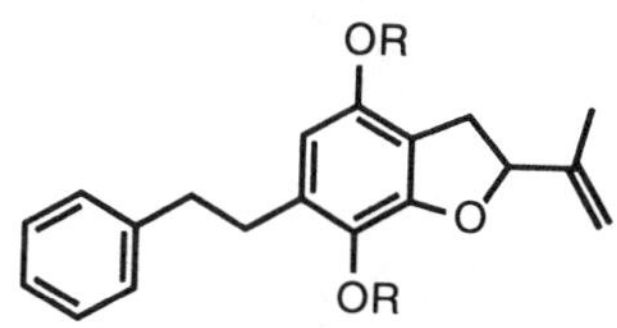

56a' R=H
56b' R=Me

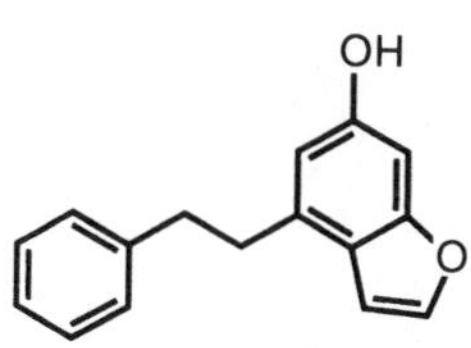

57

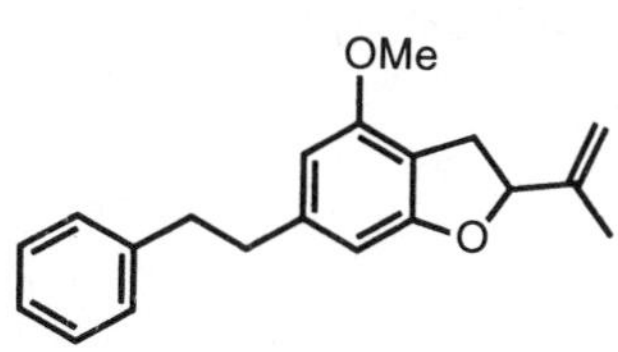

58

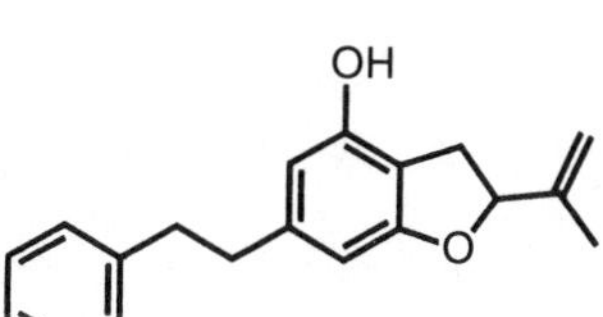

59

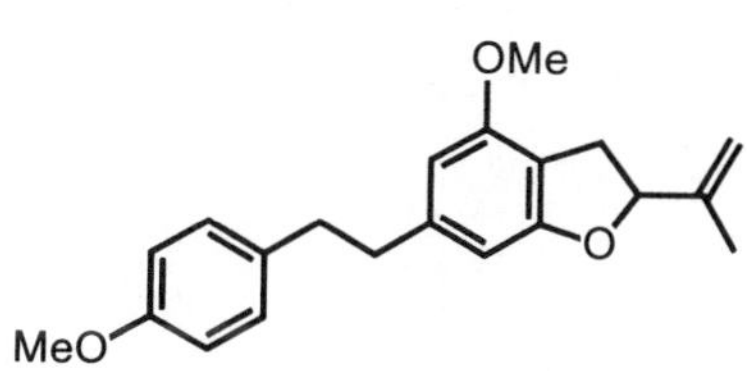

60

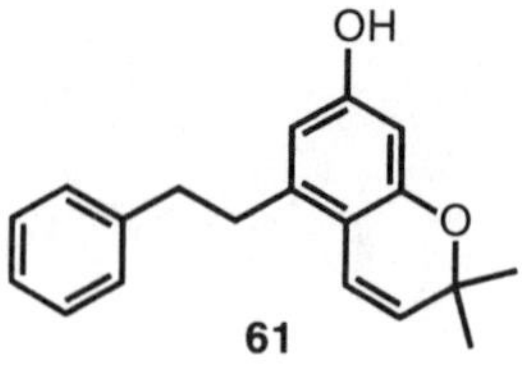

61

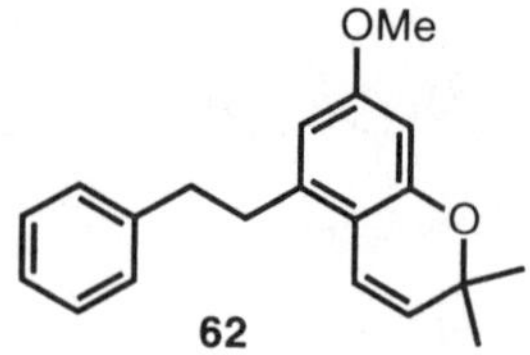

62

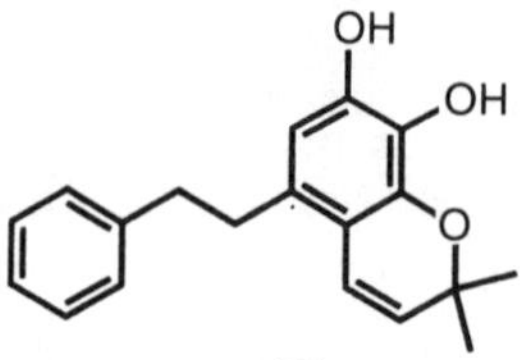

63

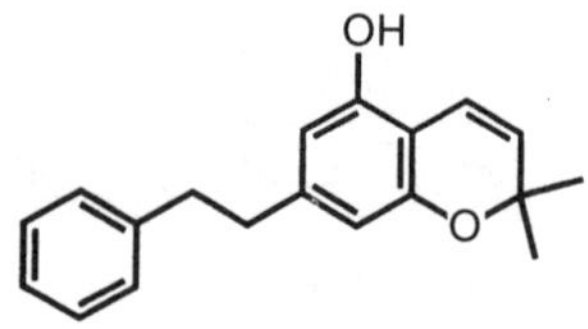

64

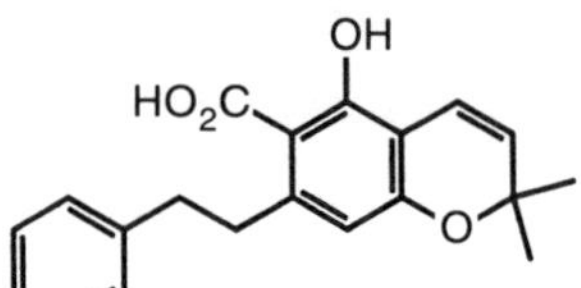

65

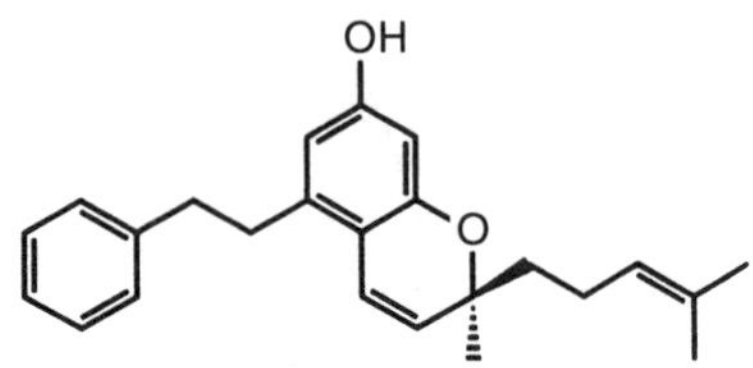

66

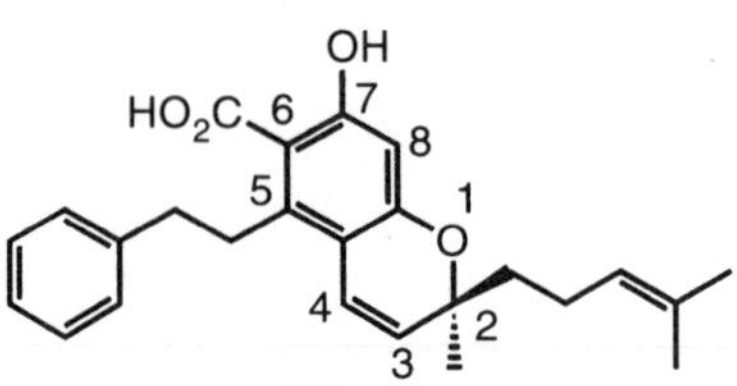

67

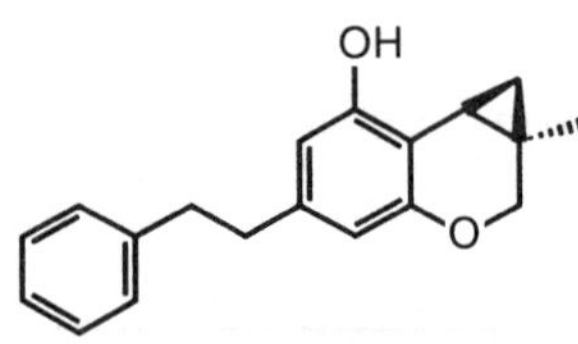

68

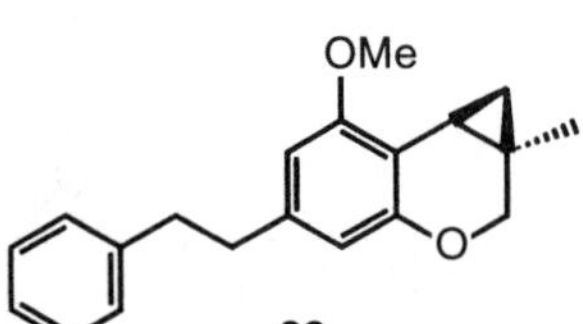

69

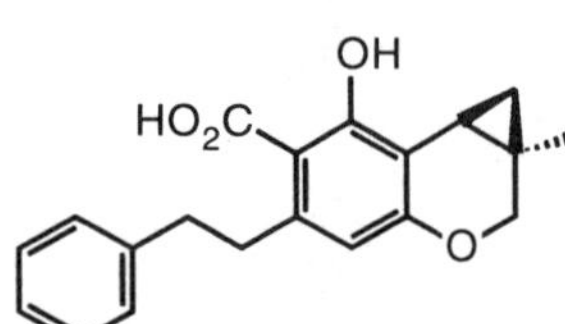

70

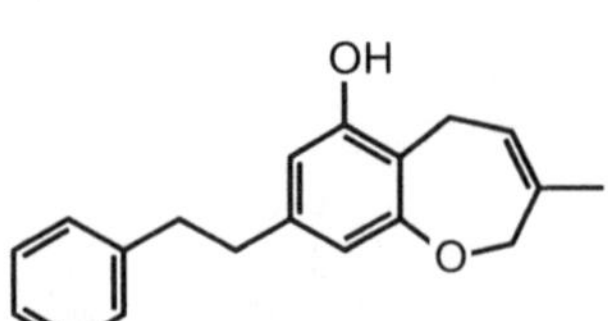

71

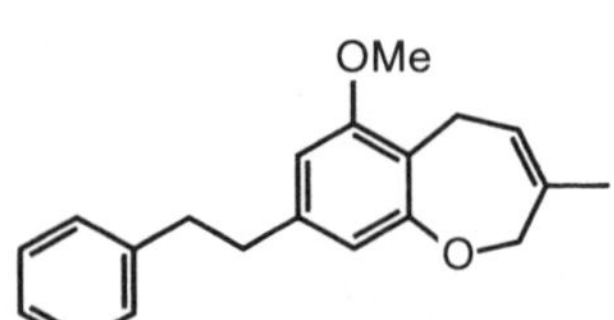

72

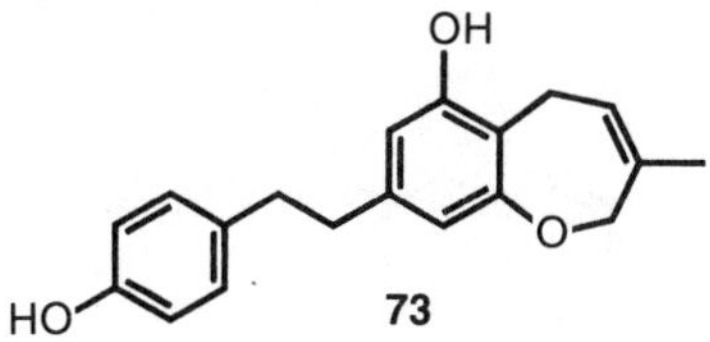

73

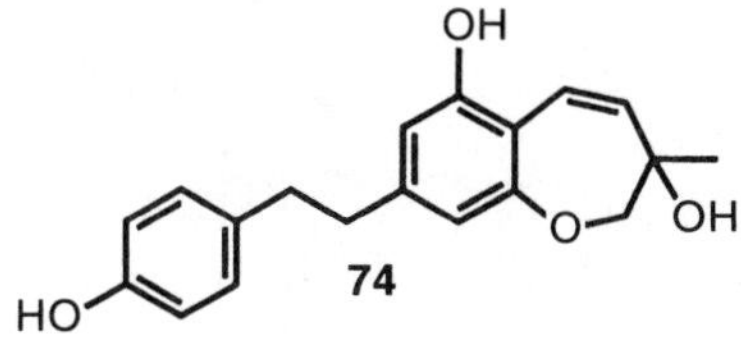

74

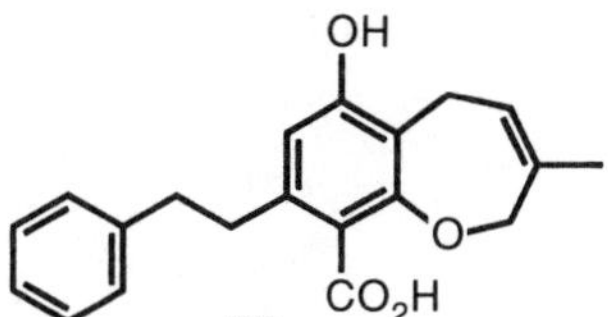

75

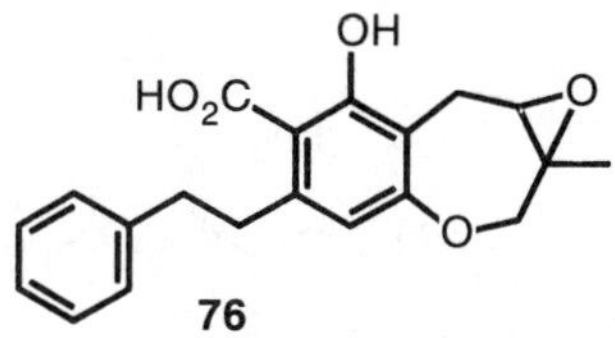

76

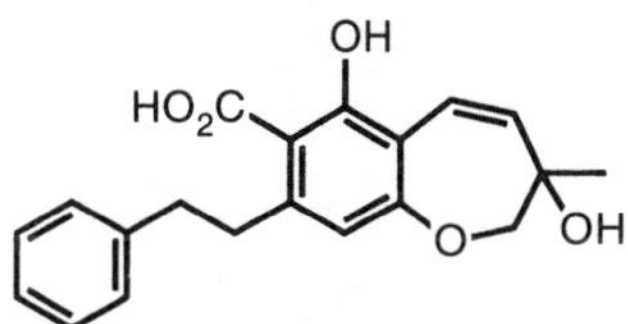

77

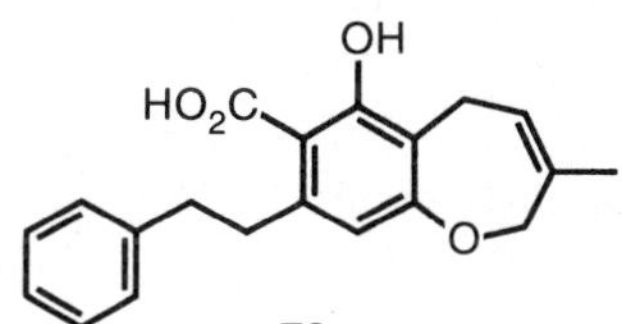

78

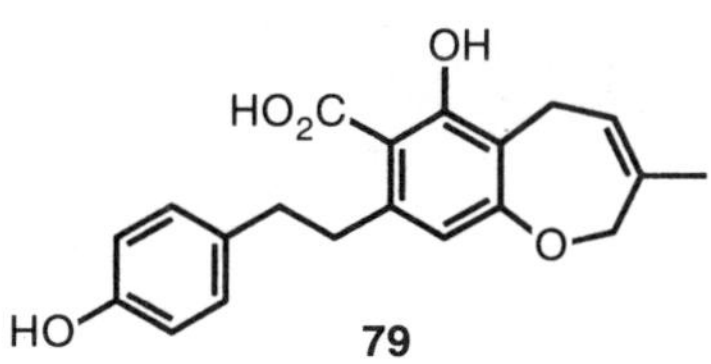

79

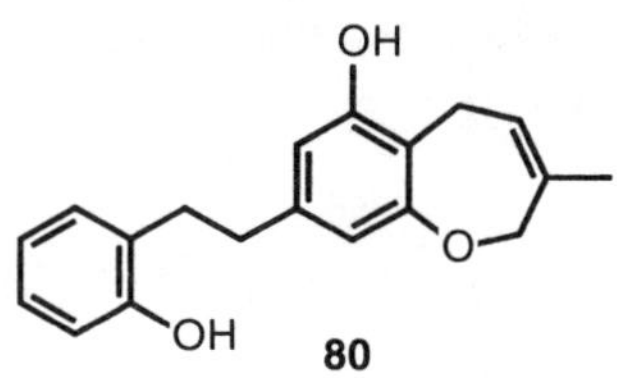

80

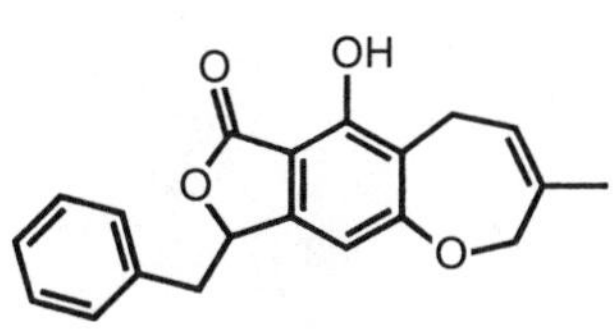

81

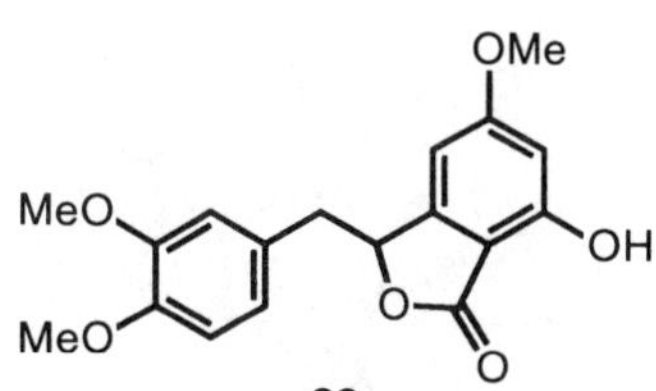

82

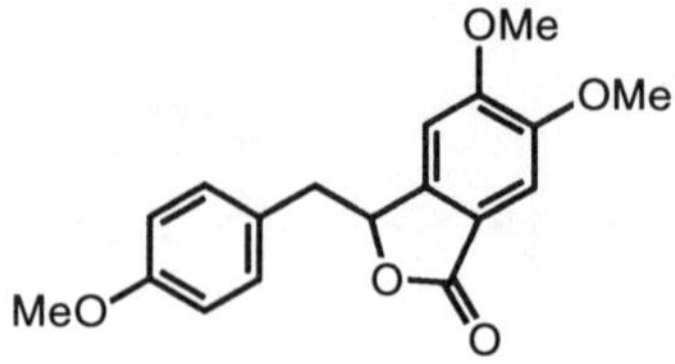

83

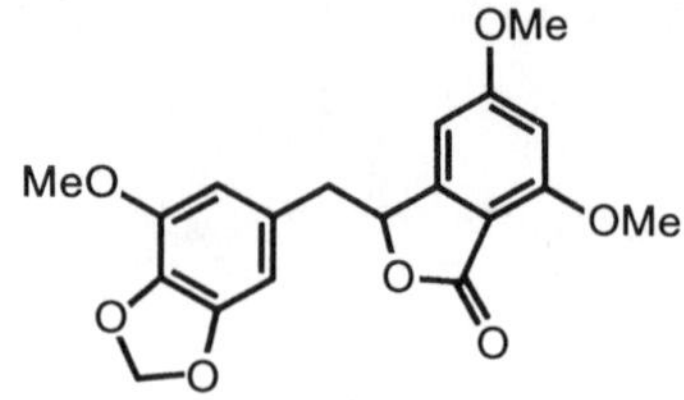

84

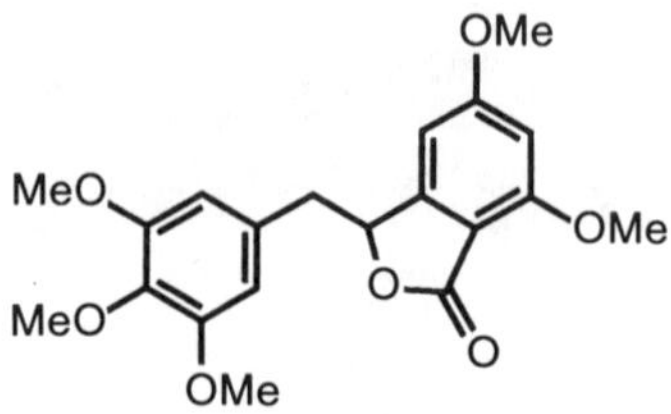

85

Group D (others)

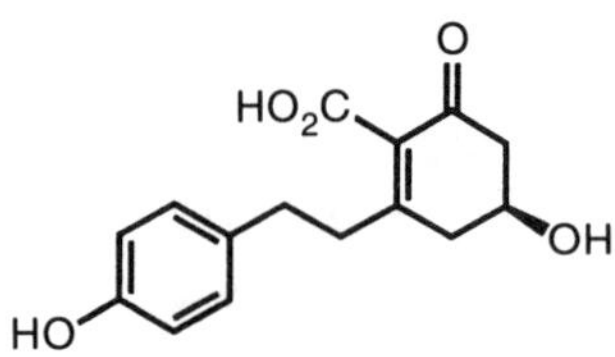

86

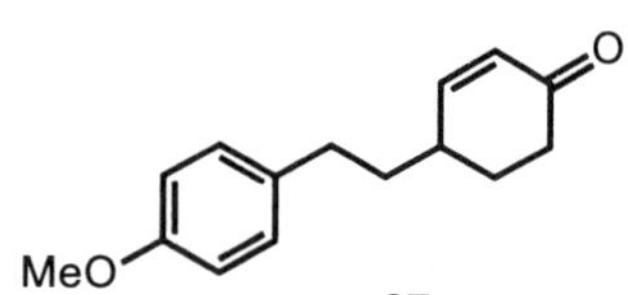

87

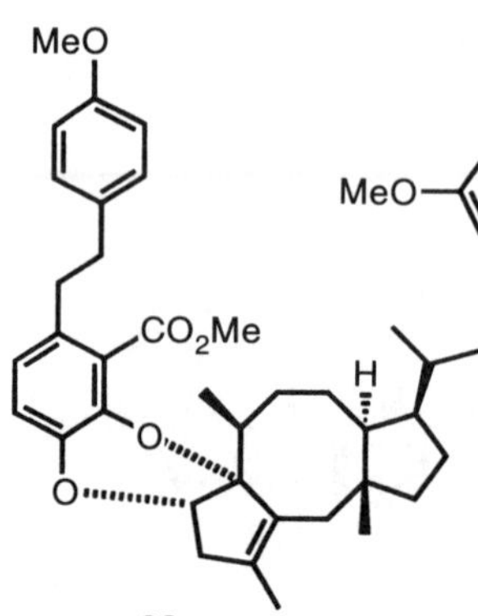

88

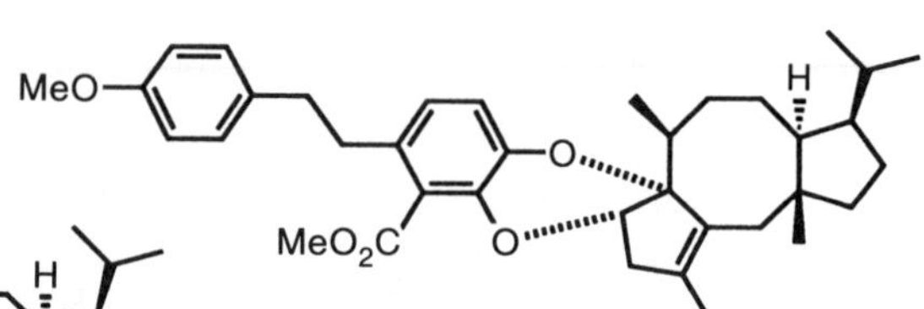

89

Type A

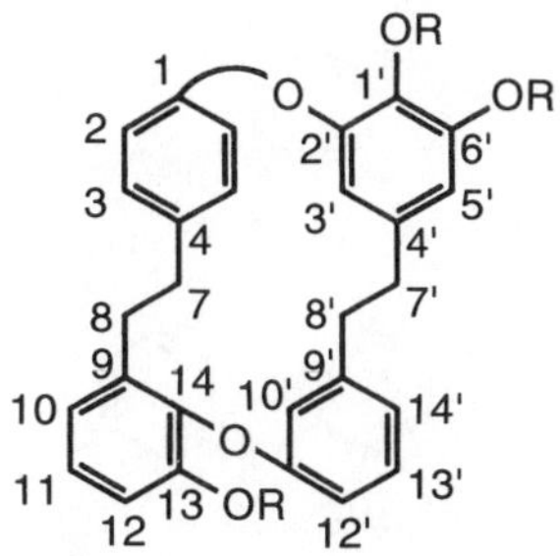

marchantin A **90a** R=H
90b R=Me
90c R=Ac
90d R=1',6'-CH_2-

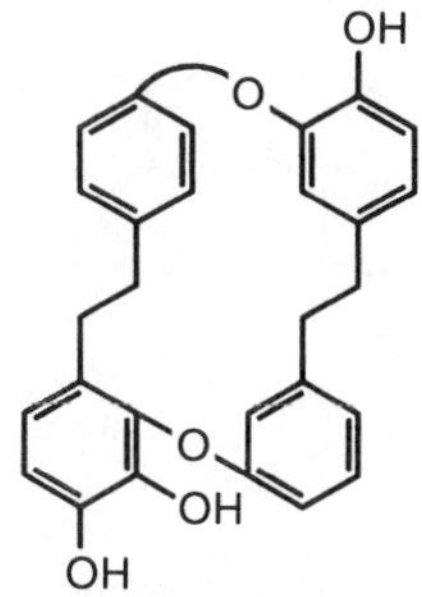

marchantin B **91**

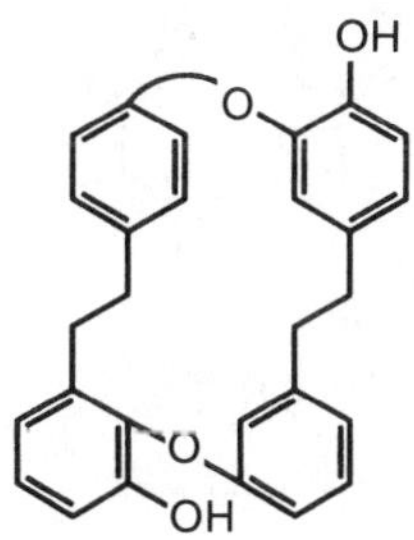

marchantin C **92**

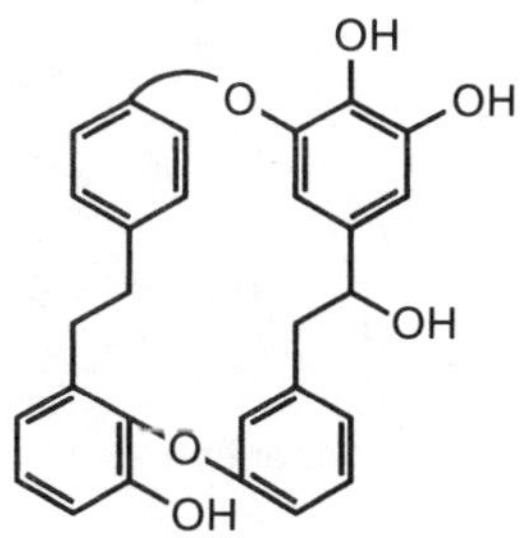

marchantin D **93**

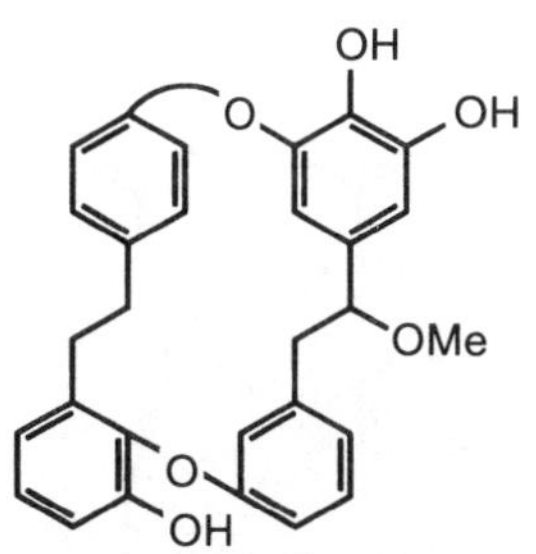

marchantin E **94**

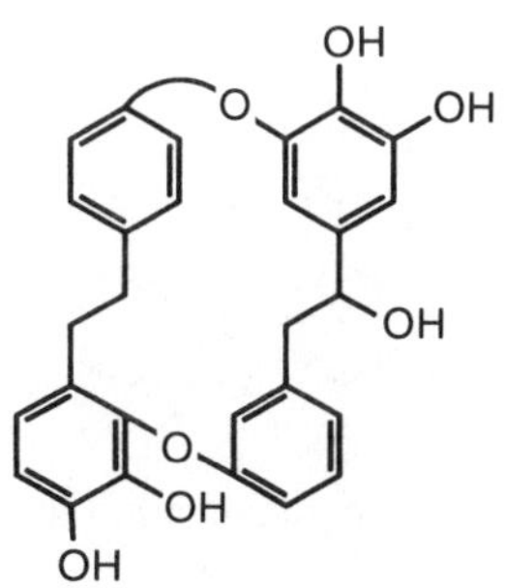

marchantin F **95**

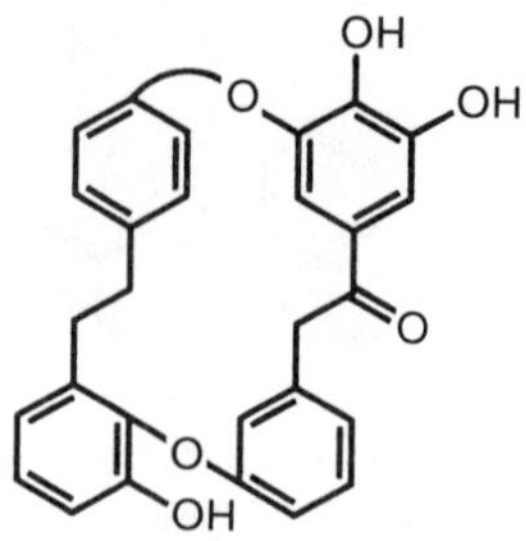

marchantin G **96**

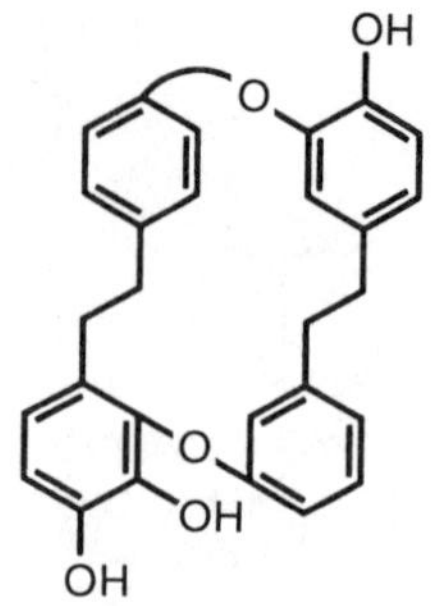

marchantin H **97**

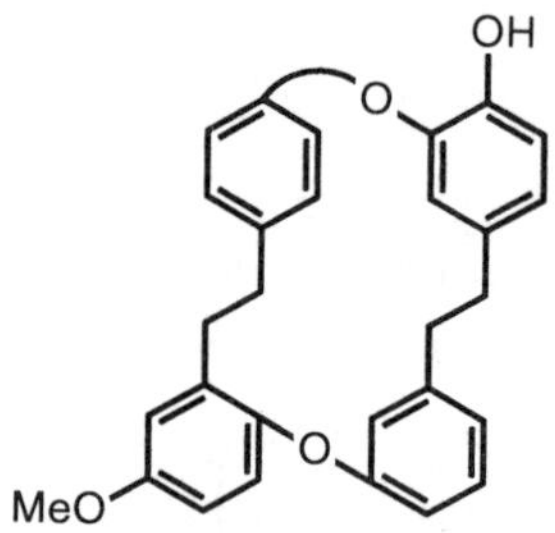

marchantin I **98**

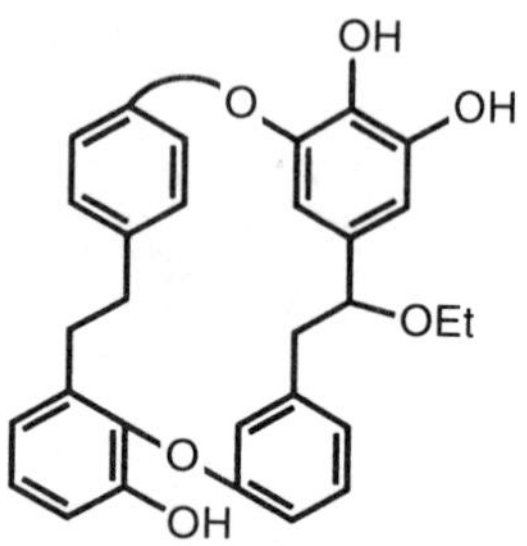

marchantin J **99**

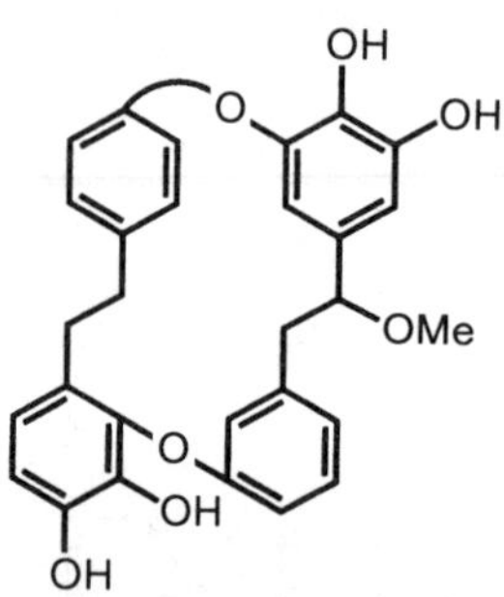

marchantin K **100**

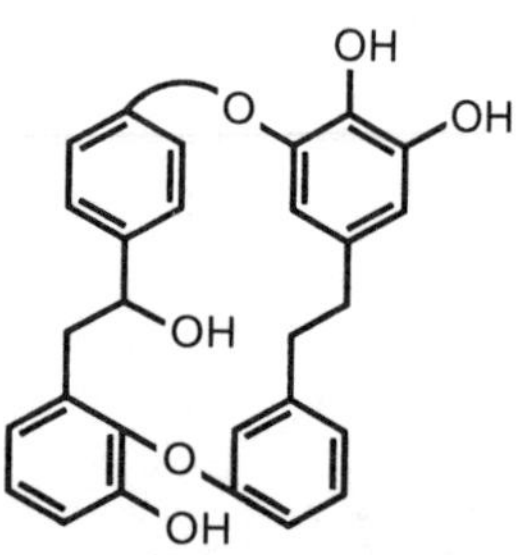

marchantin L **101**

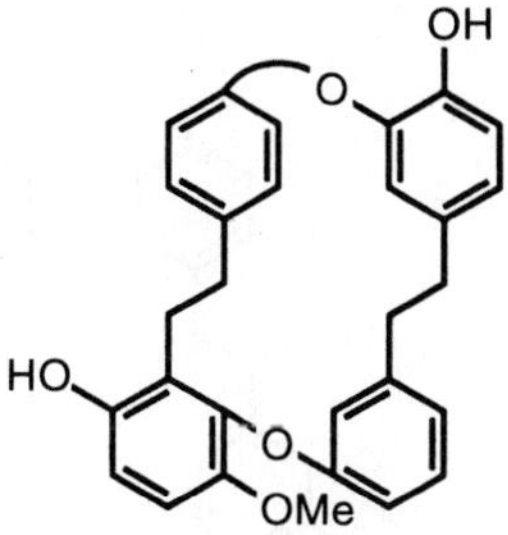

marchantin M **102**

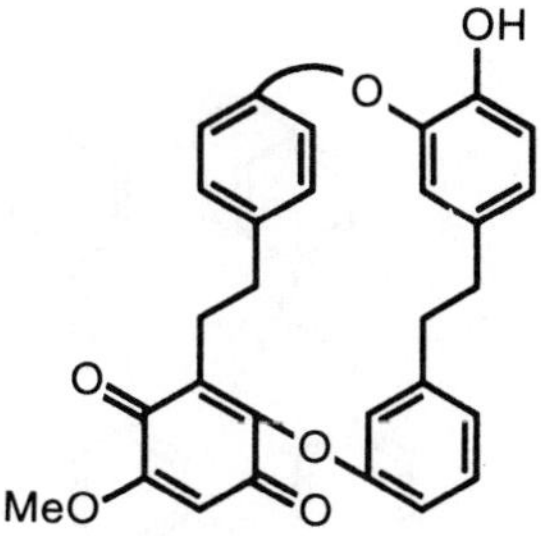

marchantin N **103**

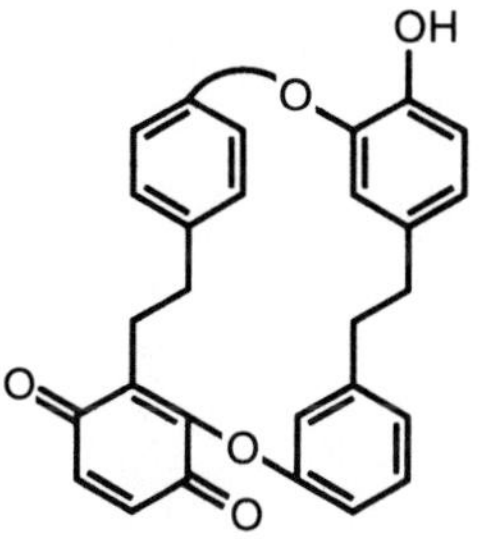

marchantinquinone **104**

Type B

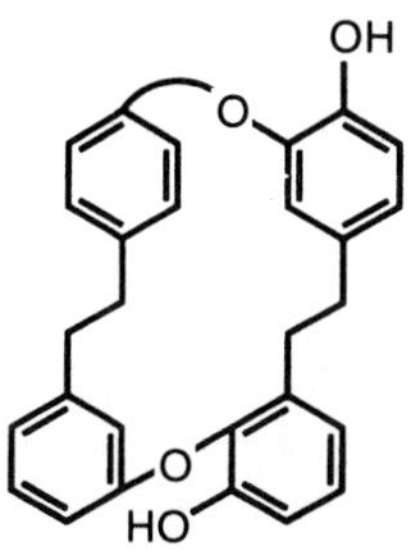

isomarchantin C **105**

Type C

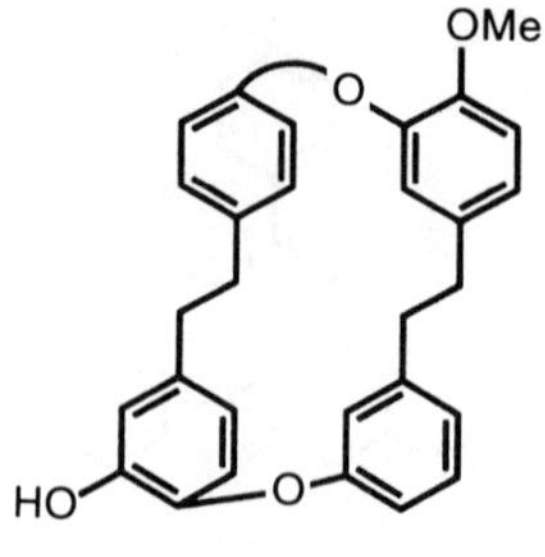

pakyonol **106**

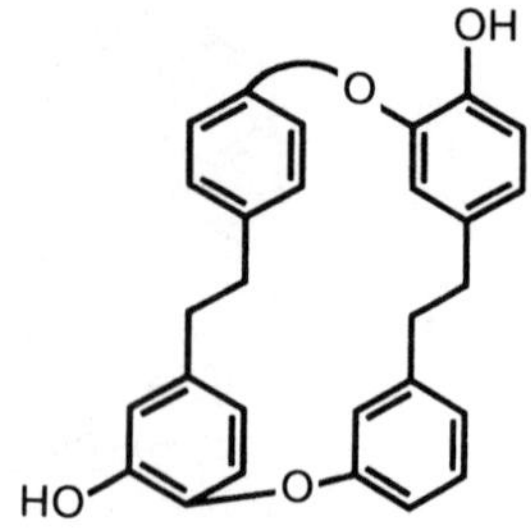

neomarchantin A **107**

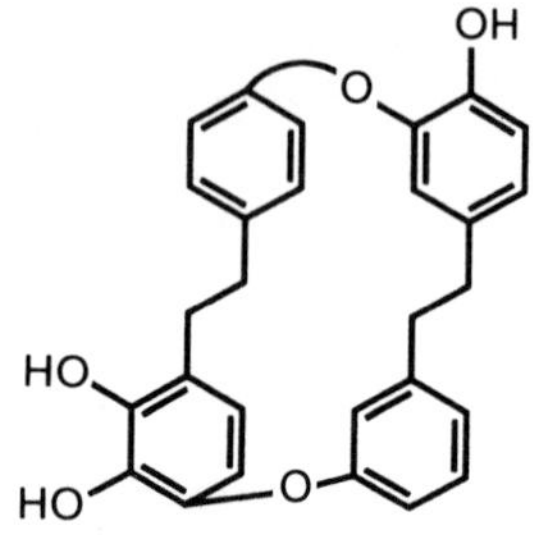

neomarchantin B **108**

Type D

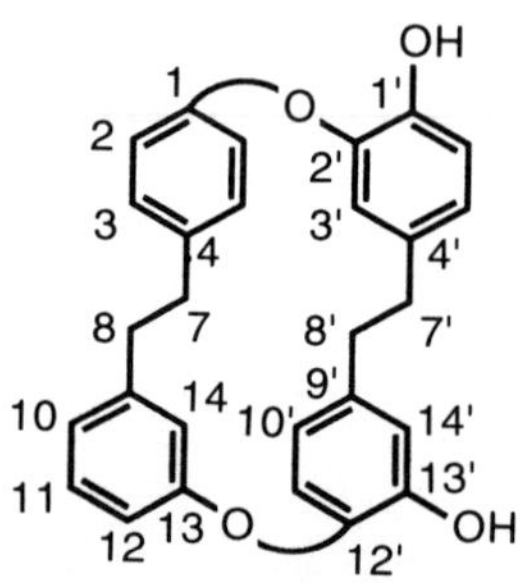

riccardin B **109a** R=H
109b R=Me
109c R=Ac

Type E

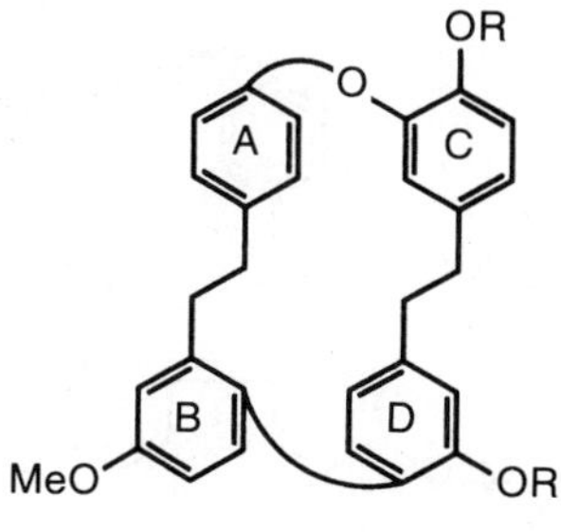

riccardin A **110a** R=H
110b R=Me
110c R=Ac

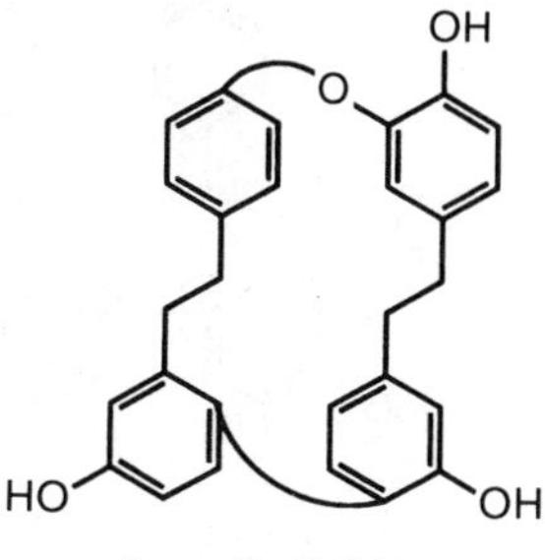

riccardin C **111**

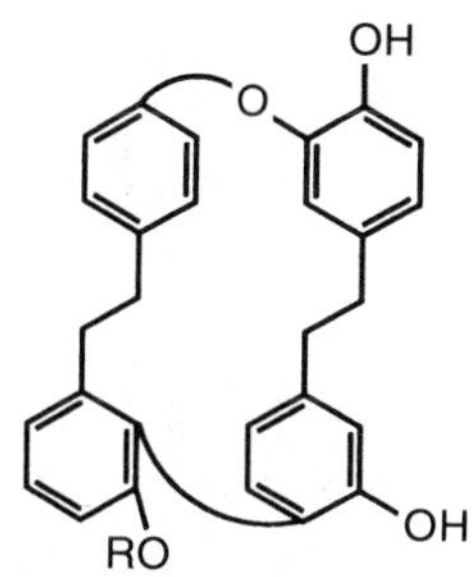

riccardin D **112** R=H
riccardin E **113** R=Me

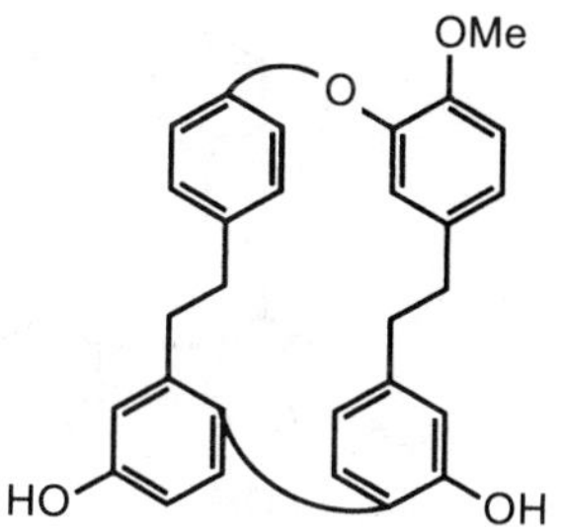

riccardin F **114**

Type F

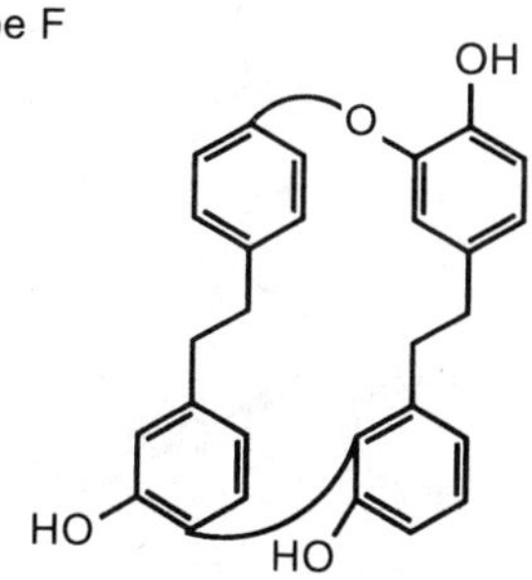

isoriccardin C **115**

Type G

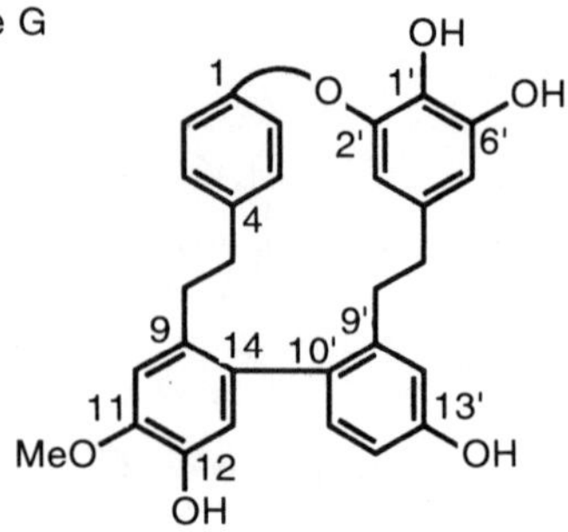

plagiochin A **116**

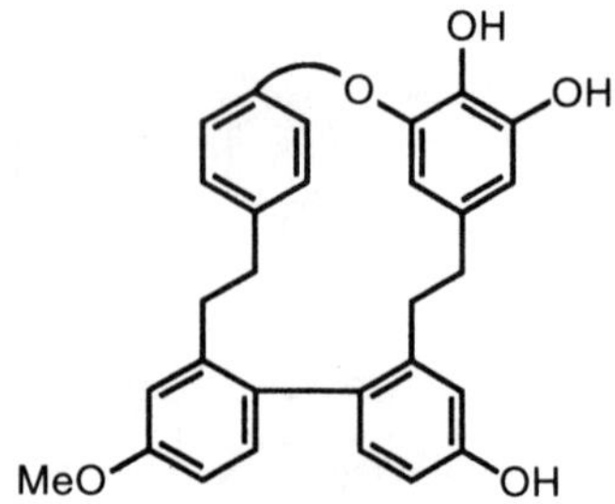

plagiochin B **117**

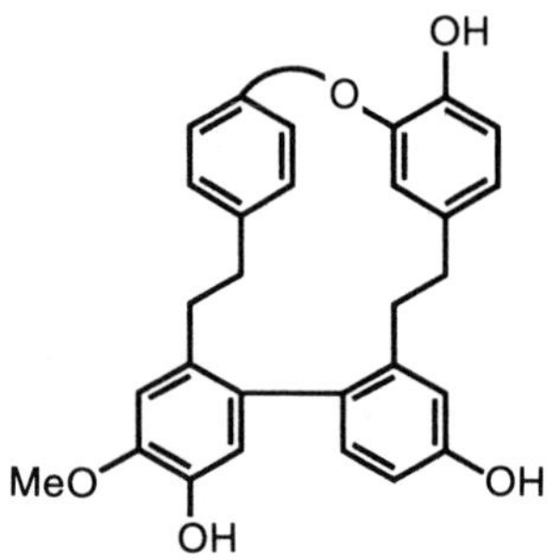

plagiochin C **118**

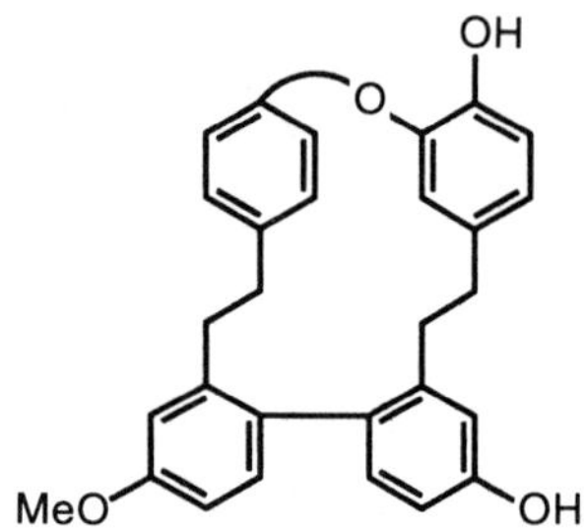

plagiochin D **119**

Type H

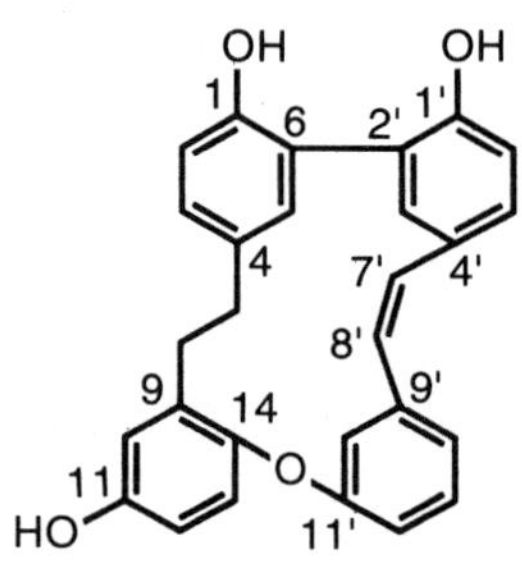

isoplagiochin A **120**

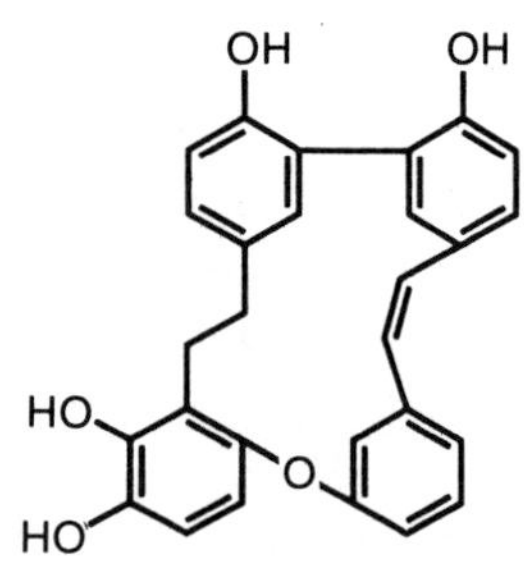

isoplagiochin B **121**

Type I

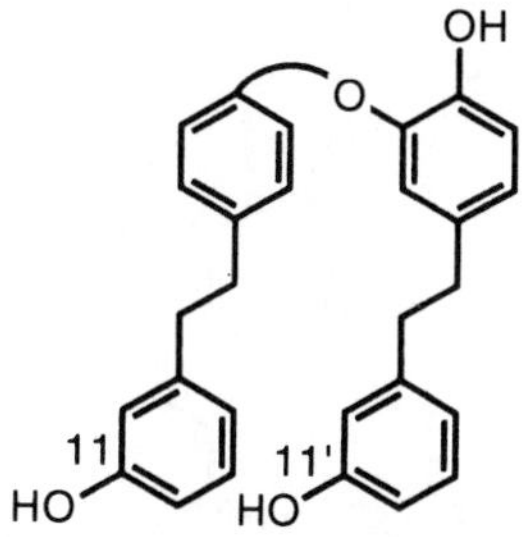

perrottetin E **122**

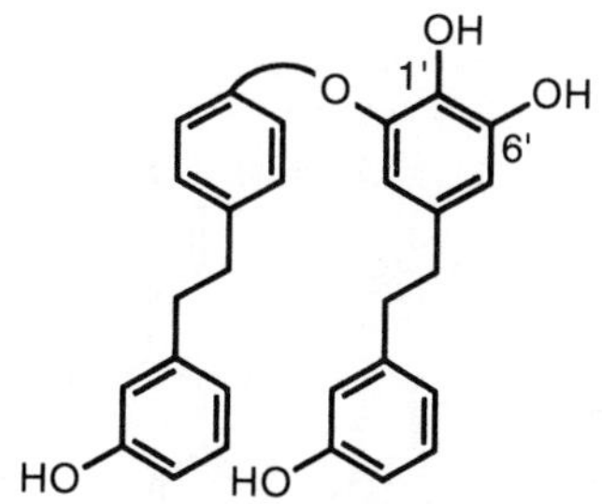

perrottetin F **123**

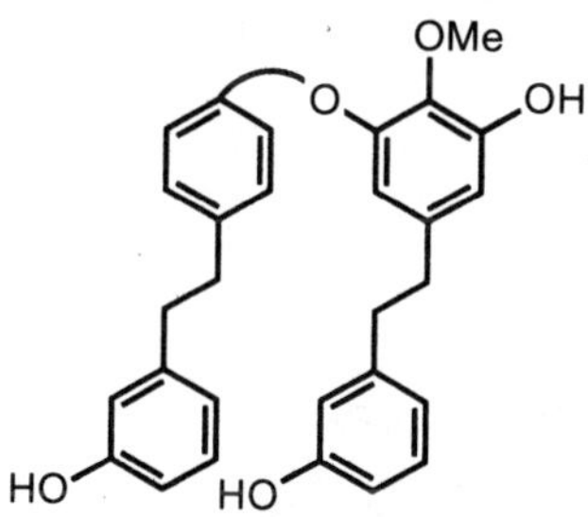

perrottetin G **124**

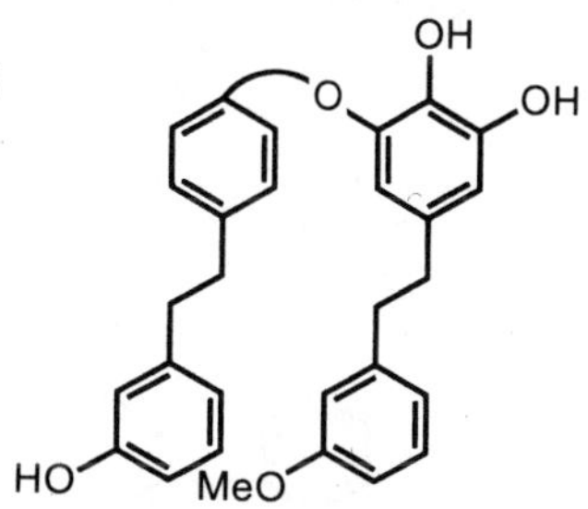

perrottetin E 11'-methyl ether **125**

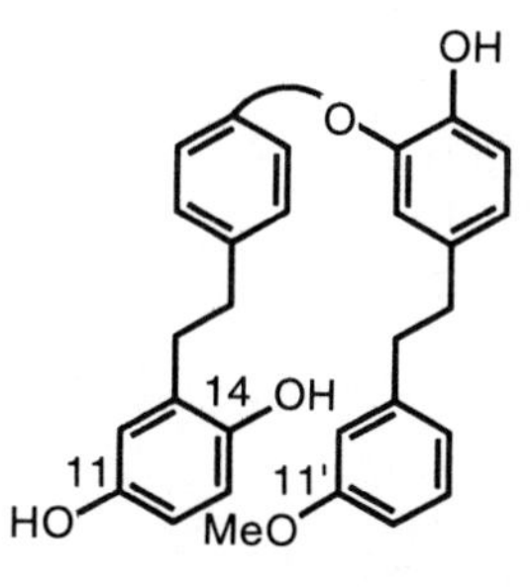

14-hydroxyperrottetin E-
11'-methyl ether **126**

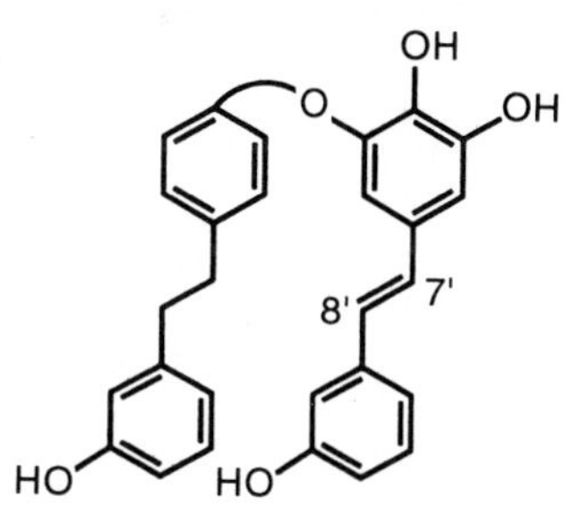

7',8'-dehydroperrottetin F **127**

Others

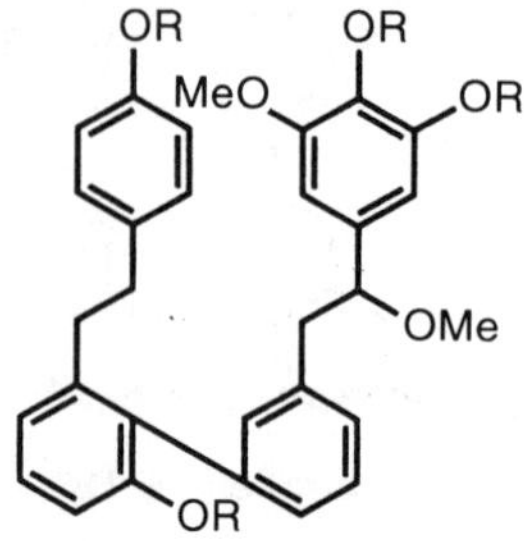

paleatin A **128a** R=H
128b R=Me
128c R=Ac

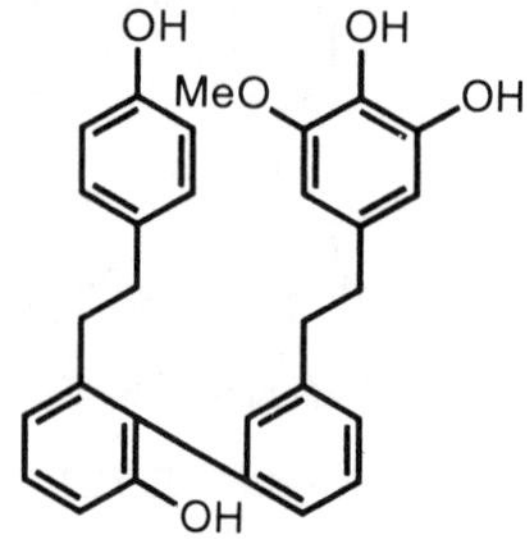

paleatin B **129**

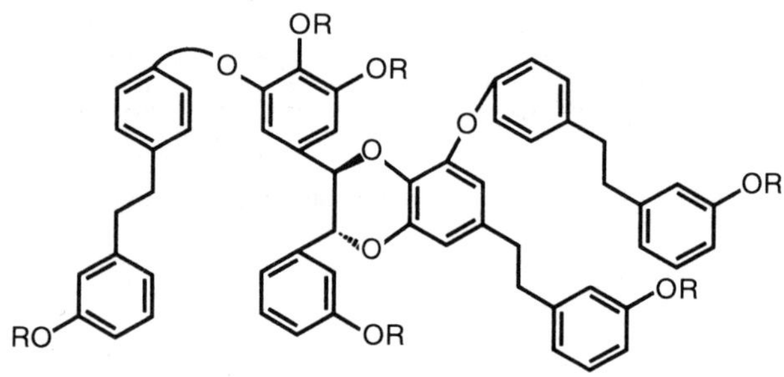

cruciatin **130a** R=H
130b R=Me
130c R=Ac

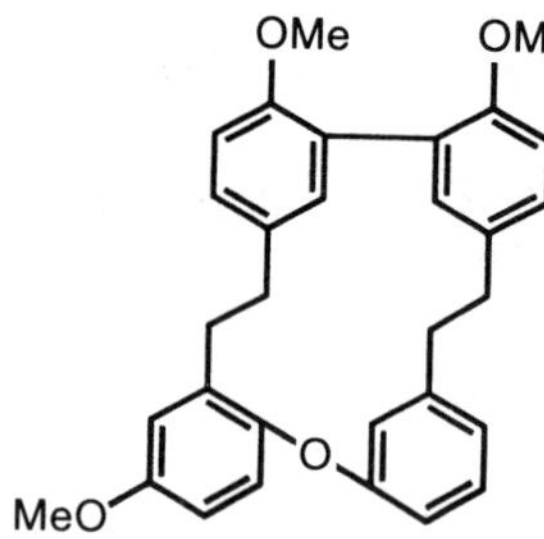

134

Phenanthrenes and dihydrophenanthrenes

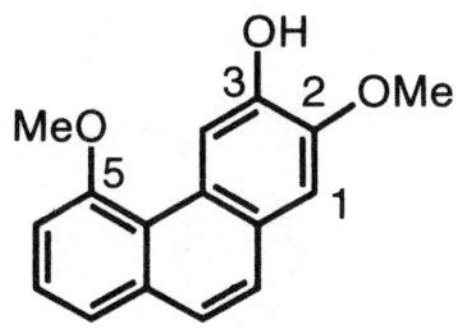

135

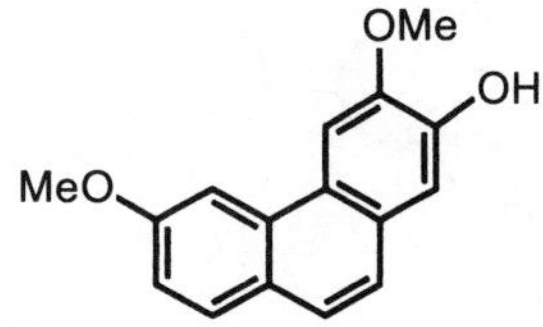

136

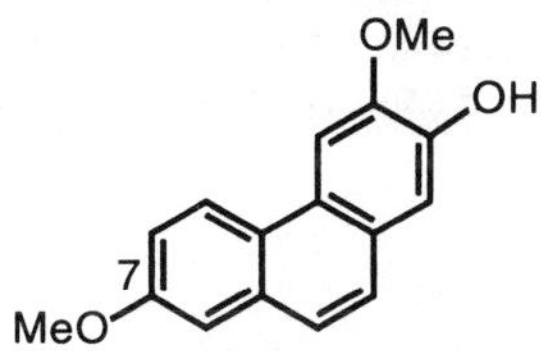

137

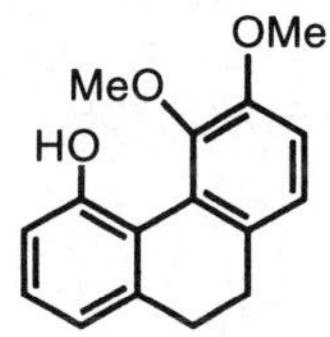

138

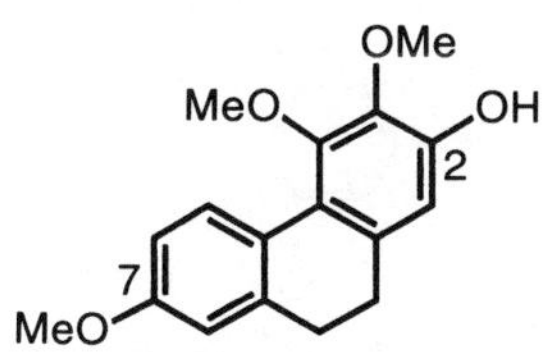

139

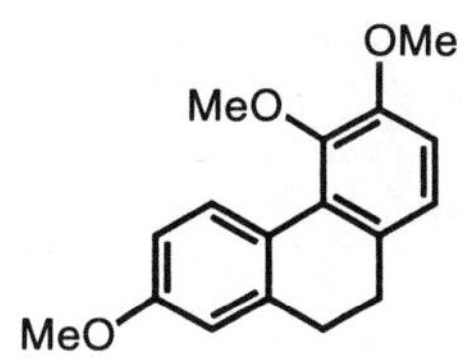

140

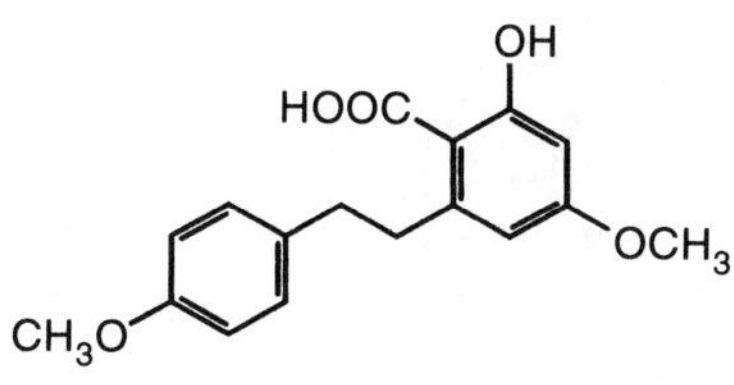

141 notholaenic acid

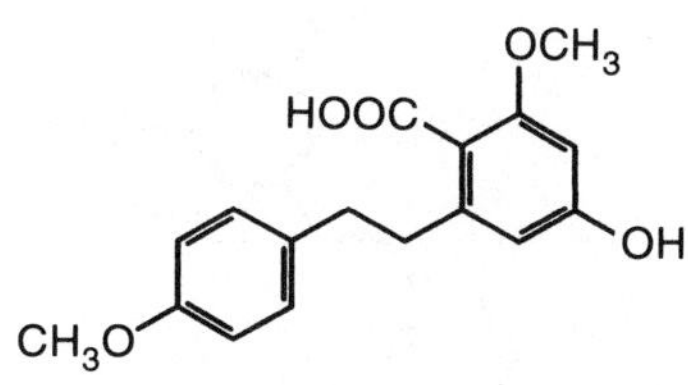

142 isonotholaenic acid

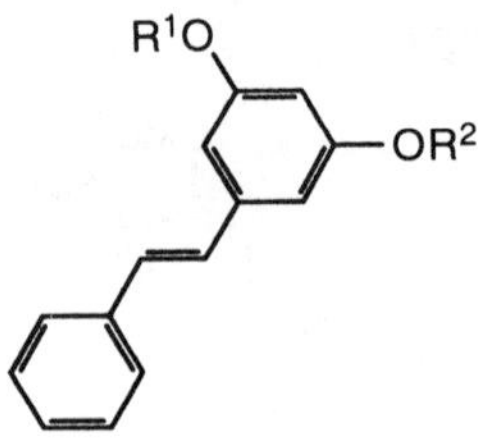

143a R^1 =H; R^2 =H pinosylvin
143b R^1 =H; R^2 =CH_3
143c R^1 =R^2 =CH_3

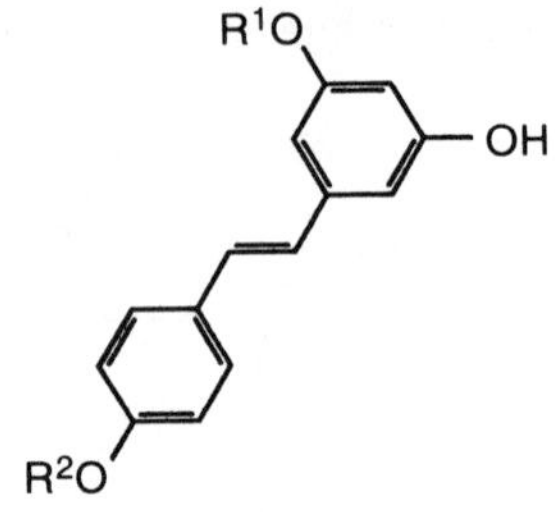

144a R^1 =R^2H resveratrol
144b R^1 =Glu, R^2 =H piceid
144c =H, R^2 =Glu resveratroloside

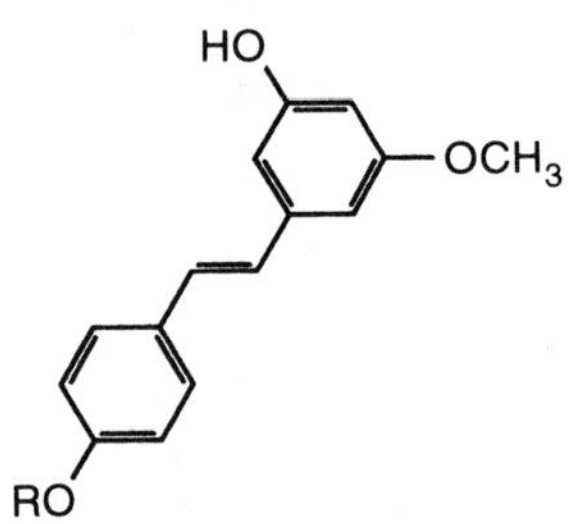

145a R =H pinostilbene
145b R^1 =R^2 =CH_3 pinostilbenoside

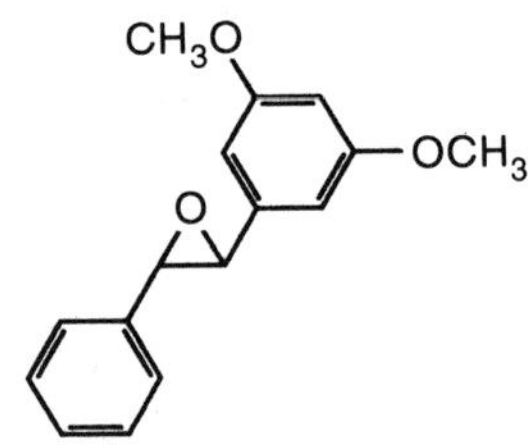

146 3,5-dimethoxystilbene oxide

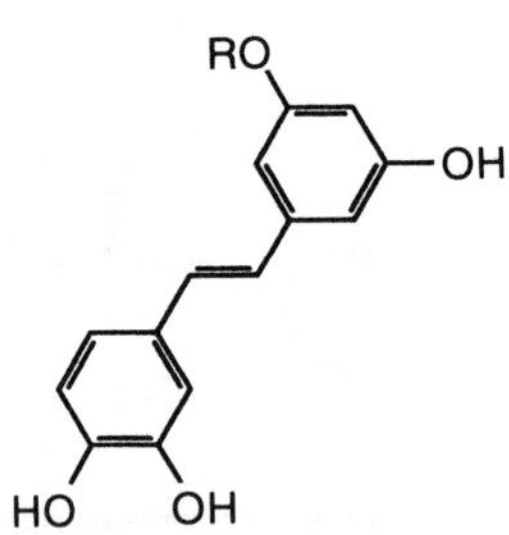

147a R =H piceatannol
147b R^1 =Glu astringin

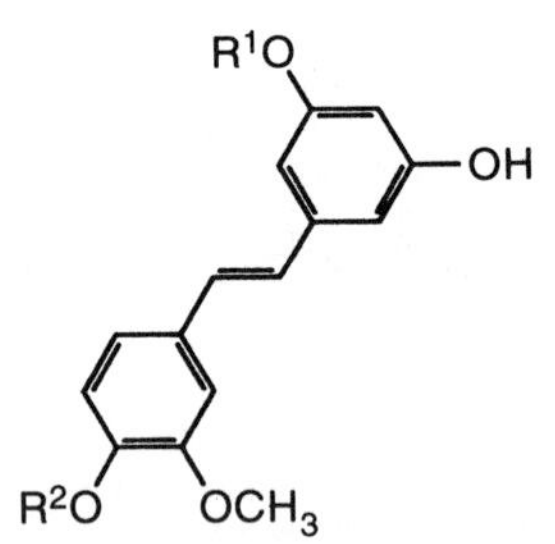

148a R^1 =H =R^2 isohapontigenin
148b R^1 =Glu; R^2 =H isorhapontin

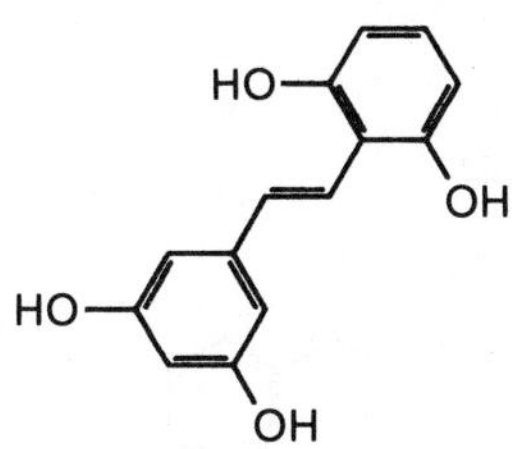

149 gnetol

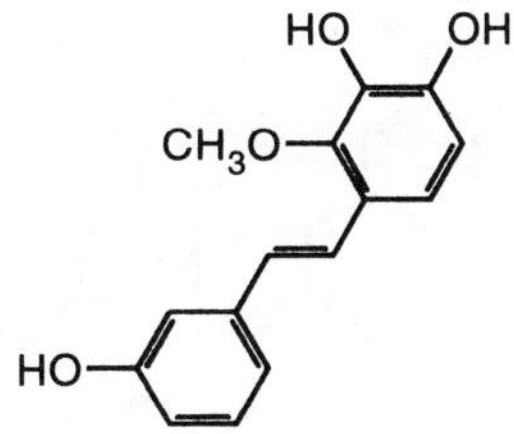

150 3'-methoxy-3,4,5'-trihydroxystilbene

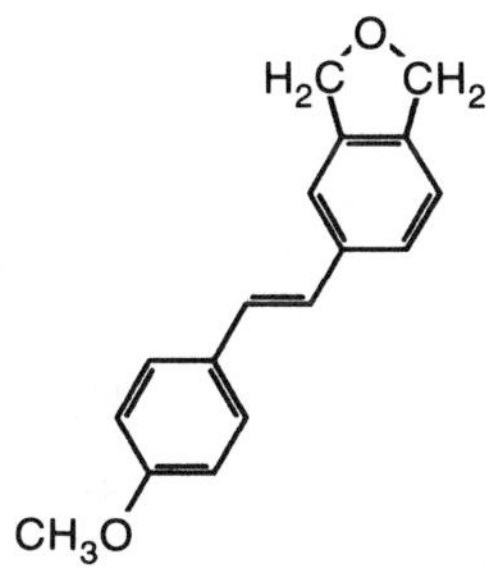

151 4'-methoxy-3,4-methylenedioxystilbene

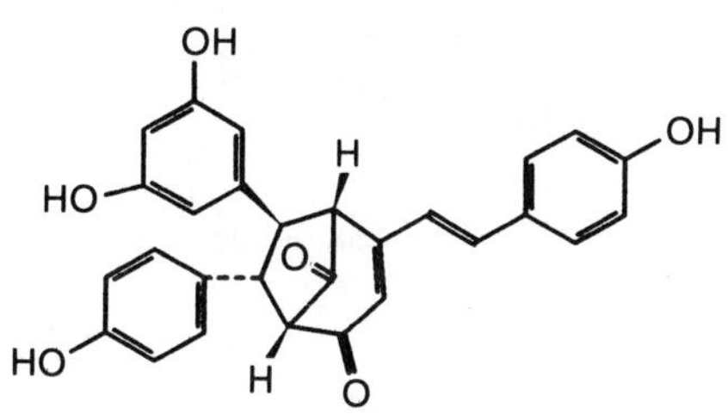

152 gnetin A

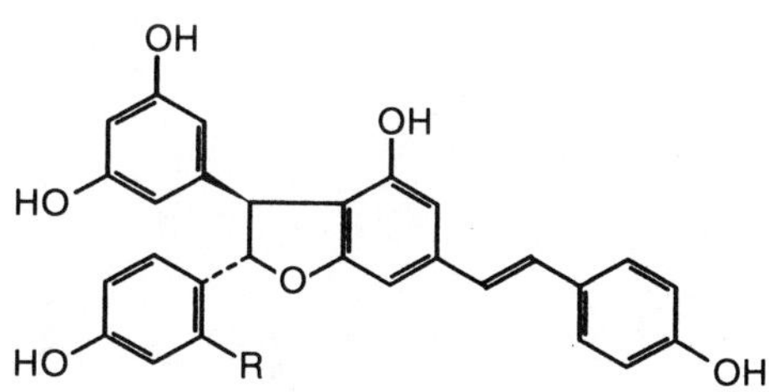

154 R =H, gnetin C
155 R =OH, gnetin D

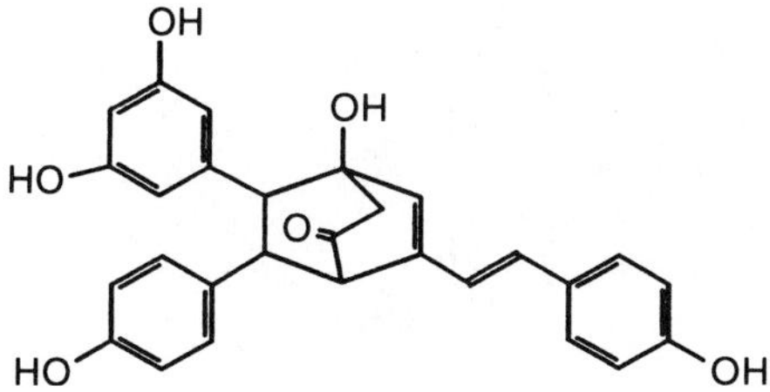

153 gnetin B

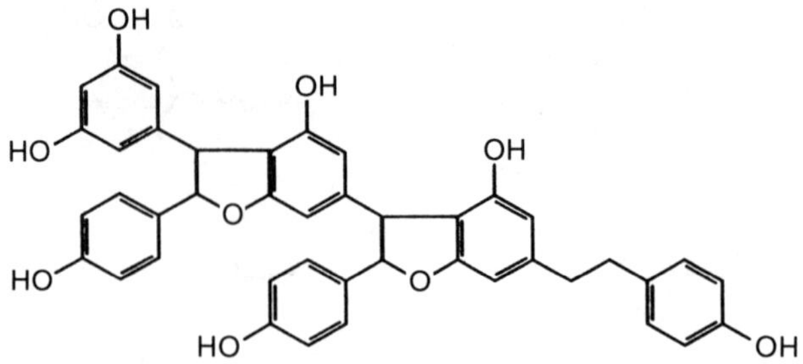

156 gnetin E

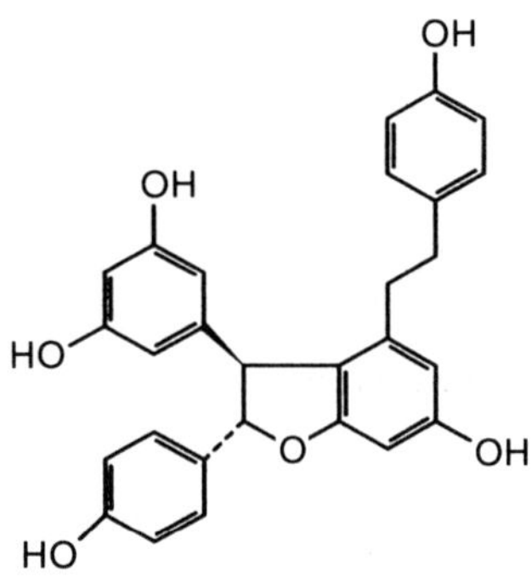

157 gnetin F

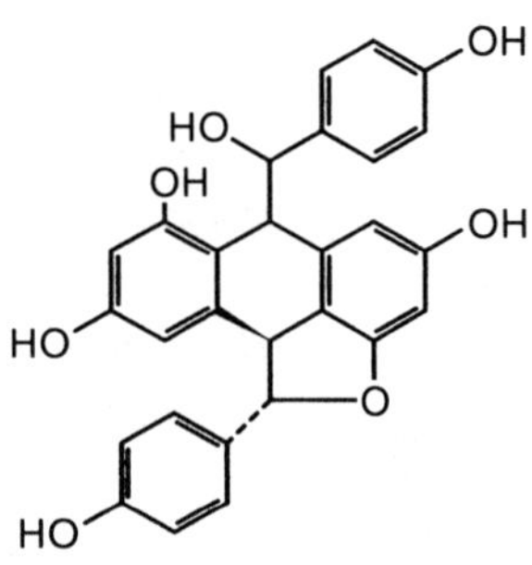

158 gnetin G

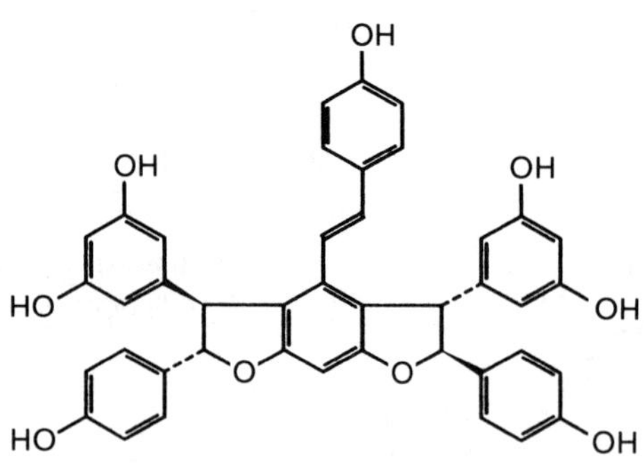

159 gnetin H

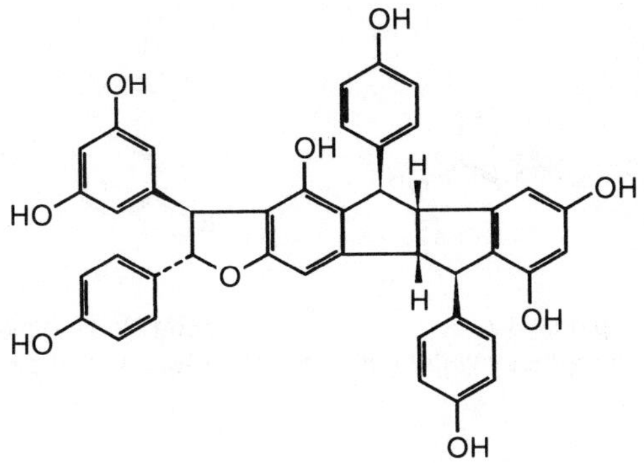

160 gnetin I

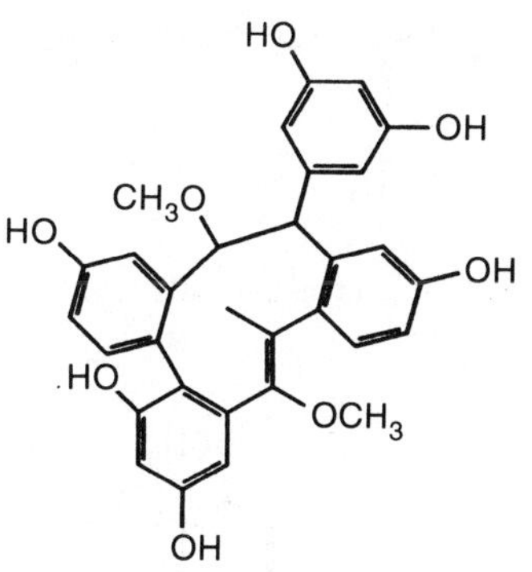

161 gnetifolin C

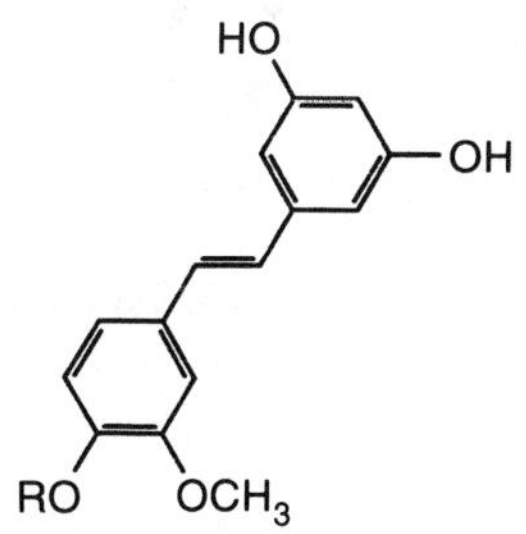

163 R =Glu, gnetifolin E

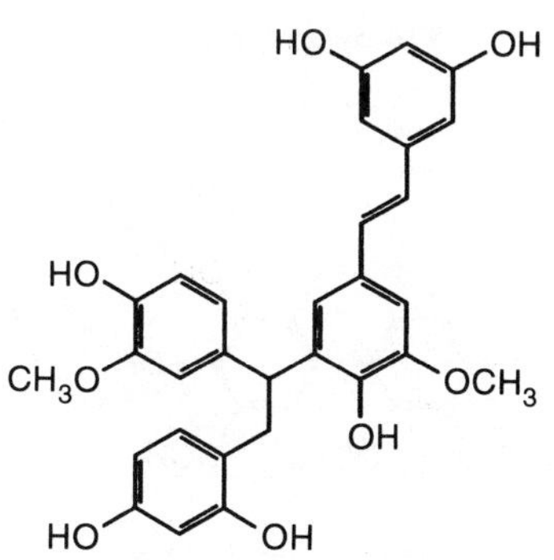

162 gnetifolin D

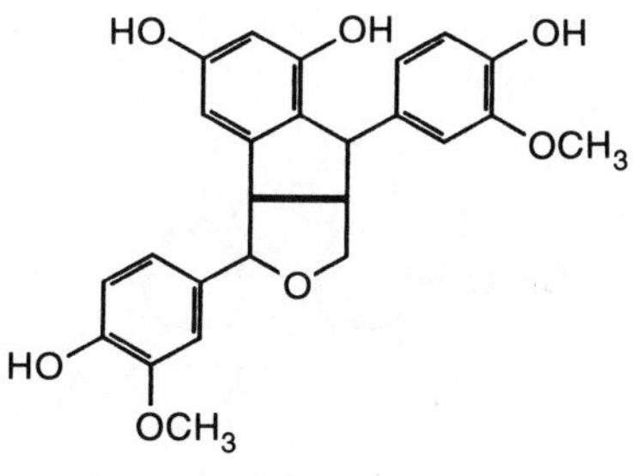

164 gnetifolin F

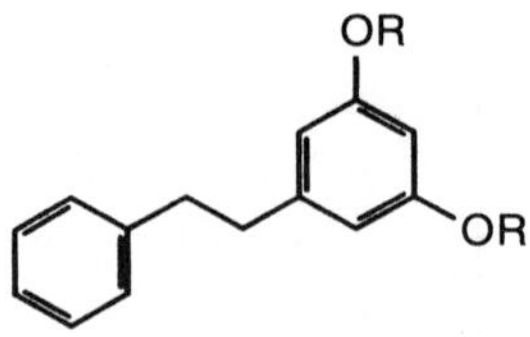

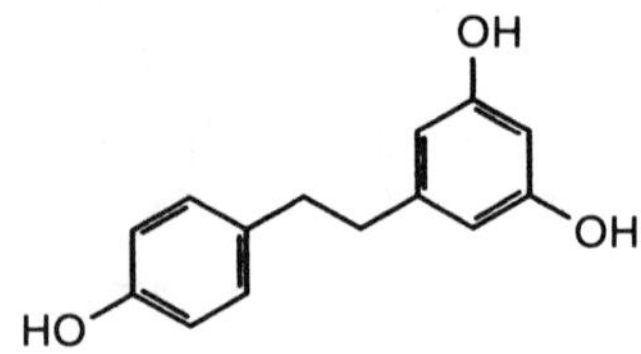

165a R =H, dihydropinosylvin
165b R =CH_3, dihydropinosylvin dimethyl ether

166a R =H, dihydroresveratrol
166b R =OH dihydrooxyresveratrol

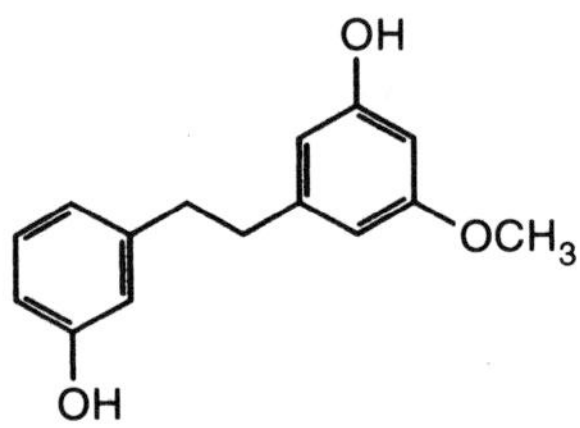

167 batatasin III

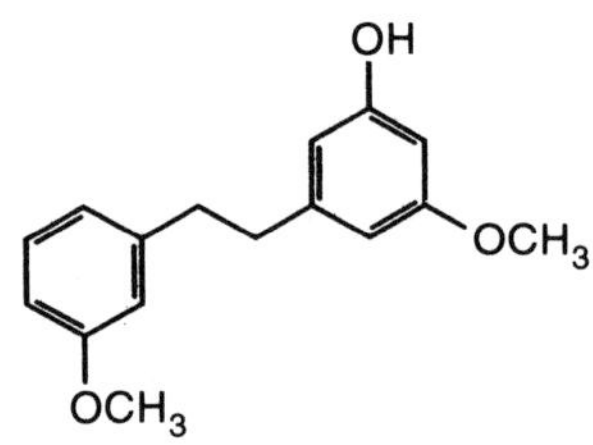

168 3′-*O*-methyl batatasin III

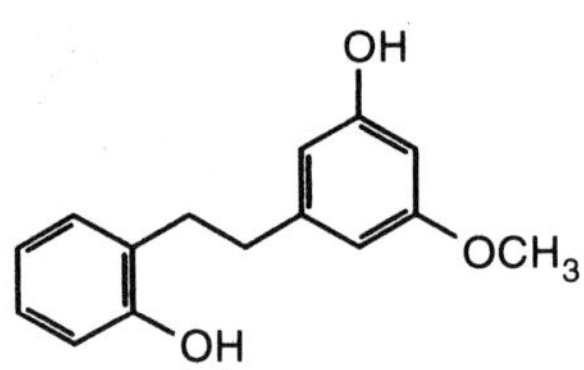

169 batatasin IV

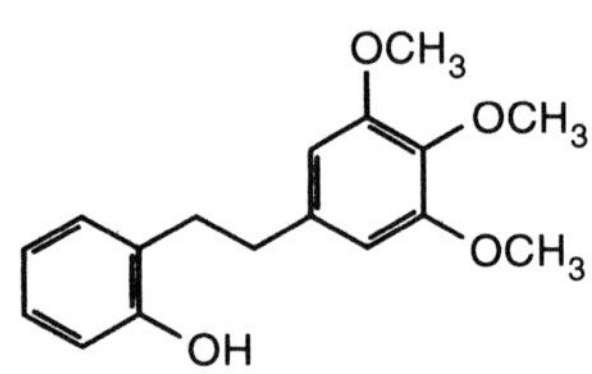

170 batatasin V

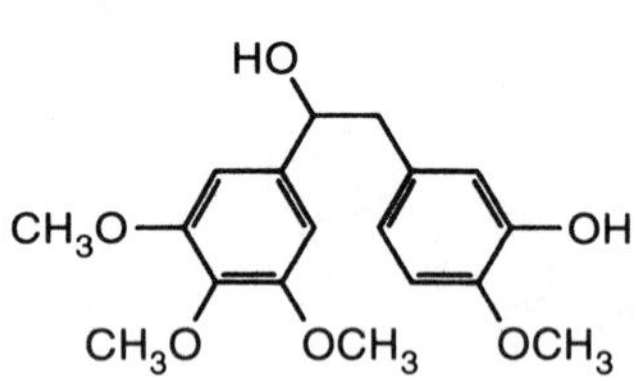

171 combretastatin

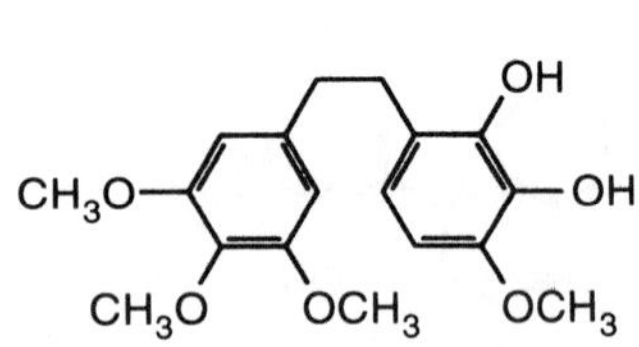

172 combretastatin B-1

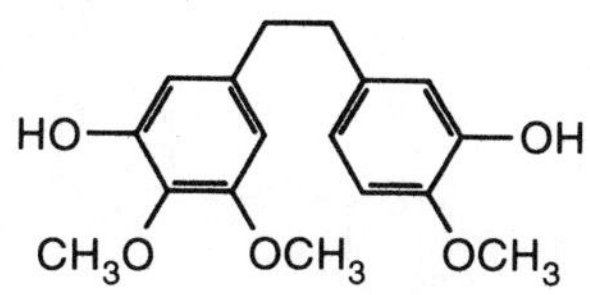

173 combretastatin B-2

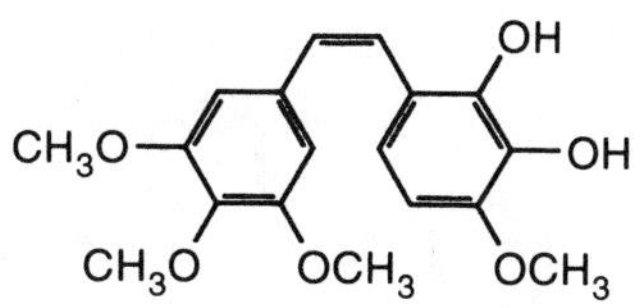

174 combretastatin A-1

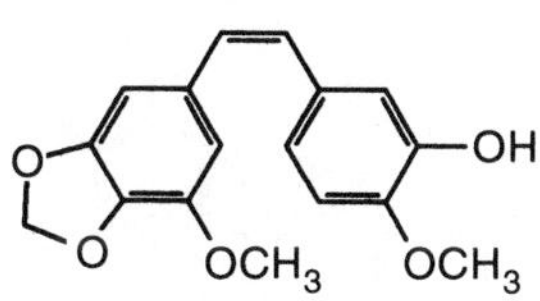

175 combretastatin A-2

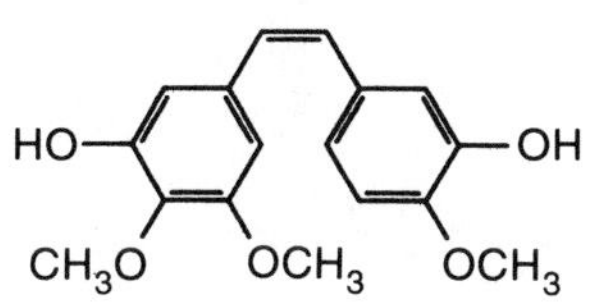

176 combretastatin A-3

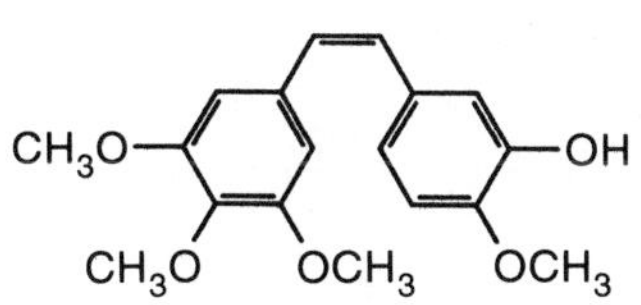

177 combretastatin A-4

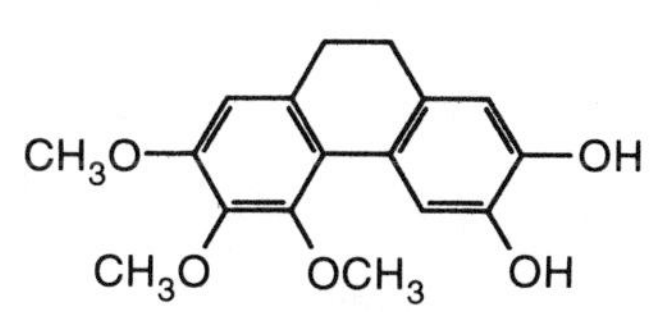

178 6,7-dihydroxy-2,3,4-trimethoxy-9,10-dihydrophenanthrene

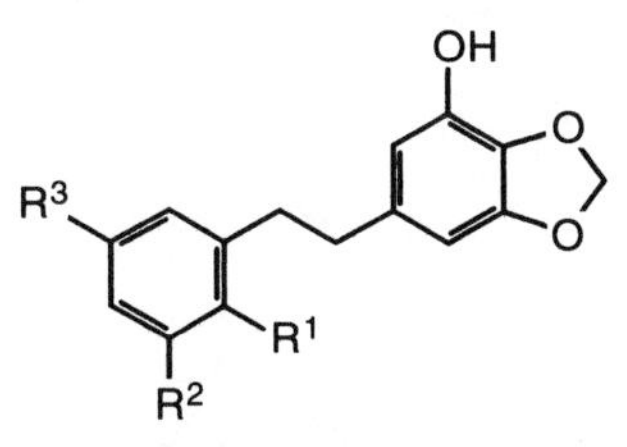

179a $R^1 = R^3 = H$, $R^2 = OH$, cirrhopetalinin

179b $R^1 = OH$, $R^2 = OCH_3$, $R^3 = H$, cirrhopetalidin

179c $R^1 = OH$, $R^2 = H$, $R^3 = OCH_3$, cirrhopetalidinin

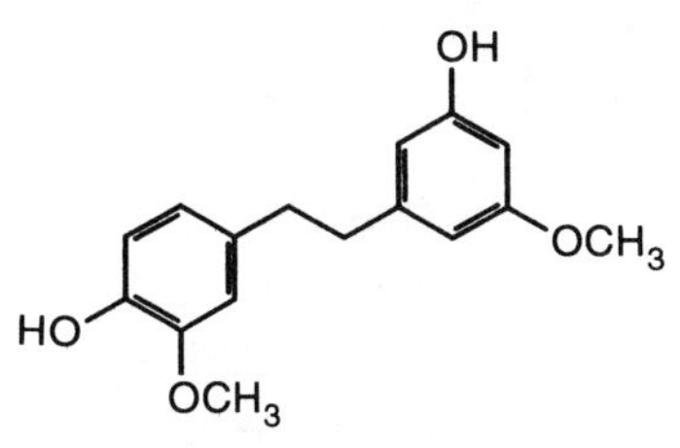

180 gigantol

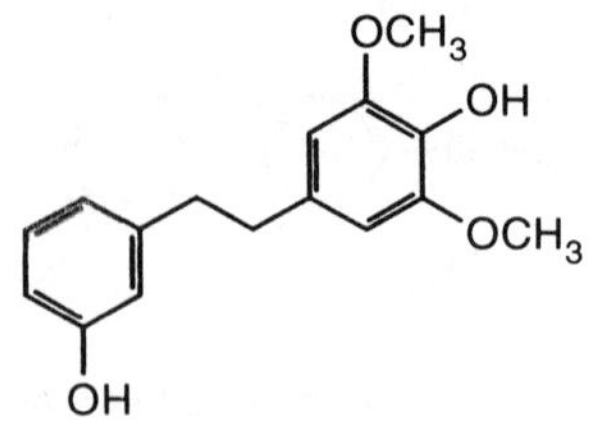

181 aloifol I

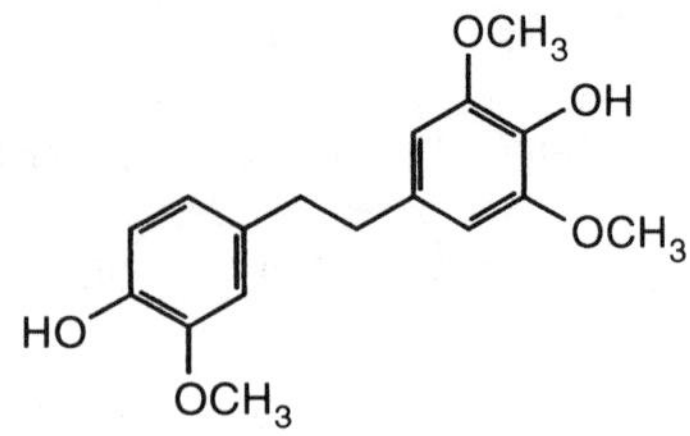

182 aloifol II

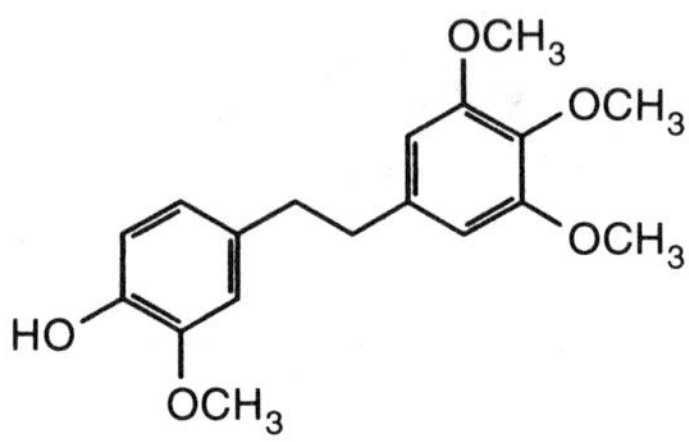

183 crepidatin

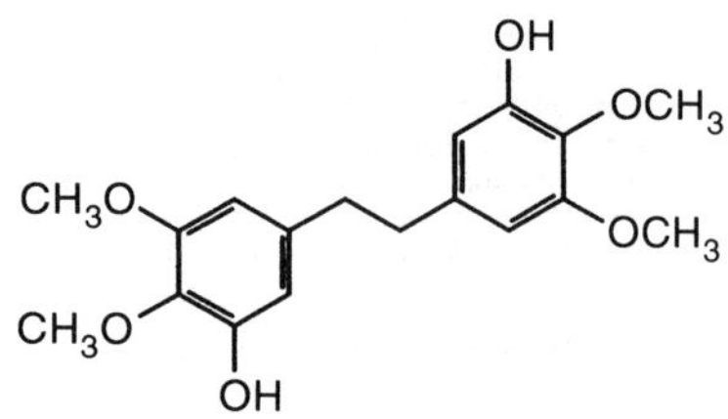

184 cumulatin

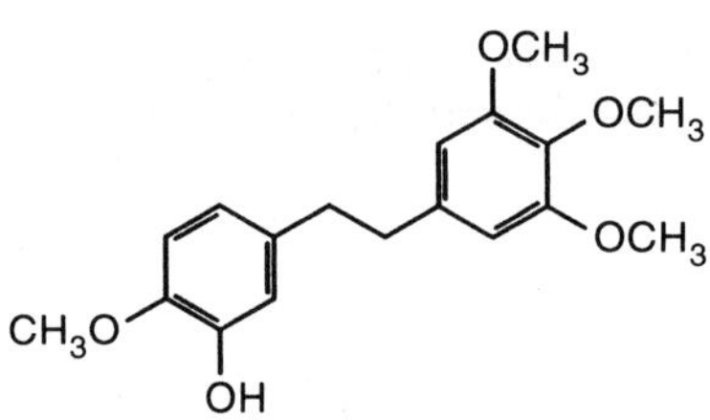

185 erianin

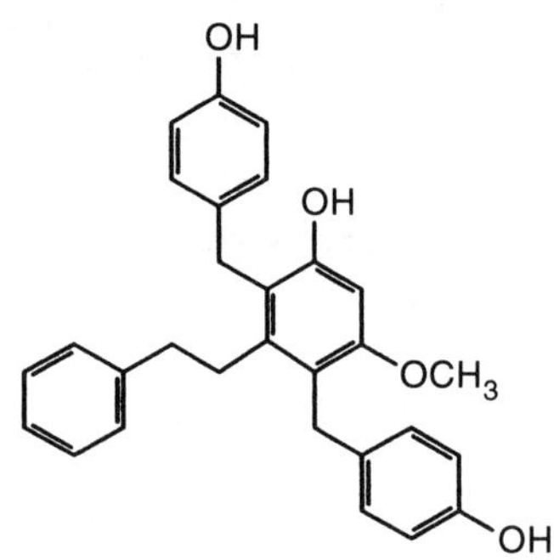

186 arundin

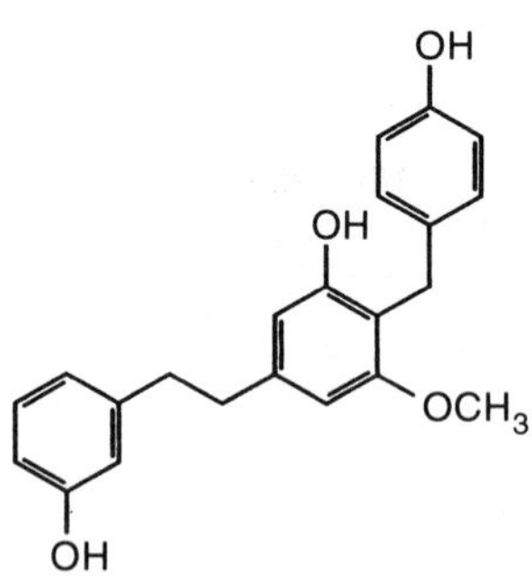

187 arundinin

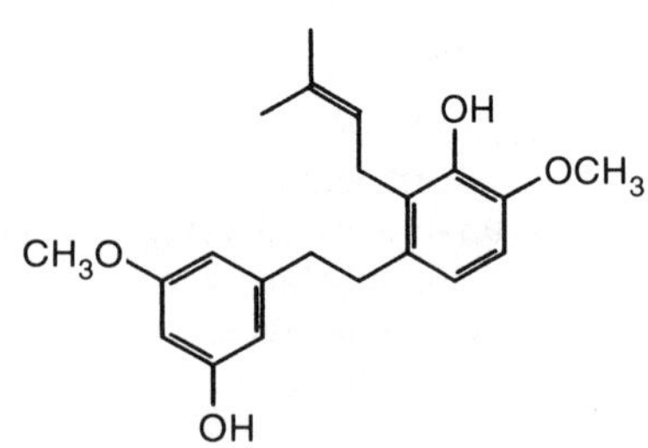

188 canniprene

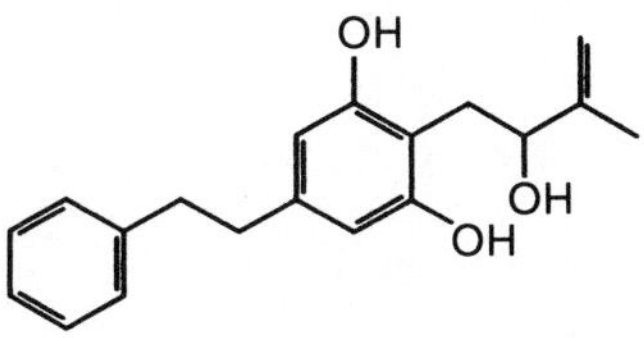

189 glepidotin C

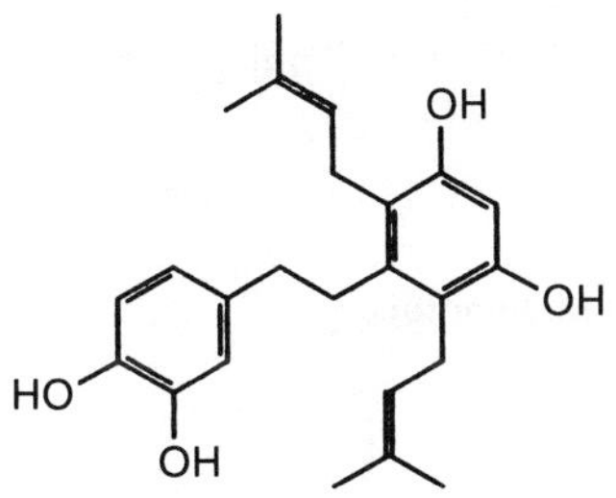

190 gancaonin R

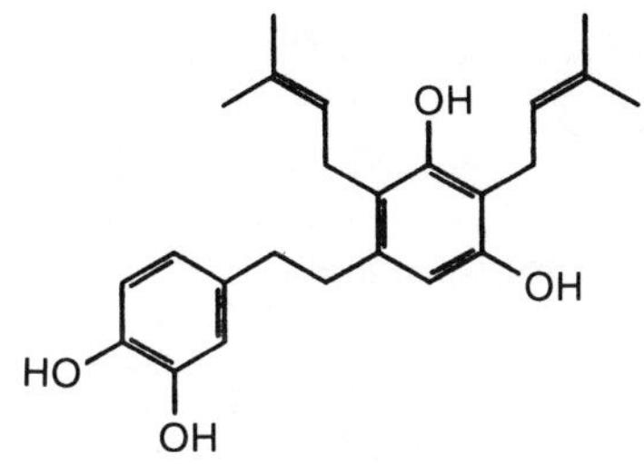

191 gancaonin S

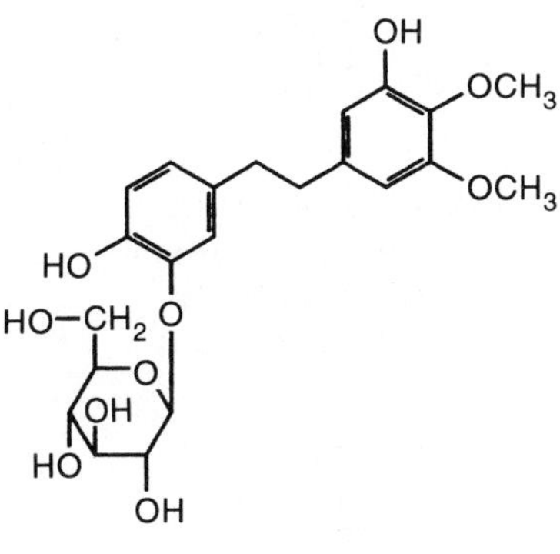

192 icariside A_4

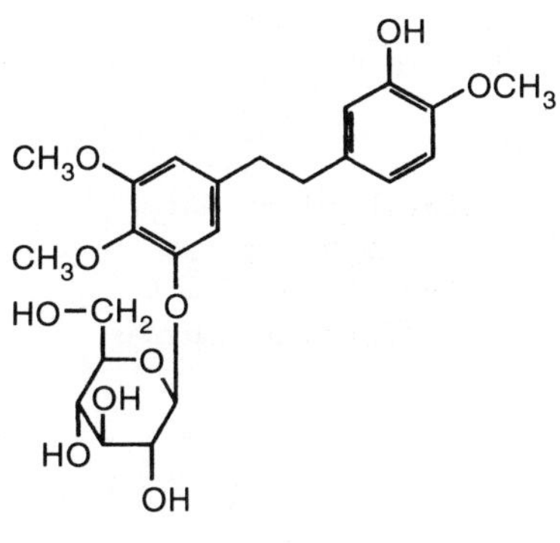

193 icariside A_6

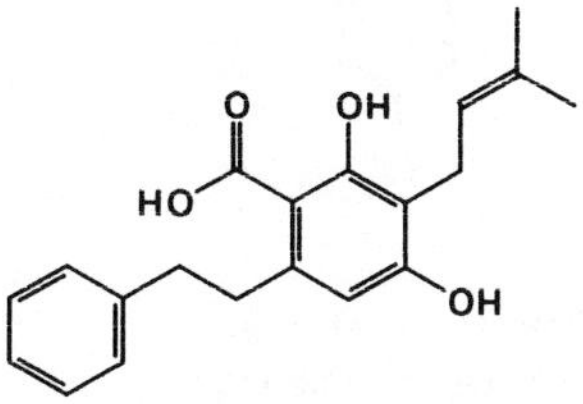

194

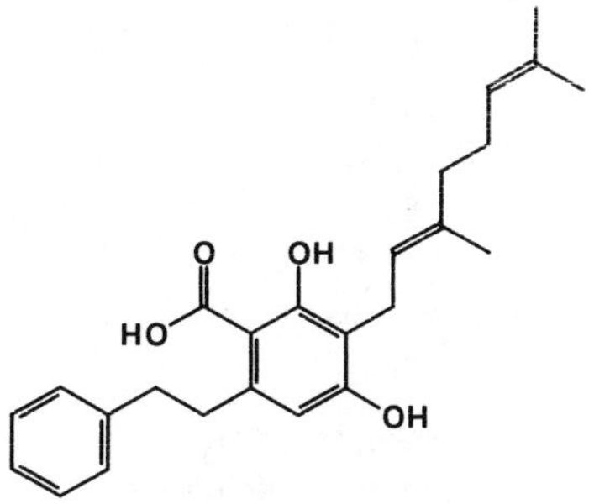

195

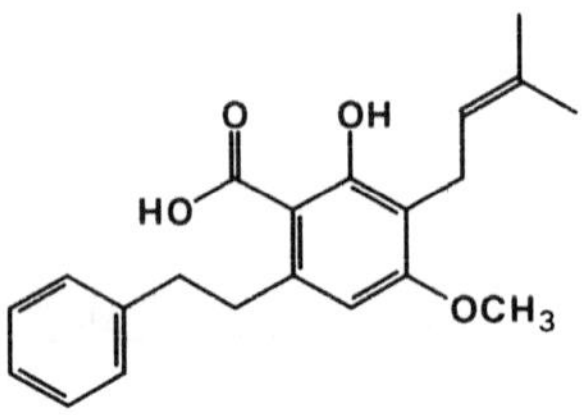

196 **amorfrutin A**

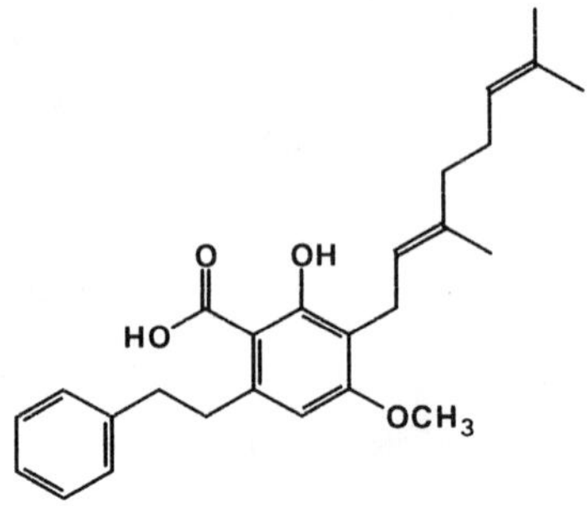

197 **amorfrutin B**

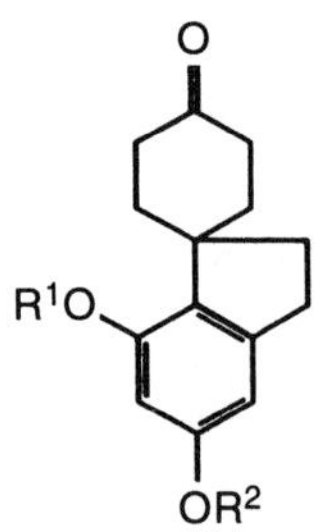

198a R^1 =H, R^2 =CH_3,
cannabispirone
198b R^1 =CH_3, R^1 =H
iso-cannabispirone

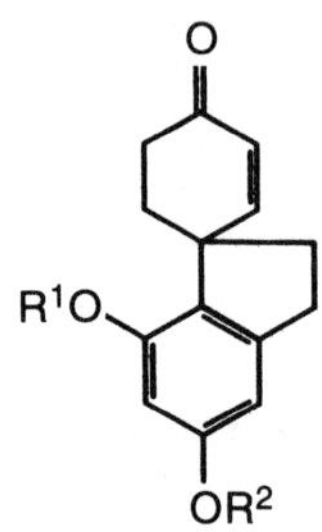

199a R^1 =H, R^2 =CH_3,
cannabispirenone-A
199b R^1 =CH_3, R^2 =H
cannabispirone-B

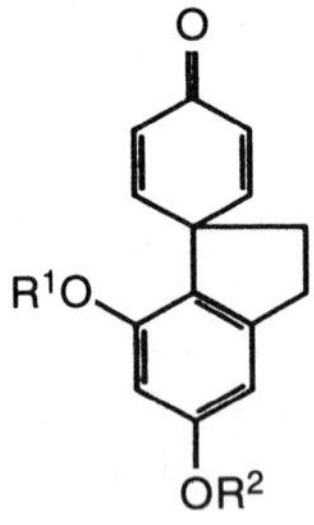

200 R^1 =H, R^2 =CH_3,
cannabispiradienone

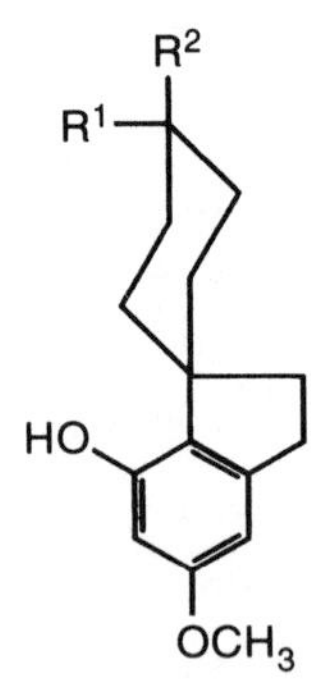

201a R^1 =H, R^2 =OH
α-cannabispiranol
201b R^1 =OH, R^2 =H
β-cannabispiranol

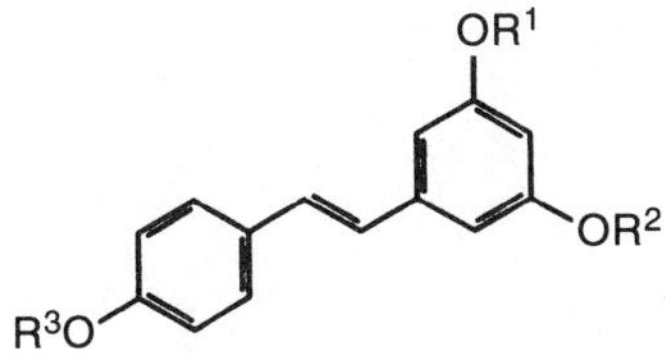

202a R^1 =CH_3, R^2 =CH_3, R^3 =H, pterostilbene
202b R^1 =H, R^2 =H, R^3 =CH_3, 3,5-dihydroxy-4'-methoxystilbene
202c R^1 =CH_3, R^2 =CH_3, R^3 =CH_3, 3,4',5-trihydroxystilbene

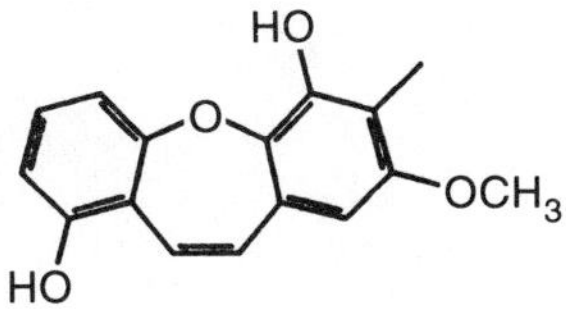

203 pacharin

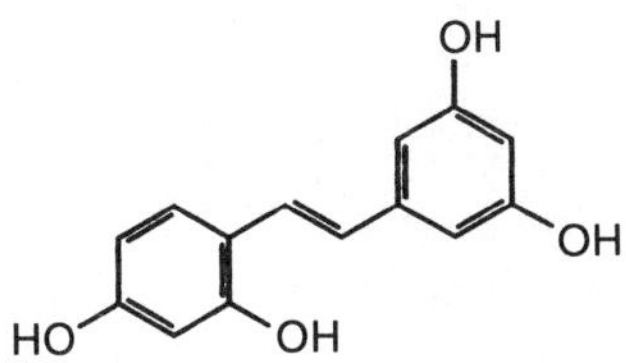

204 oxyresveratrol

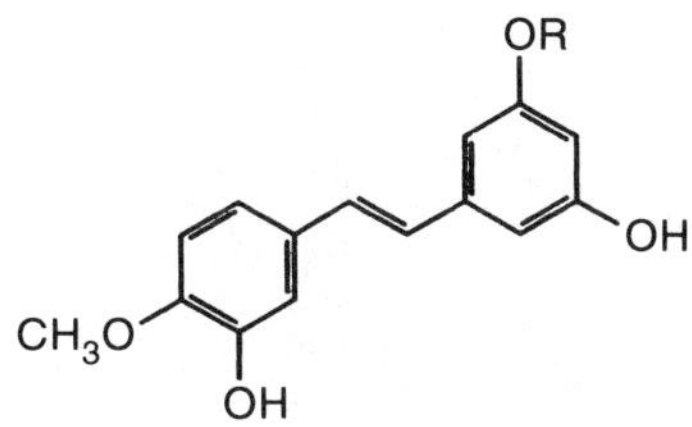

205a R =H, rhapontigenin
205b R =Glu, rhapontin

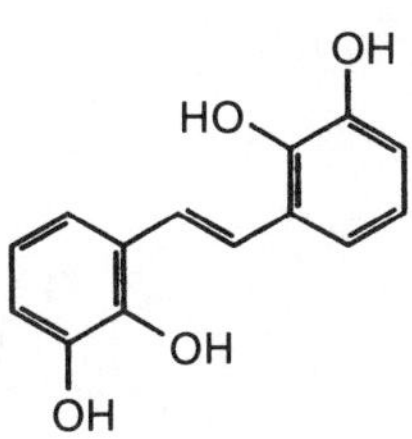

206 2,2',3,3' tetrahydroxystilbene

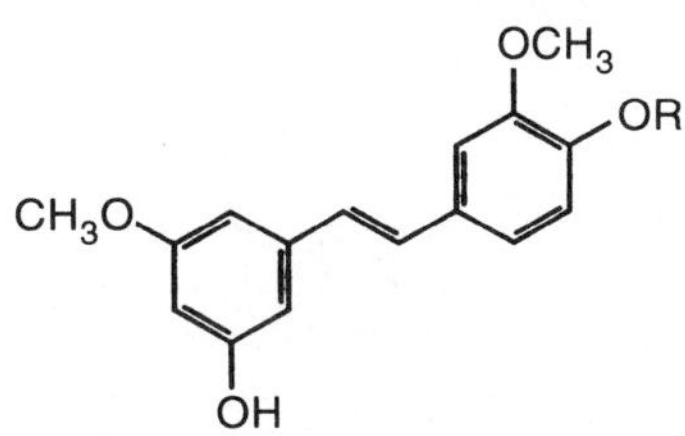

207a R =CH_3, 3'-hydroxy-3,4,5'-trimethoxystilbene
207b R =H, 3',4-dihydroxy-3,5'-dimethoxystilbene

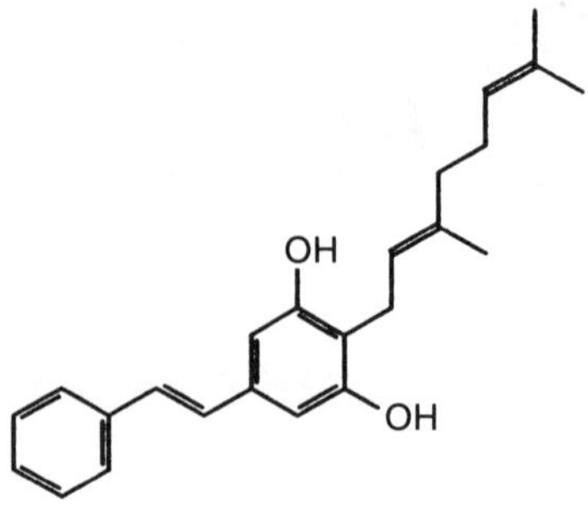

208 amorphastilbol

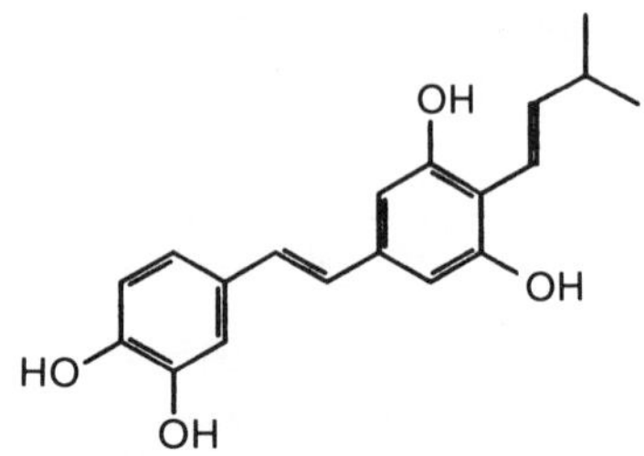

209 arachidin I

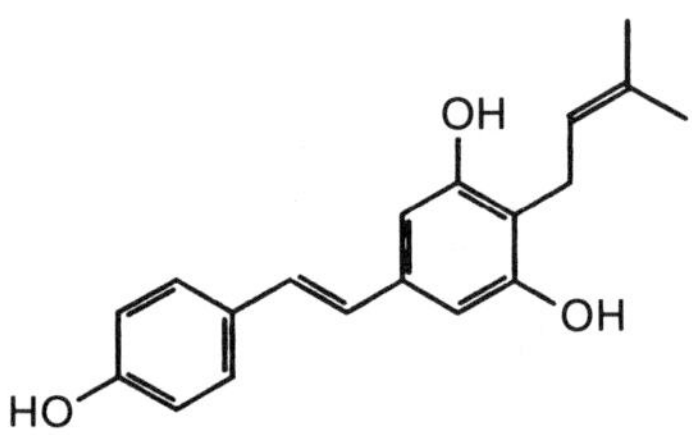

210 arachidin II

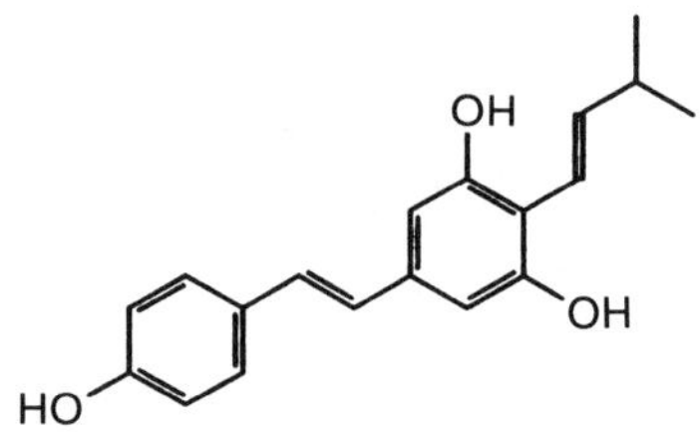

211 arachidin III

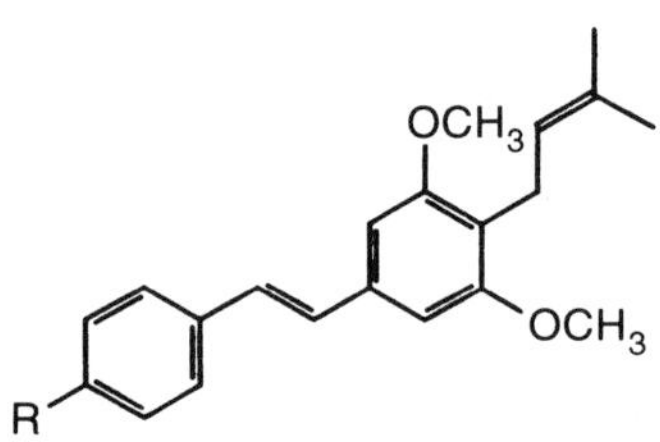

212a R =H,
3',5-dimethoxy-4-isopentenylstilbene
212b R =OCH_3,
4-isopentenyl-3,4',5'-trimethoxystilbene

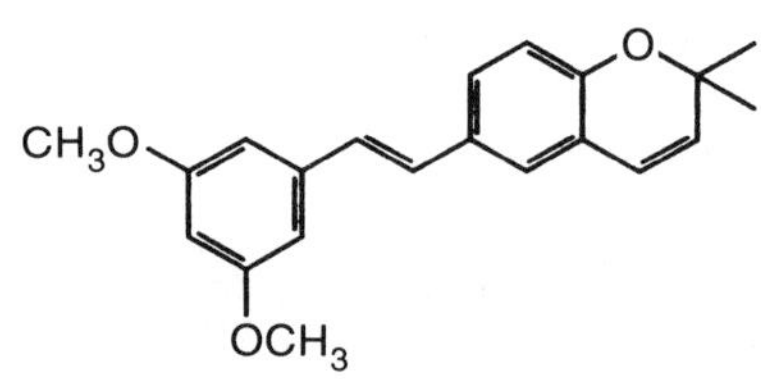

213 lonchocarpene

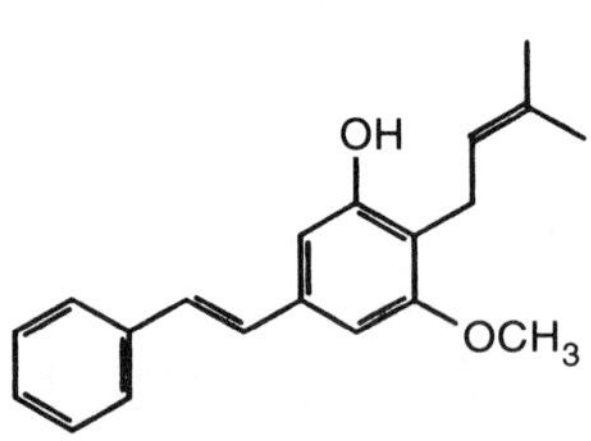

214 longistyline A

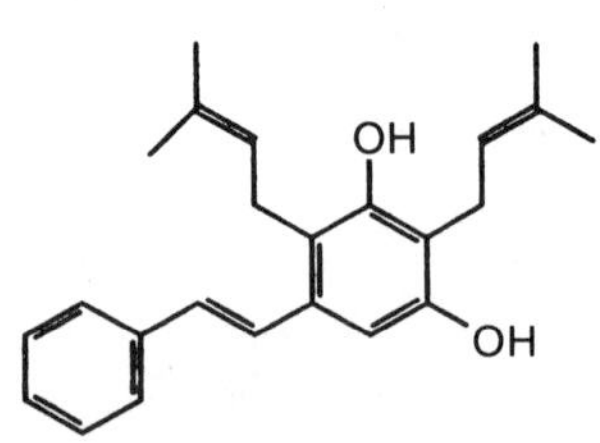

215 longistyline B

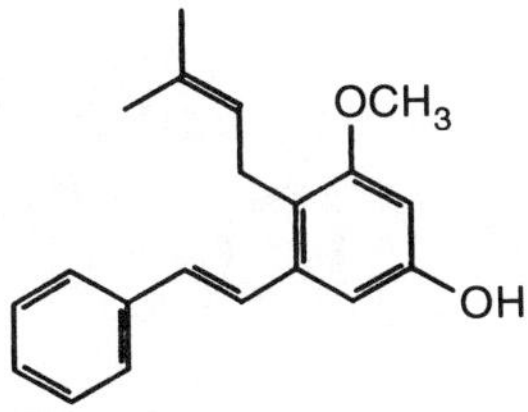

216 longistyline C

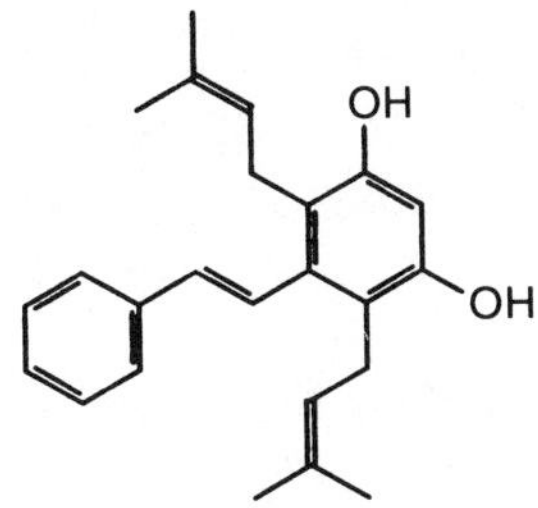

217 longistyline D

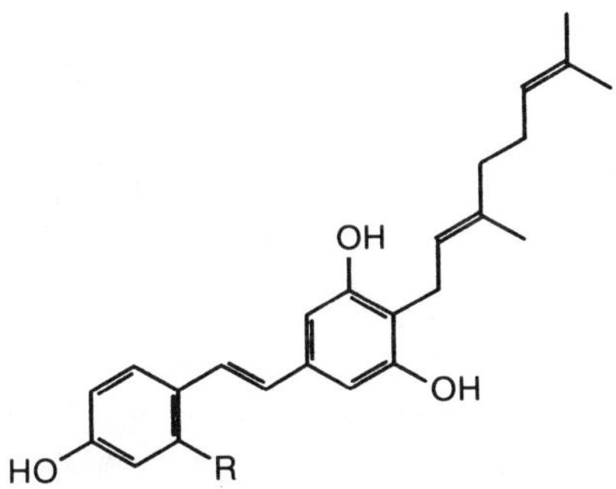

218a R =H,
4-geranyl-3,4',5-trihydroxylstilbene
218b R =OH,
chlorophorin

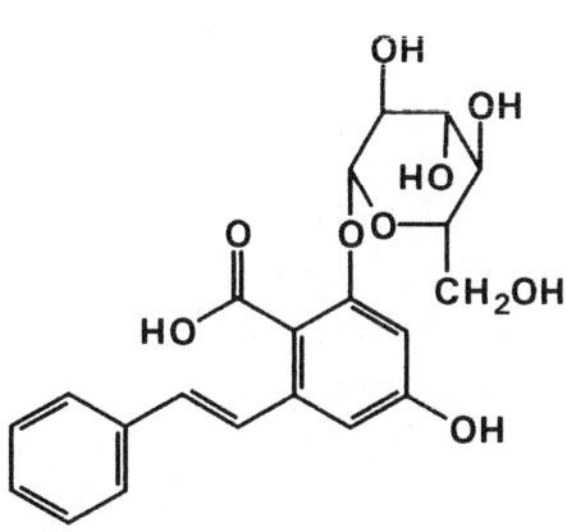

219 gaylussacin

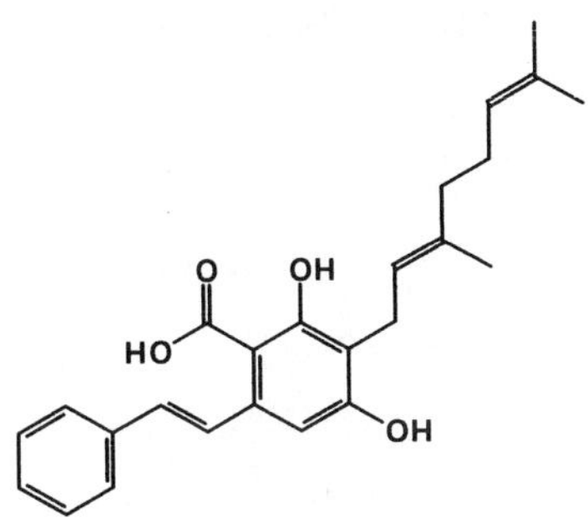

220

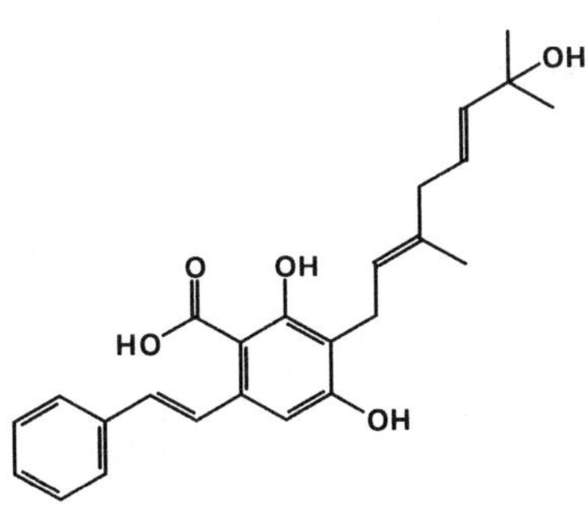

221

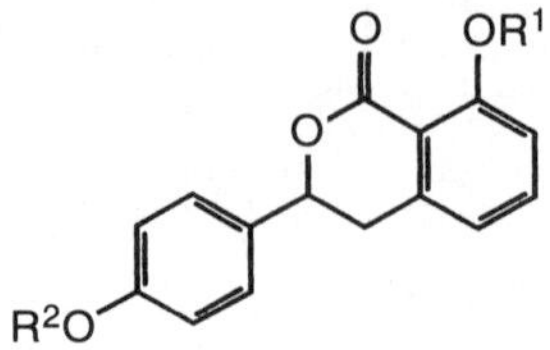

222a R^1 =H, R^2 =H, hydrangenol
222b R^1 =H, R^2 =CH_3,
hydrangenol monomethyl ether
222c R^1 =Glu, R^2 =H,
hydrangenol glucoside

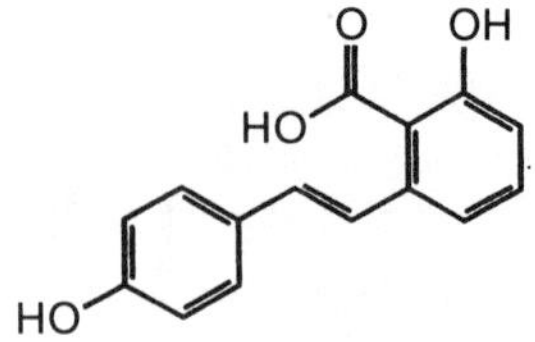

223 hydrangeic acid

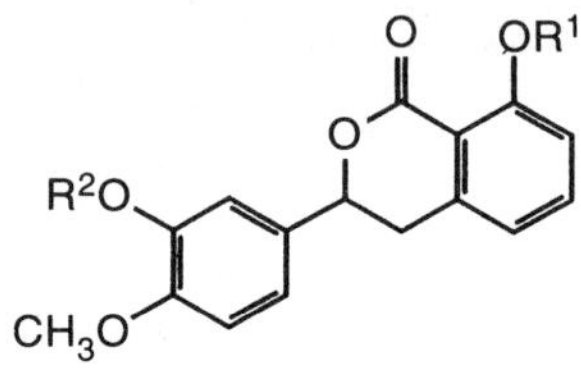

224a R^1 =H, R^2 =H, phyllodulcin
224b R^1 =H, R^2 =CH_3,
phyllodulcin monomethylether
224c R^1 =Glu, R^2 =H, phyllodulcin glucoside

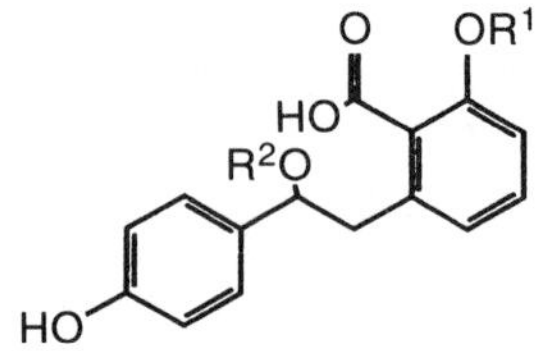

225a R^1 =Glu, R^2 =H,
hydrangea glucoside A
225b R^1 =H, R^2 =Glu,
hydrangea glucoside B

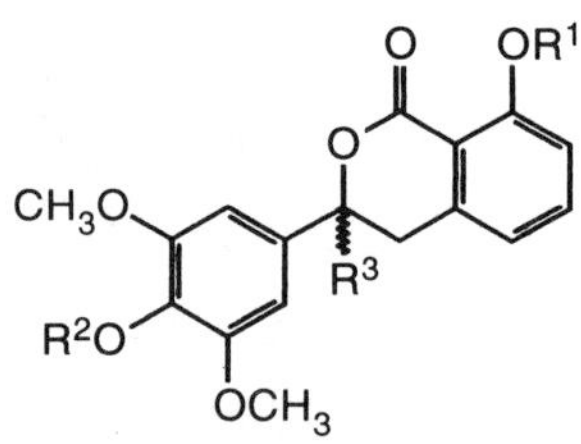

226a R^1 =Glu, R^2 =CH_3, R^3 =- - - - H,
macrophylloside A
226b R^1 =Glu, R^2 =H, R^3 =◄H,
macrophylloside B
226c R^1 =Glu, R^2 =CH_3, R^3 =- - - - H,
macrophylloside C

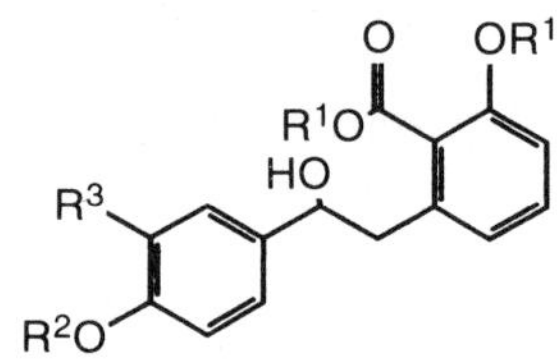

227a R^1 =Glu, R^2 =CH_3, R^3 =OH,
stilbene glucoside I
227b R^1 =Glu, R^2 =H, R^3 =H,
stilbene glucoside II

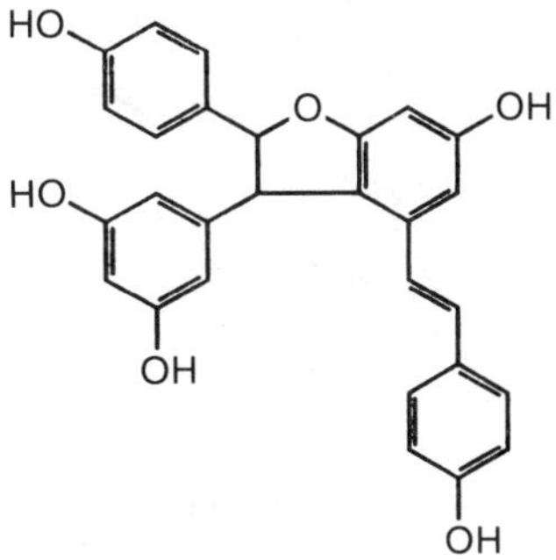

228 ε-viniferin

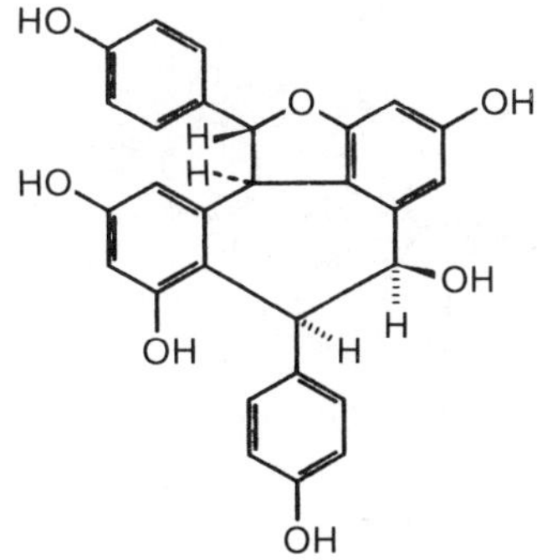

229 balanocarpol

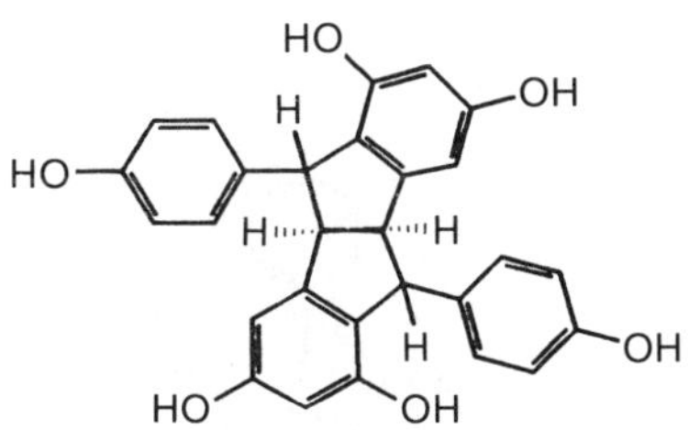

230 pallidol

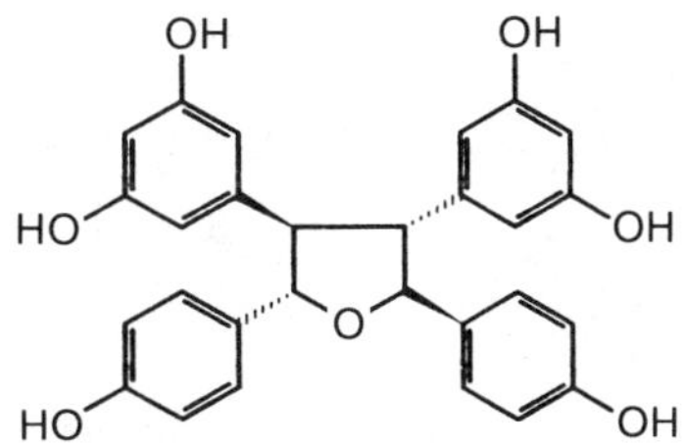

231 tricuspidatol A

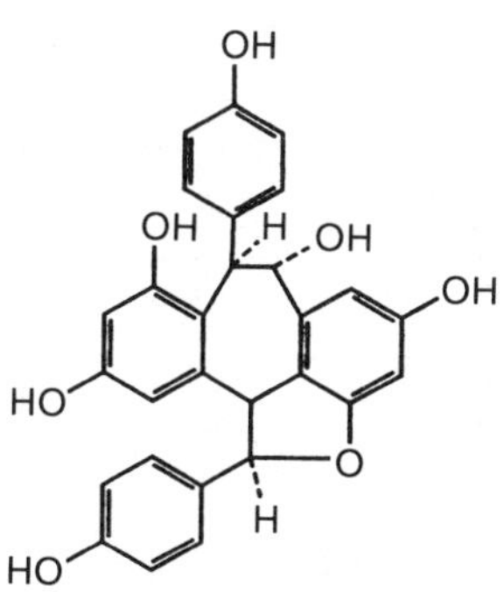

232 ampelopsin A

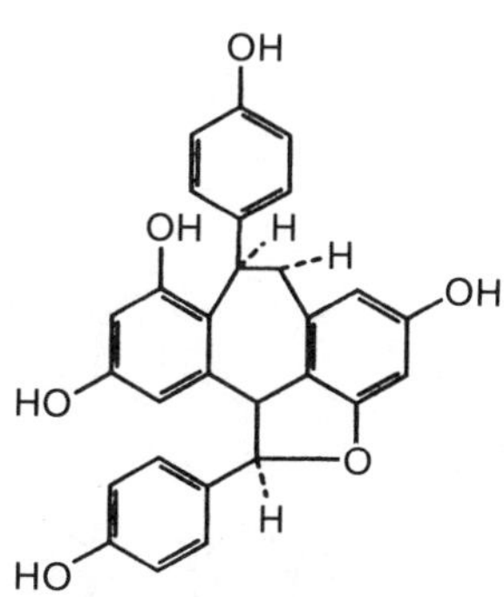

233 ampelopsin B

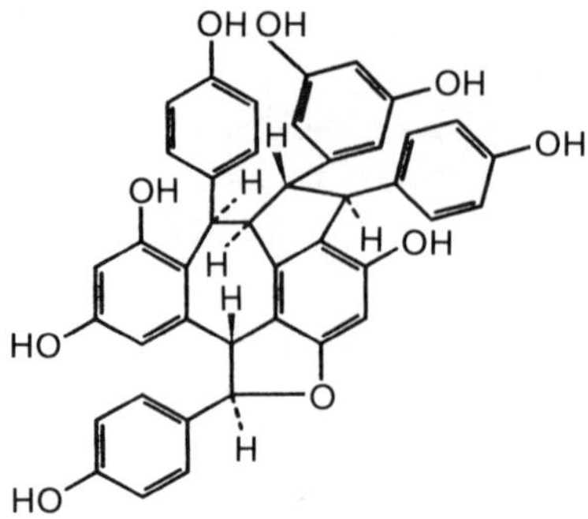

234 ampelopsin C

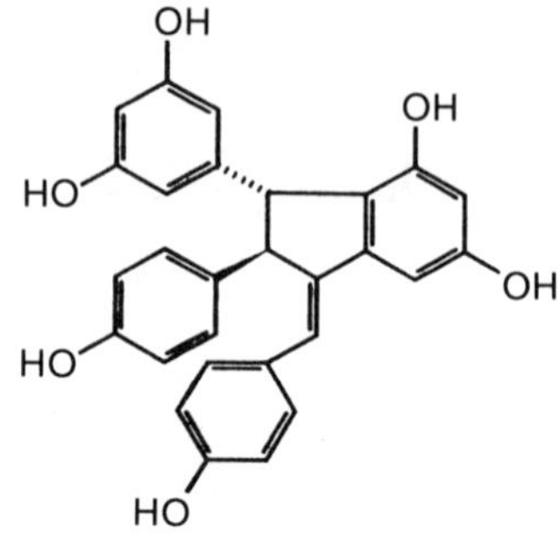

235 ampelopsin D

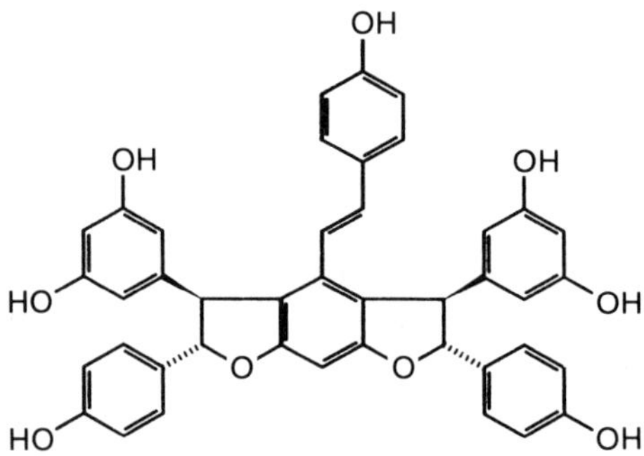

236 *trans*-ampelopsin E

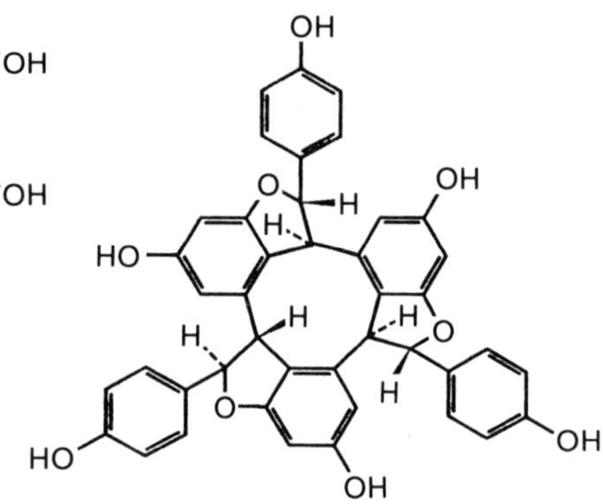

238 α-viniferin

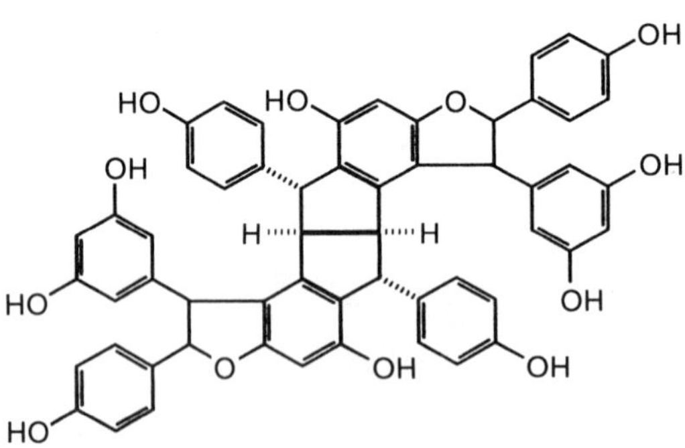

237 ampelopsin H

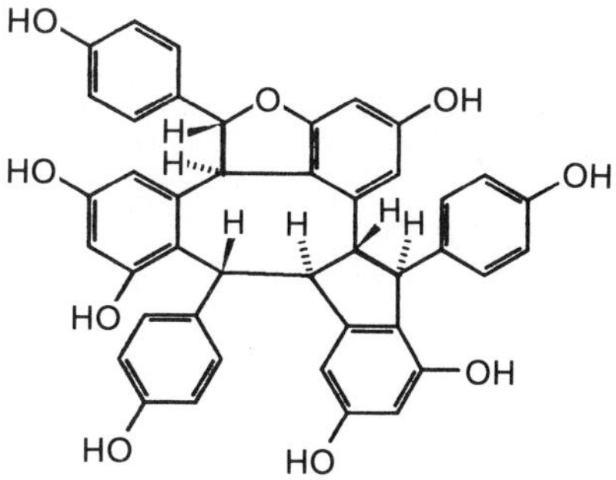

239 distichol

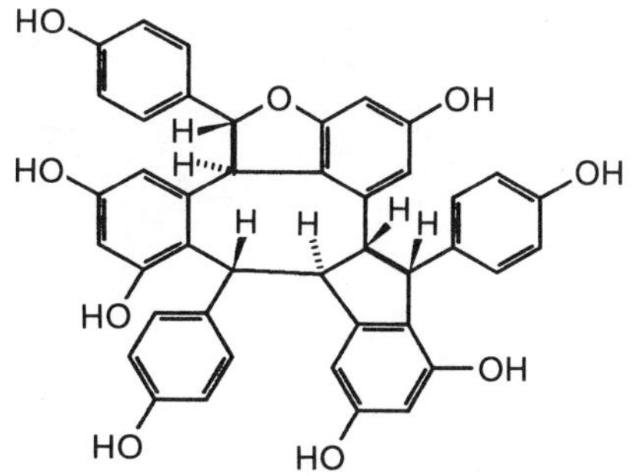

240 canaliculatol

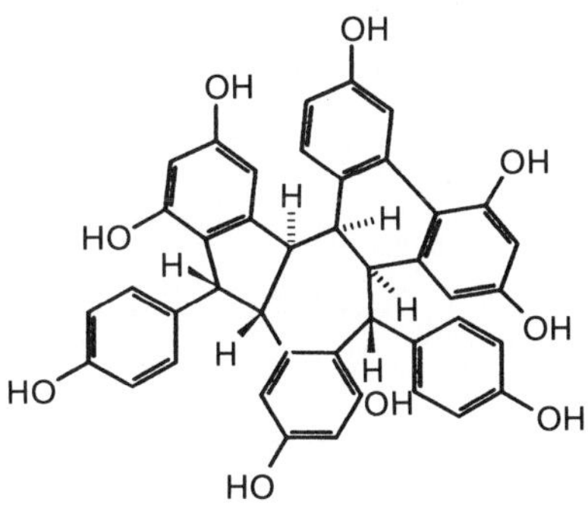

241 copalliferol A

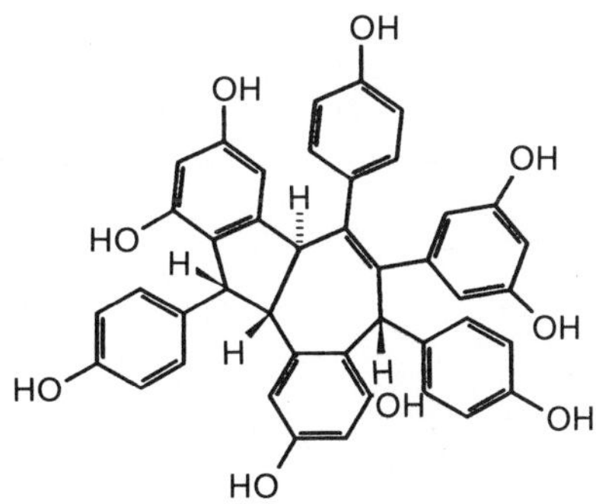

242 copalliferol B

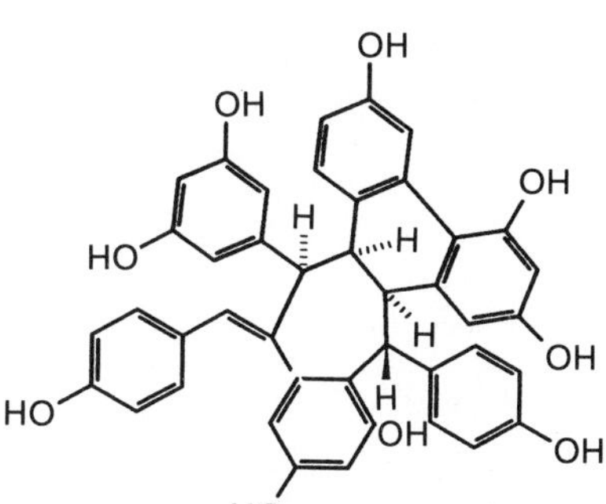

243 stemonoporol

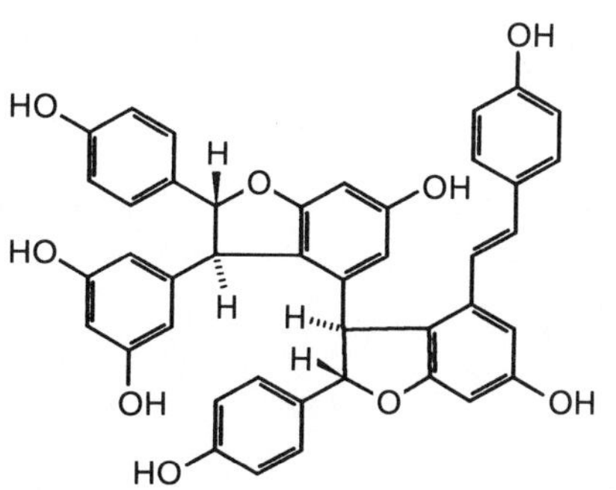

244 miyabenol C

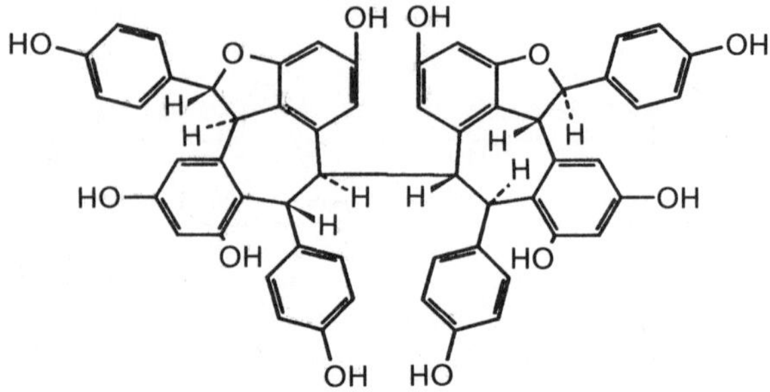

245 hopeaphenol

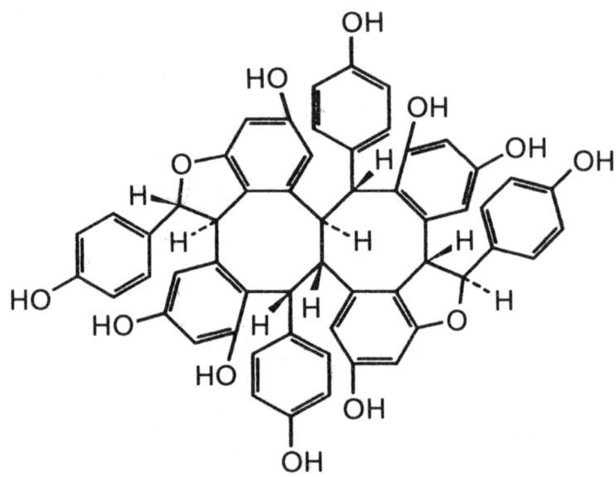

246 vaticaffinol

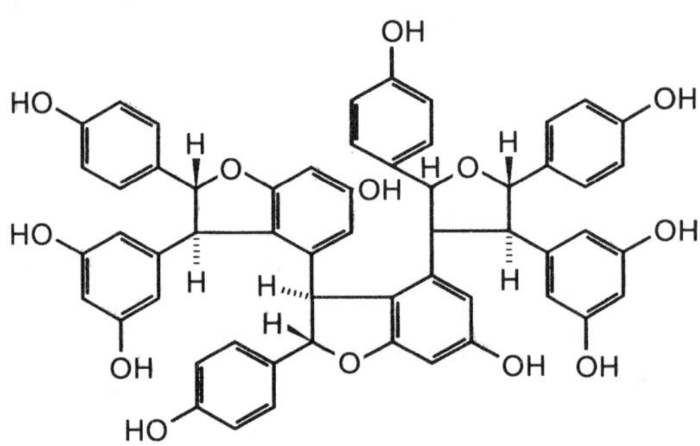

247 kobophenol A

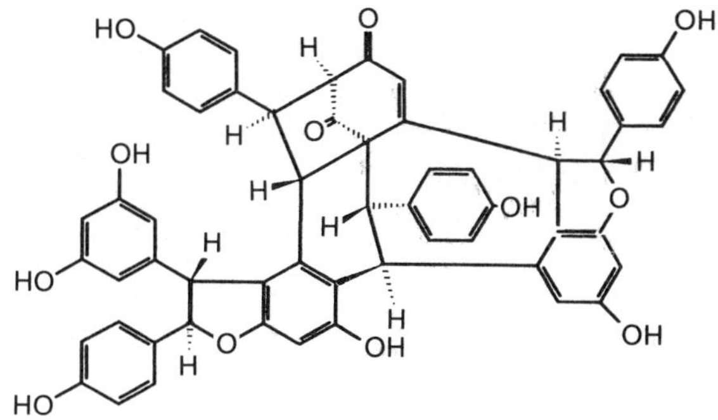

248 kobophenol B

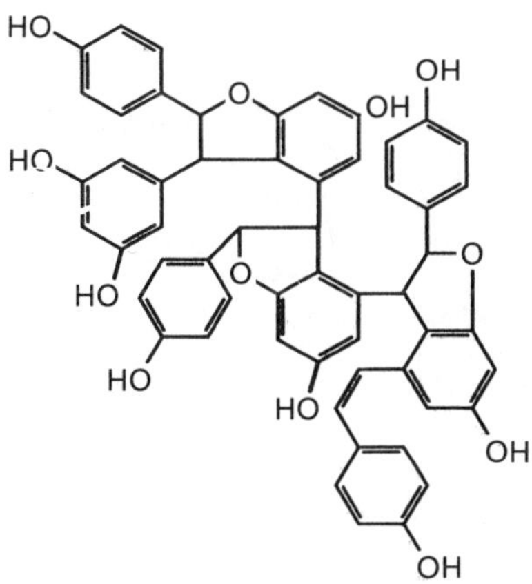

249 miyabenol A

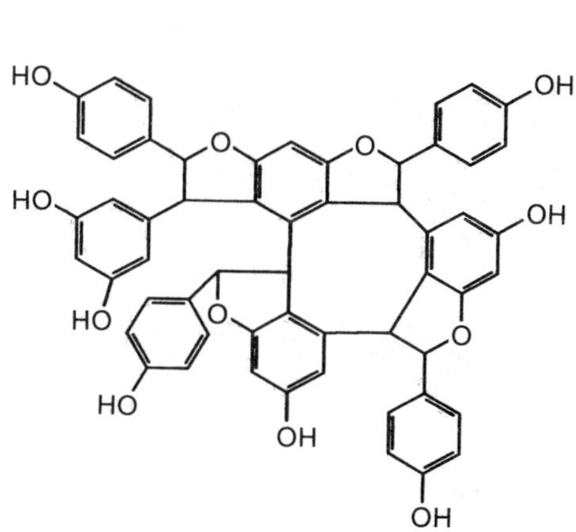

250 miyabenol B

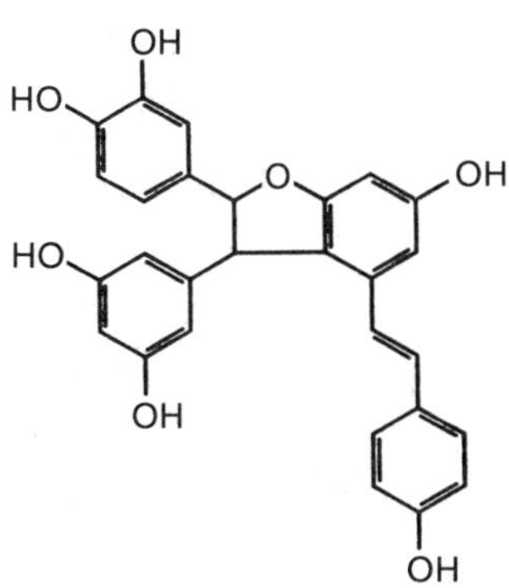

251 scirpusin A

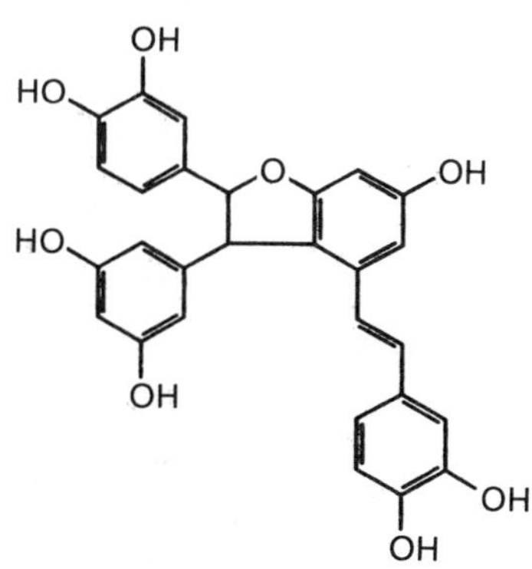

252 scirpusin B

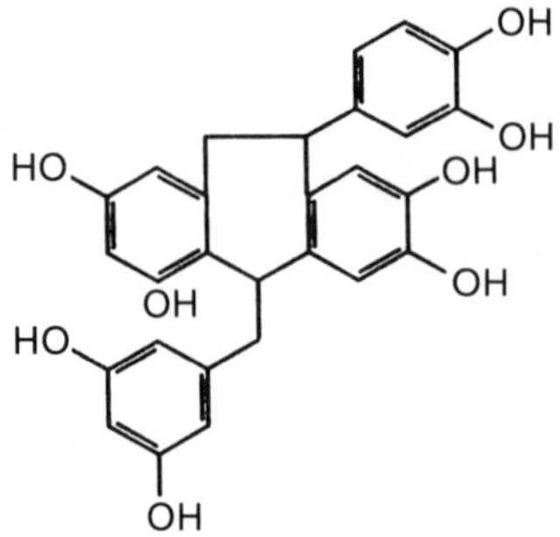

253 cassigarol A

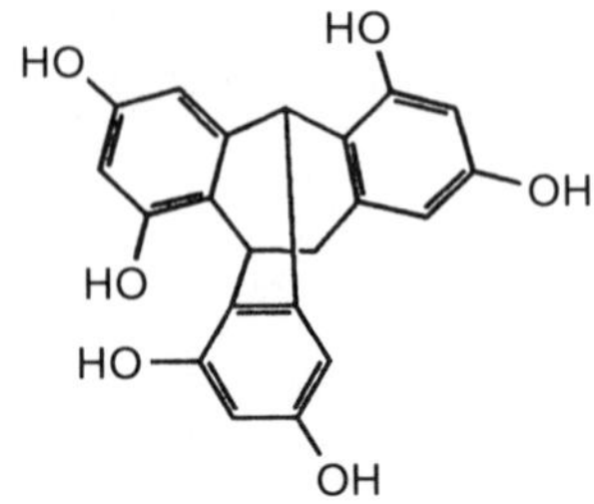

254 cassigarol B

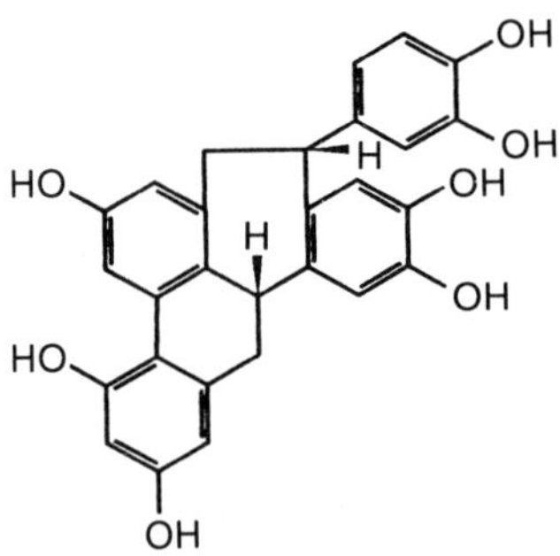

255 cassigarol C

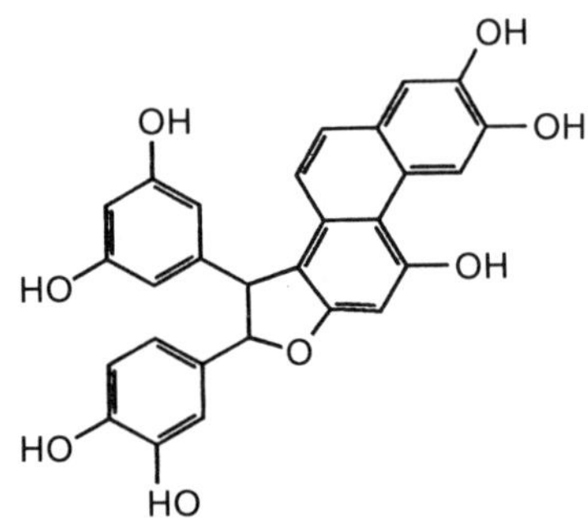

256 cassigarol D

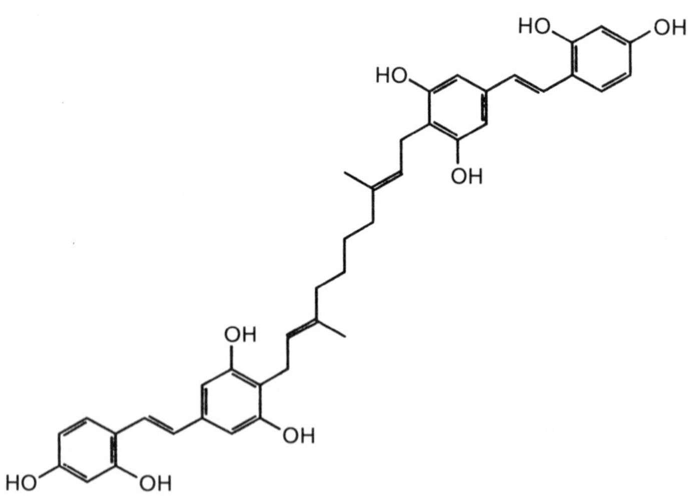

257 excelsaoctaphenol

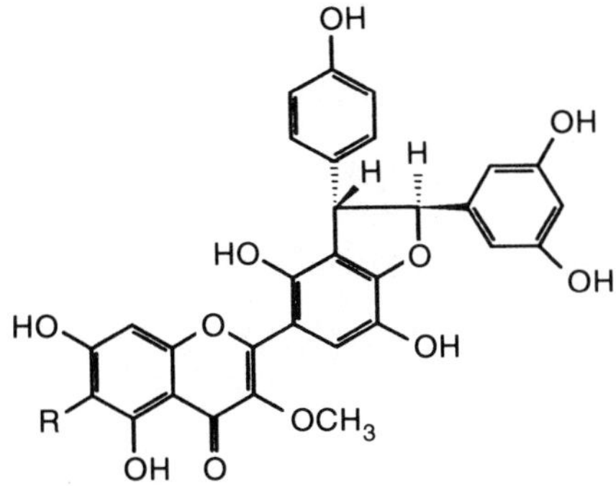

258a R =H, androyol
258b R =CH_3, methylandroyol

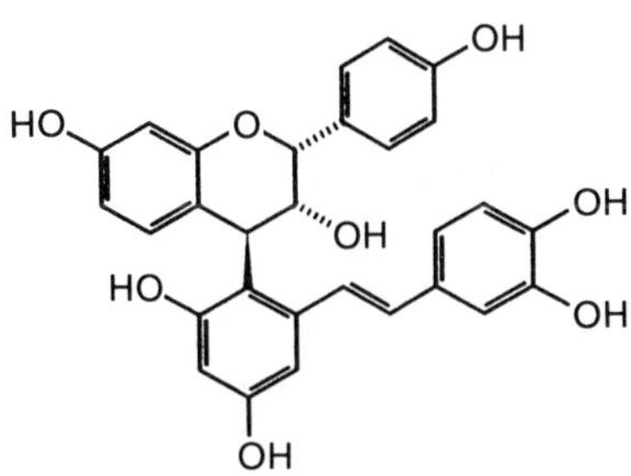

259

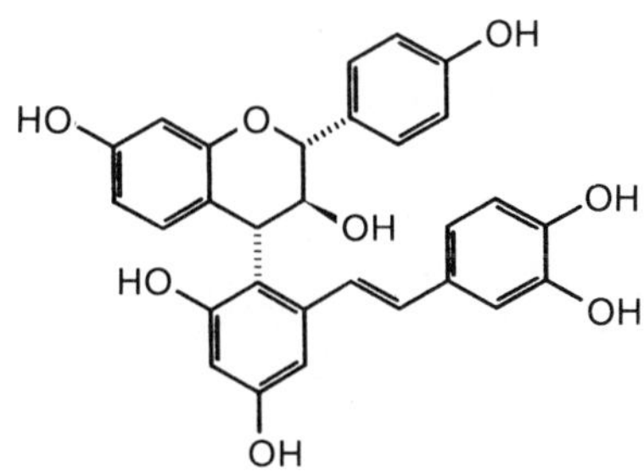

260

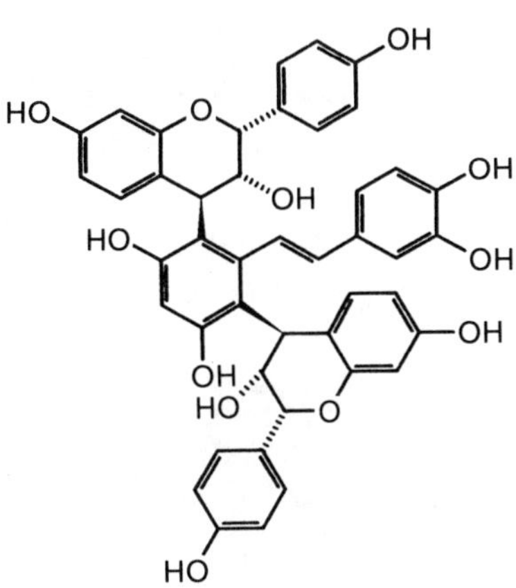

261

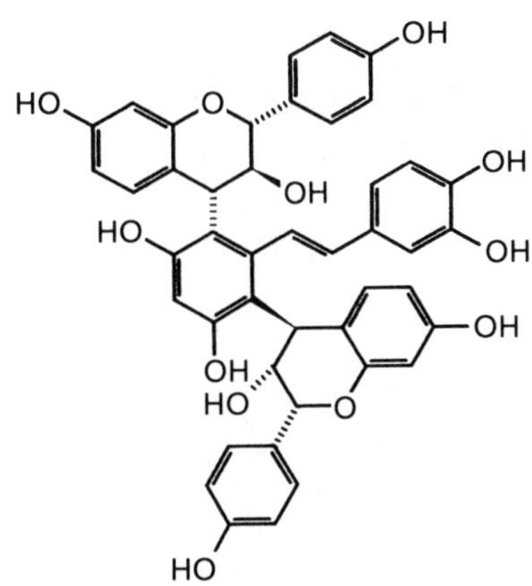

262

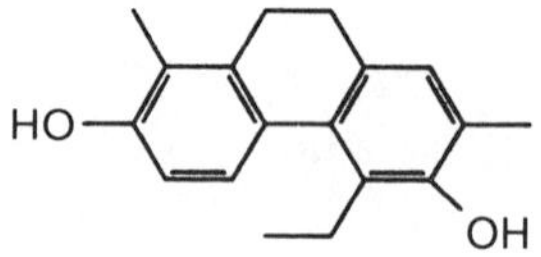

263 juncunone

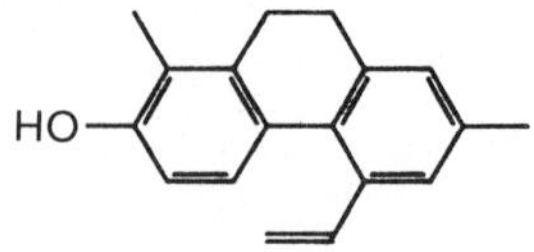

264 juncunol

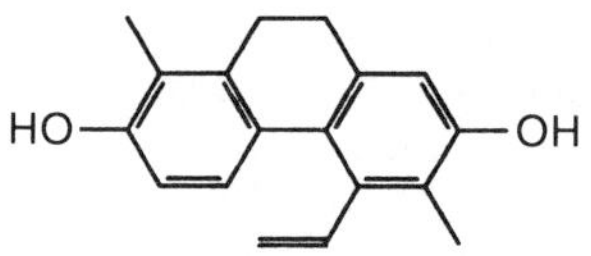

265 juncusol

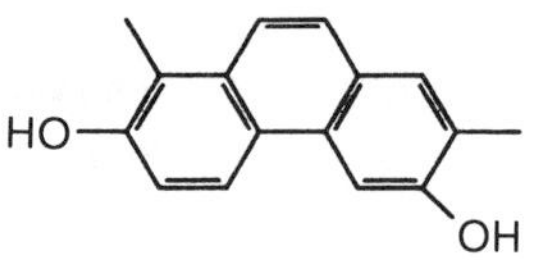

266 micrandrol A

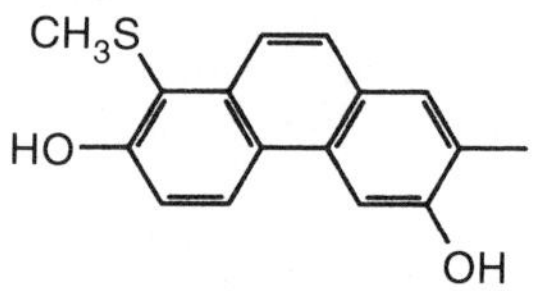

267 micrandrol C

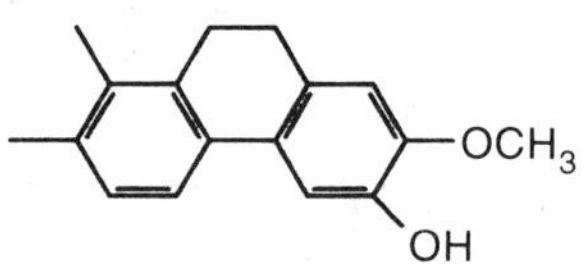

268 micrandrol F

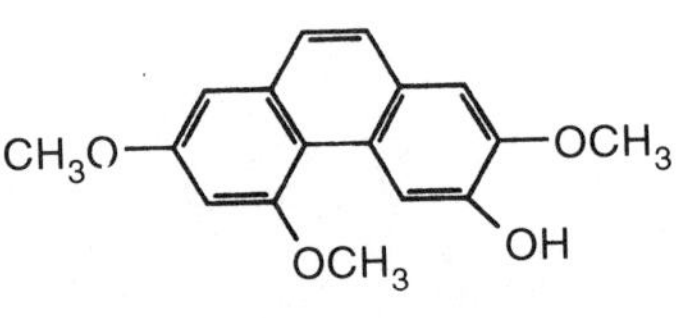

269 batatasin I

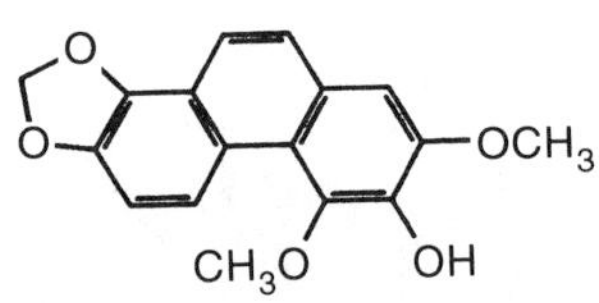

270 Ta IV

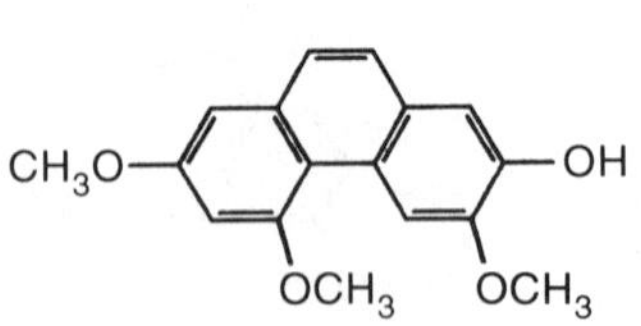

271 Ta VI

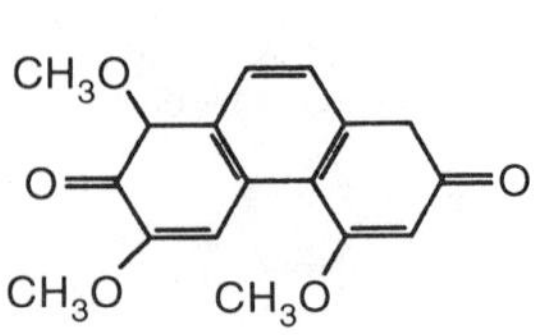

272 Ta II

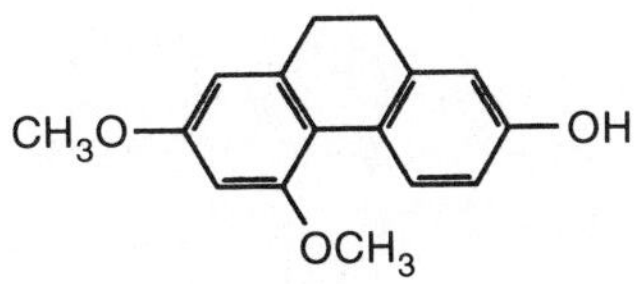

273 orchinol

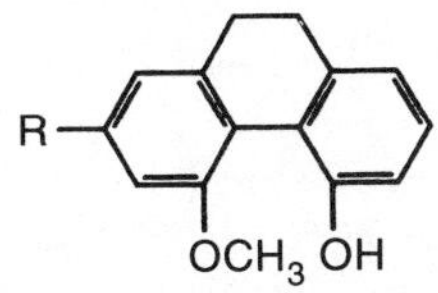

274a R =OCH$_3$, loroglossol
274b R =OH, hircinol

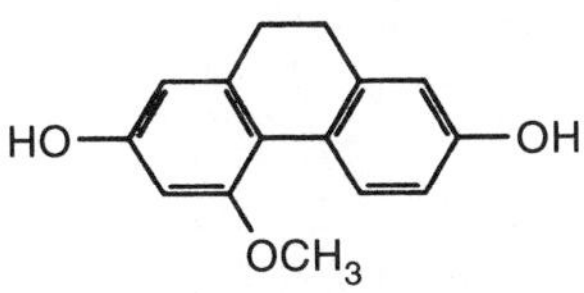

275 coelonin

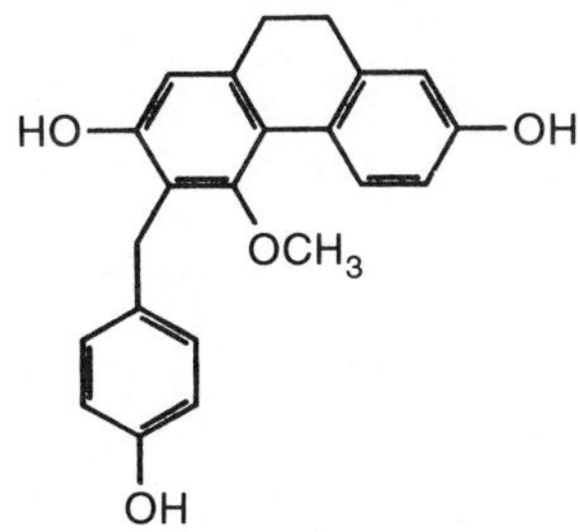

276 3-(4-hydroxybenzyl)-coelonin

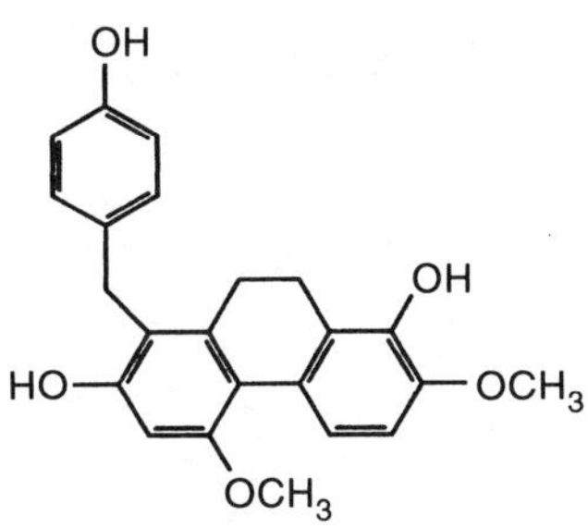

277

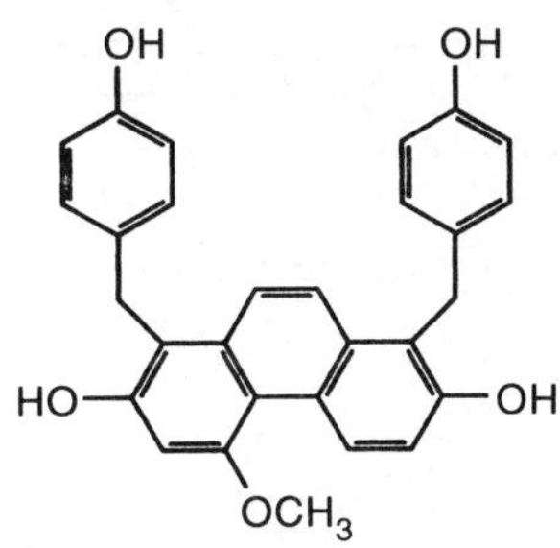

278

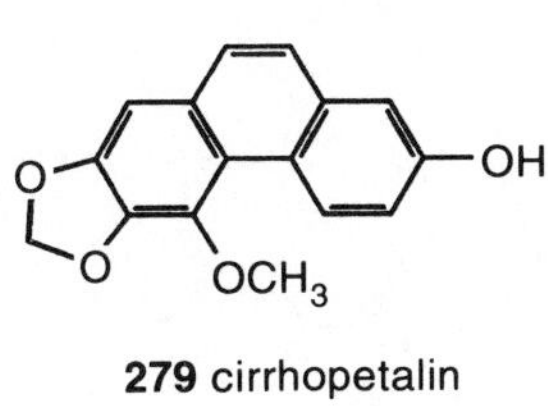

279 cirrhopetalin

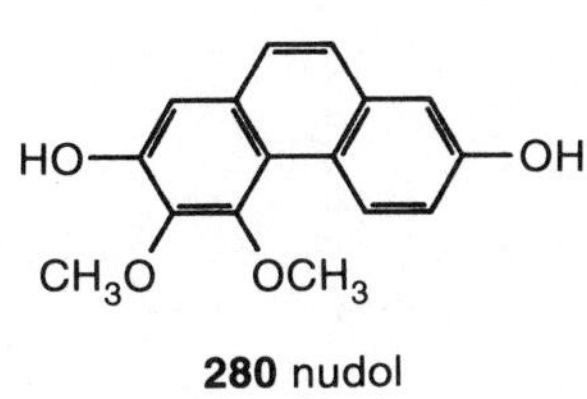

280 nudol

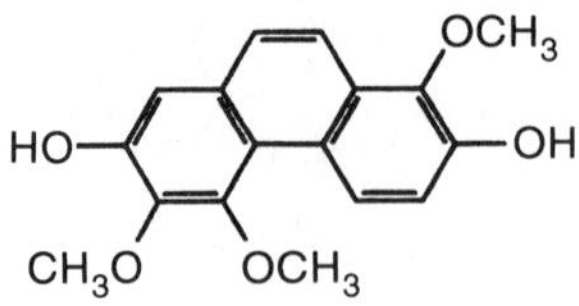

281 confusarin

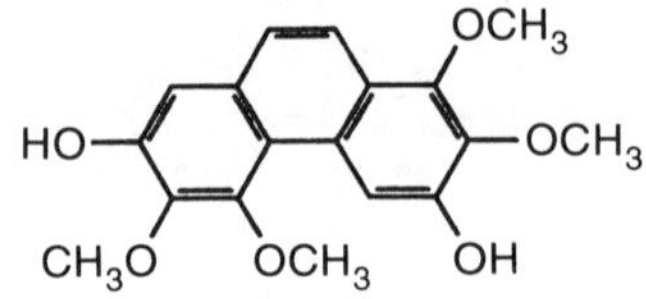

282 confusaridin

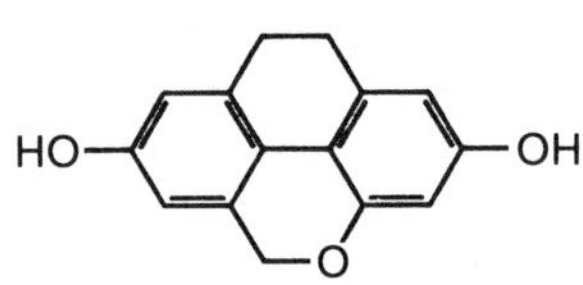

283 flavidin

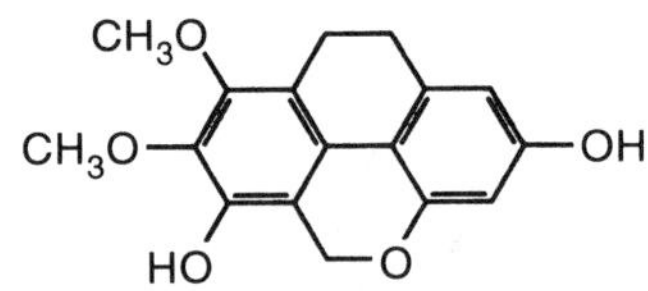

284 coelogin

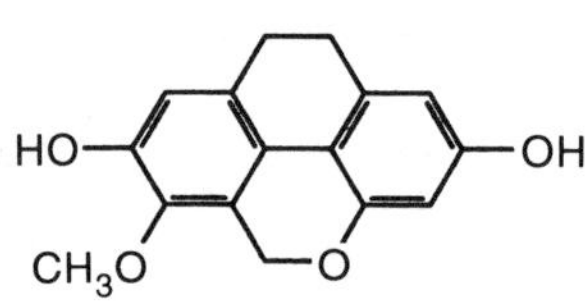

285 imbricatin

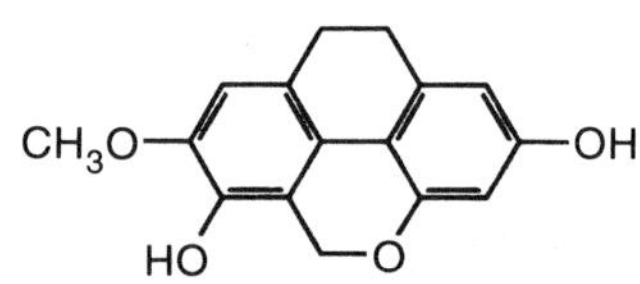

286 amoenumin

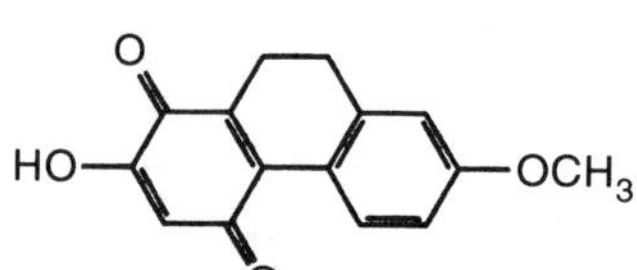

287 ochrone A

288 ochrone B

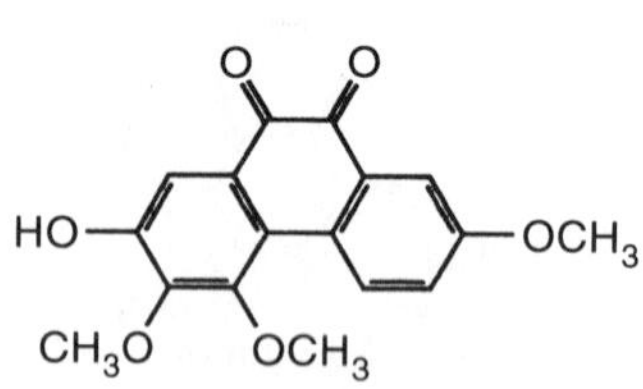

289 bulbophyllanthrone

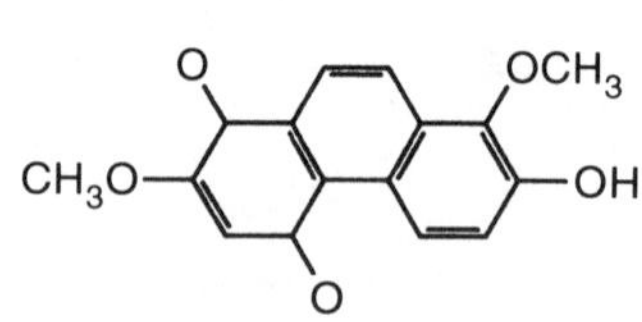

290 cypripedin

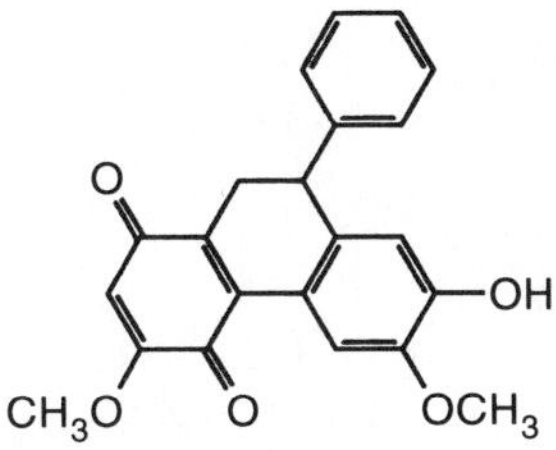

291 latinone

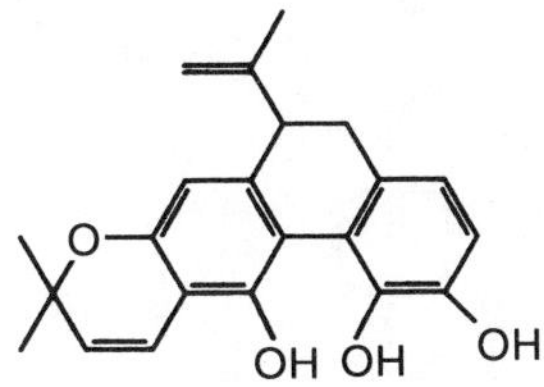

292 paralycolin

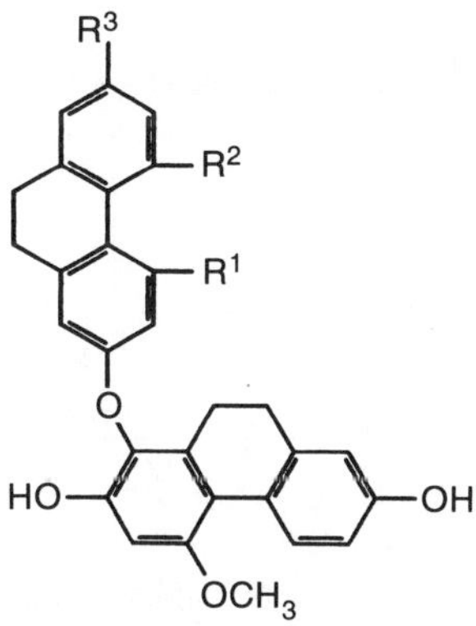

293a R^1 =OCH_3, R^2 =H, R^3 =OH
blestrin A
293b R^1 =H, R^2 =OH, R^3 =OCH_3,
blestrin B

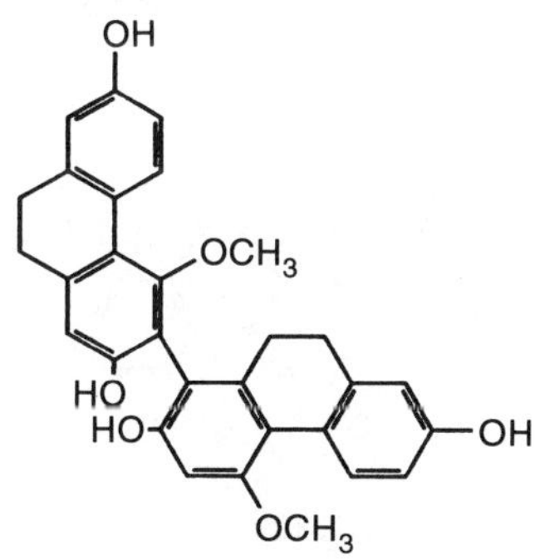

294 blestrianol A

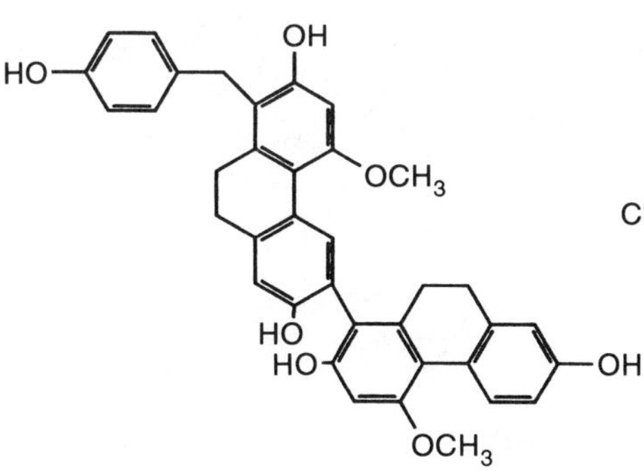

295 blestrianol B

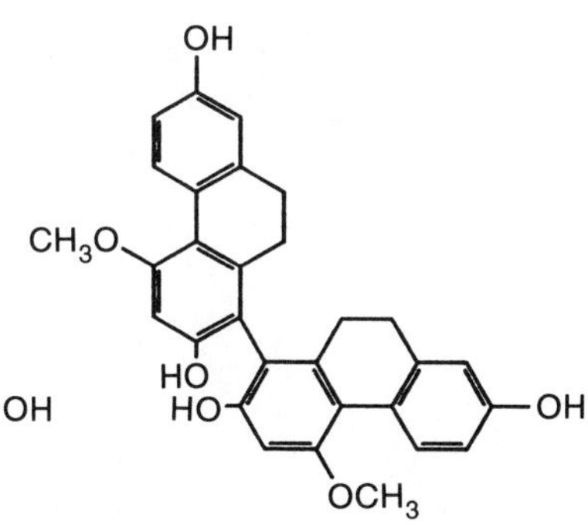

296 blestriarene A

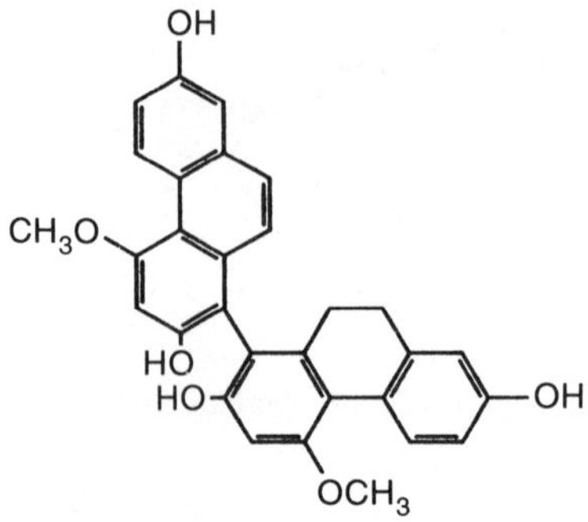

297 blestriarene B

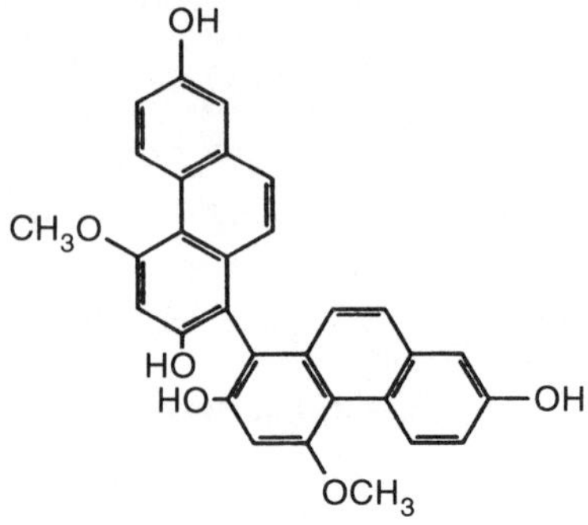

298 blestriarene C
=cirrhopetalanthrin

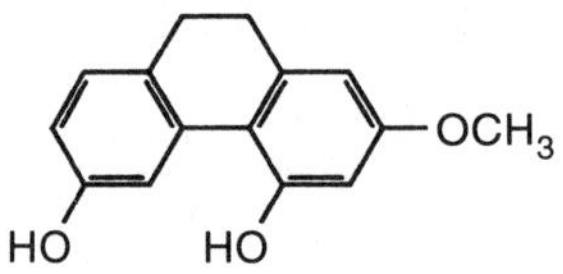

299 cannithrene 1

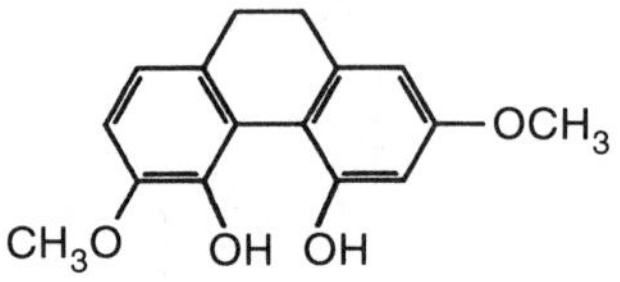

300 cannithrene 2

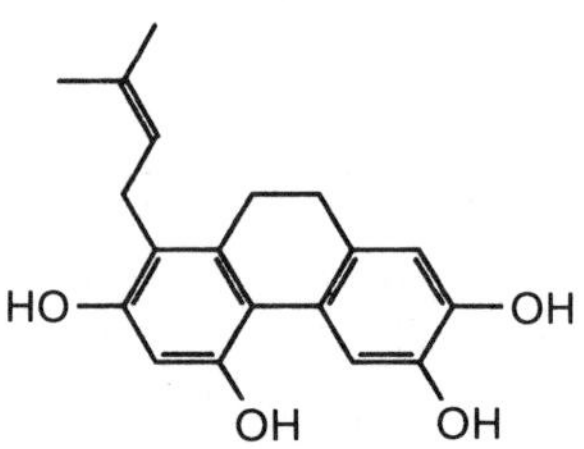

301 gancaonin V

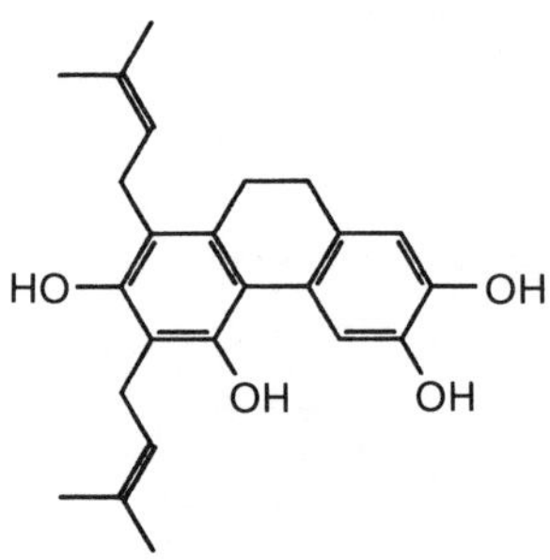

302 gancaonin U

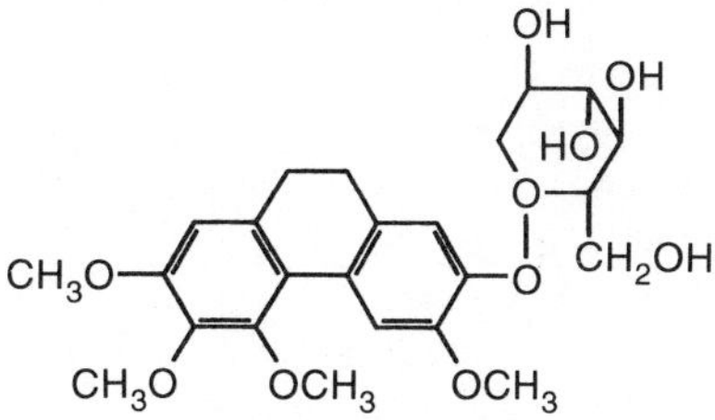

303 icariside A_1

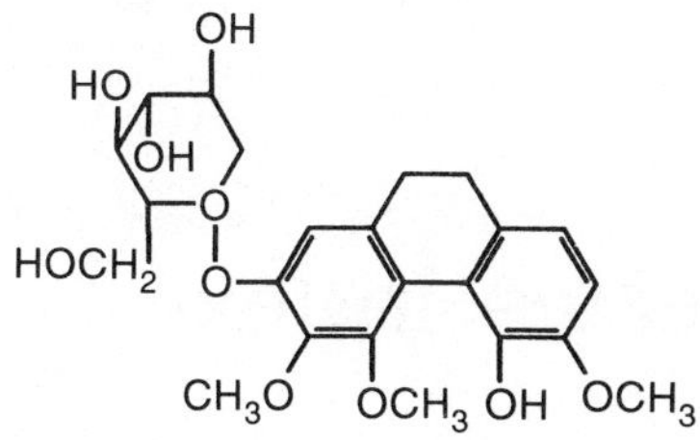

304 icariside A_2

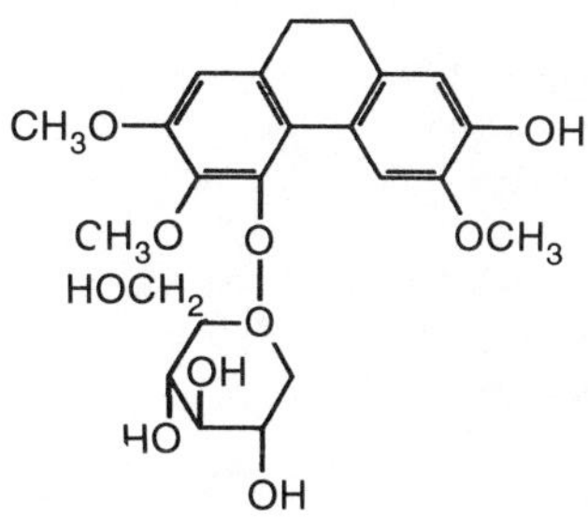

305 icariside A_3

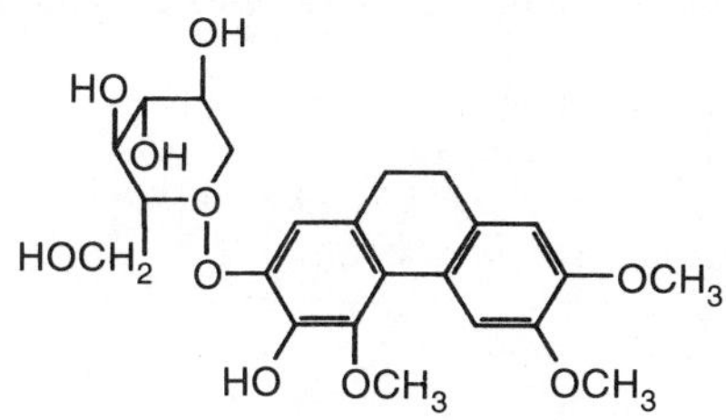

306 icariside A_5

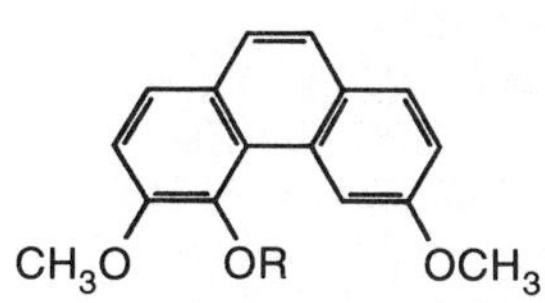

307a R = α-thebaol,
307b R =CH_3, O-methyl-α-thebaol

Species index

Page numbers in *italics* refer to tables

Subject index

Page numbers in *italics* refer to tables, and page numbers in **bold** refer to chemical structures